Biotechnology and
Bioactive Polymers

Biotechnology and Bioactive Polymers

Edited by

Charles G. Gebelein

Lionfire, Inc.
Edgewater, Florida
and Florida Atlantic University
Boca Raton, Florida

and

Charles E. Carraher, Jr.

Florida Atlantic University
Boca Raton, Florida

Plenum Press ● New York and London

05892193

CHEMISTRY

Library of Congress Cataloging-in-Publication Data

Biotechnology and bioactive polymers / edited by Charles G. Gebelein
  and Charles E. Carraher.
      p.   cm.
    Includes bibliographical references and index.
    ISBN 0-306-44629-4
    1. Biopolymers--Biotechnology--Congresses.   I. Gebelein, Charles
  G. II. Carraher, Charles E.
  TP248.65.P62B565 1994
  610'.28--dc20   93-48722
                                                             CIP

Proceedings of an American Chemical Society Symposium on Biotechnology and Bioactive Polymers,
held April 5–10, 1992, in San Francisco, California

ISBN 0-306-44629-4

©1994 Plenum Press, New York
A Division of Plenum Publishing Corporation
233 Spring Street, New York, N.Y. 10013

Printed in the United States of America

PREFACE

Some have predicted that the coming several decades will be the decades of "biotechnology," wherein cancer, birth defects, life span increases, cosmetics, biodegradation, oil spills and exploration, solid waste disposal, and almost every aspect of our material life will be affected by this new area of science. There will also be an extension of emphasis on giant molecules: DNA, enzymes, polysaccharides, lignins, proteins, hemoglobin, and many others.

Biotechnology has been defined in various ways. In one sense, this field is older than human history and references to the human use of biotechnology-derived materials can be found in the oldest human writings, such as the Bible. In this book, biotechnology refers to the direct usage of naturally occurring materials or their uses as a feedstock, including the associated biological activities and applications of these materials. Bioactive polymers, on the other hand, are polymers which exert some type of activity on living organisms. These polymers are used in agriculture, controlled release systems, medicine and many other areas. The papers in this book describe polymers which essentially combine features of biotechnology and bioactivity.

The International Symposium on Biotechnology and Bioactive Polymers, jointly sponsored by the Biotechnology Secretariat and the Division of Polymeric Materials, was presented at the Spring, 1992 American Chemical Society, meeting in San Francisco. This book contains chapters derived from this symposium representing the spectrum of applications, both demonstrated and exploratory, that this combined realm of Biotechnology and Bioactive Polymers encompasses. This book is divided into four sections, with a some overlap in both concepts and materials. Thus, Section 1 deals mainly with polysaccharide-based applications, while Section 2 is concerned with both polypeptide applications and biodegradable polymers. Section 3 covers some diagnostic applications of biotechnological polymers, while Section 4 considers controlled release and other medical applications of these systems.

The specific topics in Section 1 cover: the elimination of yeast-causing microorganisms via modified polysaccharides (Carraher, Butler, Naoshima, Sterling & Saurino); a novel bacterial-derived polysaccharide (Morris); the blood coagulation properties of some modified heparins (Stemberber, Bader, Haas, Walenga and Blümel); polysaccharide biomaterials for use in soft tissue augmentation (Larsen, Pollack, Reiner, Leshchiner & Balazs); the use of polysaccharides to capture metal ions [hydrometallurgy] (Inoue, Yoshizuka & Baba); the range of applications of chitin and chitosan (Hirano, Inui, Kosaki, Uno & Toda); the recovery of proteins from whey using chitosan (Kennedy, Paterson & Silva); and microorganism inhibition by lignin-based polymers (Carraher, Sterling, Butler & Ridgeway).

v

Section 2 contains the following topics: a review of actin and tubulin proteins (Dirlikov); the bioactive roles of elastin-like materials (Nicol, Gowda, Parker & Urry); the biotechnological expression of synthetic polypeptide adhesives in a common bacteria (Salerno & Goldberg); medical treatments using modified collagens (Stemberber, Ascherl, Scherer, Bader, Erhardt, Adelmann-Grill, Sorg, Thomi, Stoltz, Jeckle, Seaber & Blümel); the construction of polymeric phosphocholine bilayers (Singh, Markowitz & Tsao); the synthesis and properties of semi-IPNs derived from veronia oil (Sperling & Barrett); an analysis of the actual environmental biodegradibility of polymers (Swift) and the absorption of polymers when used as bone plates (Jadhav & Deger).

In Section 3 the following topics are covered: the self-monitoring of blood glucose levels using nonaqueous diagnostic tests (Kennamer, Burke & Usmani); protein purification via selective adsorption (Ngo); the diagnostic properties of some dry polymeric systems (Kennamer, Burke & Usmani); the application of bacteria-absorption polymers as biosensors (Tsushima, Kondo, Sakata & Kawabata); and the polymeric immobilization of adenosines for the separation of nucleic acid oligomers (Inaki, Matsukawa & Takemoto).

In the final Section 4, the specific topics treated include: antimicrobial polymers (Vigo); the synthesis and characterization of a polymeric ß-lactam (Ghosh); the controlled delivery of vaccines and antigens using biodegradable polymers (Schmidt, Flanagan & Linhardt); the polymeric controlled release of cardiovascular agents (Levy, Golomb, Trachy, Labhasetwar, Müller & Topol); the use of polyICLC for treatment of AIDS patients (Levy, Salazar, Morales & Morgan); a trio of papers describing various aspects of the release of 5-fluorouracil including: the 5-FU release profiles from several polysaccharides (Gebelein, Williams, Marshall & Slaven), the anticancer activity of 5-FU released from chitosan microspheres (Ouchi, Shiratani, Kobayashi, Takei & Ohya), and the release profile from a 5-FU prodrug (Gebelein, Gardner & Ellis); the bioactivity of some synthetic polyphosphates and polyphosphonates (Carraher, Powers & Pandya); and some novel anti-infective polymeric biomaterials (Kohnen, Jansen, Ruiten, Steinhauser & Pulverer).

The Editors thank the authors for their excellent contributions to this book. The partial financial support of the Symposium from the Division of Polymeric Materials, Allied-Signal, Inc., Biomatrix, Inc., Boehringer-Mannheim Corp., Johnson & Johnson Orthopaedics and the Rohm & Haas Company is gratefully acknowledged.

Charles G. Gebelein and Charles E. Carraher, Jr., Editors
Florida Atlantic University

CONTENTS

FACTORS AFFECTING THE BACTERIAL ACTIVITY OF SACCHARIDES AND

POLYSACCHARIDES MODIFIED THROUGH REACTION WITH ORGANOSTANNANES

Charles E. Carraher, Jr., Cynthia Butler, Yoshinobu
Naoshima, Dorothy Sterling and Vincent Saurino

Florida Atlantic University
Departments of Chemistry & Biological Sciences
Boca Raton, FL 33431
and
Okayama University of Science
Department of Biochemistry
Ridaicho, Okayama, 700 Japan

The biological activities of tin-containing materials
derived from numerous saccharides and polysaccharides is
discussed. Emphasis is placed on bacterial results. In
general, inhibition is only secondarily dependent on the
percentage of organostannane incorporation. Inhibition
follows the general inhibition order for alkyl-substitution
where general activity decreases as the length of the alkyl
chain increases. Finally, general activity is dependent on
the nature of the saccharide.

INTRODUCTION

Polysaccharides and saccharides are essential materials of nature,
important in the basic structural aspects of plants and animals as
foodstuffs, transmission of fluids, providing mechanical strength and
lubrication, physiological information-carrying agents, involved in the
active and passive transport of materials, control of tissue growth
through contact inhibition, etc. Polysaccharides offer quite dramatic, as
well as subtle, changes in their biological activity through changes in
their intimate molecular structure including shape and molecular weight.[1]
Often it is not only their molecular shape that is critical, but also
their supermolecular architecture may play an important role. Thus, in
addition to the usual parameters of size and shape, items such as source,
time of year of collection, isolation procedure are important. With such
an array of variables it is important to clearly describe the
polysaccharides employed in studies and a clear description of any
treatment of the materials is also important for many applications.

Use of polysaccharides as control release agents is widespread and
diverse. For simplicity we can consider two broad categories of use of
polysaccharides involved with control release of drugs. These two
divisions are:

(a) as simple encapsulation materials; and

(b) as complexing agents (through chemical and secondary bonding to the drug).

Schuerch has briefly reviewed these approaches.[1] Focusing on the second mode of control release, a number of iron-containing polysaccharide complexes are in clinical application.[2] Combination of oxytocin and carboxymethyl dextran has been reported.[1,2] More recently, our group has employed numerous polysaccharides as carriers of tin-containing moieties (see, for instance, references 3-11). It is the purpose of the present report to detail findings aimed at comparing the activities of some of these materials as a function of polysaccharide or saccharide.

EXPERIMENTAL

Synthetic and source details are described elsewhere (see, for instance, references 3-11). Structure and other physical characterizations are also described in these references.

Bacterial studies were conducted in the usual manner. For instance, plates containing a suitable growth medium, such as tryptic soy agar or Mueller Hinton agar, were seeded with suspensions of the test organism after 24 hours incubation at 37°C. Shortly after the plates were seeded, the test compounds were introduced as solids directly or as solids in an emulsion or in solution employing sterile disks (standard 5mm). The plates were incubated and the inhibition noted. Compounding was accomplished through grinding the products into powders and then adding the powder to the latex paste with mixing and, finally, allowing the mixture to dry to a film.

RESULTS AND DISCUSSION

GENERAL STRUCTURES

General structures for some of the polysaccharides and dextran tin-containing materials are given below in Figures 1-3. As usual, those materials derived from monohalo organostannanes do not have increased crosslinking, whereas crosslinking is increased when dihalo and trihalo organostannanes are employed.

Numerous saccharides have been modified through reaction with orga-nostannane halides. A major reason for desiring such materials is the biological properties that might be exhibited by such materials. Follow-ing is a brief study relating the inhibition of selected modified products, shown in Figure 4, as a function of the nature of the organos-tannane, nature of the saccharide and percentage tin. These results are preliminary and should be considered as such.

CONCENTRATION

Concentration studies are difficult to carry out from products derived from reaction with dihaloorganostannanes since the products are generally crosslinked and insoluble. (Even so, compounding of the materi-

2

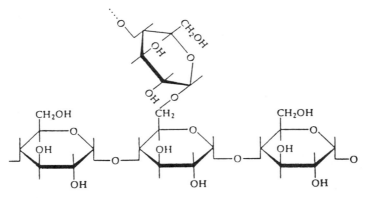

Figure 1. The structure of amylopectin, with details of a branch point.

als into caulks and paints and with powders such as talc has been done and such samples can be used to evaluate concentration affects but such affects are also moderated by factors associated with the compounding material.) The exception to this insolubility is products derived from reaction with xylan. Xylan has only two hydroxyls per ether unit with the alcohols situated such as to encourage the formation of five-membered cyclic rings through reaction with the dihalostannanes. These products are soluble and show the expected concentration responses with inhibition decreasing as concentration decreases.[12]

Concentration studies are also difficult to carry out since the amount of substitution on a specific polysaccharide is generally largely independent of the particular reaction conditions (when employing the interfacial route of synthesis). This is believed to happen because once modification begins on a chain, it continues with steric and other factors being the limiting agent. Even so, some variation was achieved through the use of differing molar ratios of reactants and different reaction systems for dextran. Results appear in Tables 1 and 2. Here, general inhibition appears to be independent of the amount of incorporated tin. This is not in opposition to the solution results where the amount of tin present was varied about 100 fold whereas in Tables 1 and 2 the concentration of tin varied by only about two fold.

Table 1. Results of disc assays for dextran modified through reaction with dibutyltin dichlorides.

	Growth Inhibition			
% Sn	P.aueriginosa	E.coli	S.Aureus	C.albicans
26	0	C	C	S (0.6)
23	0	C	C	S (2.5)
20	0.1	C	C	S (2.0)
17	0	C	C	S (0.8)
14	0	C	C	S (1.0)
Monomer	C	0.1	C	S (0.4)
				D (0.1)

where: C = Cidal, S = Statis

3

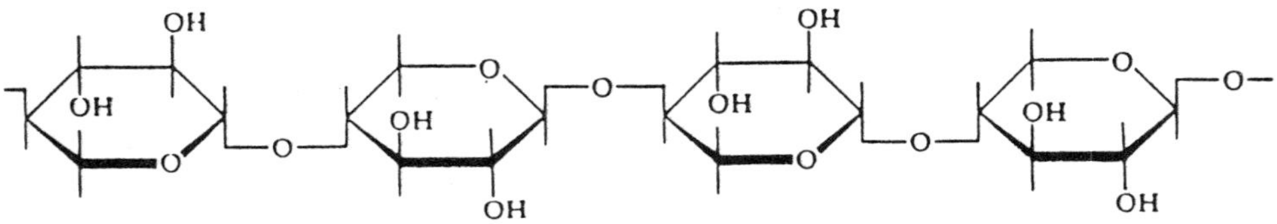

Figure 2. The structure of xylan.

TOXICITY OR STARVATION

Are the organostannane-modified polysaccharides toxic themselves or are the microorganisms simply not able to digest them and die due to starvation? In general, the plates containing the microorganisms contain nutrient agar such that there should be little question with respect to starvation. Even so, studies were carried out using two microorganisms that are known to digest saccharides such as cellulose and dextrose. For one set of experiments, dextrose was added to one set of test tubes containing dibutyltin-modified cellulose. The second set of test tubes was identical except no dextrose was added. The growth decrease for the test tubes containing dibutyltin-modified cellulose and dextrose was similar to that of the test tubes that did not contain dextrose (Table 3). This is consistent with the butyltin-containing cellulose being toxic to the microorganisms and not simply having the modified cellulose material inhibiting or preventing saccharide metabolism.

1. NATURE OF ALKYLSTANNANE

The precise structural-biological property relationships of saccharides and polysaccharides has been an area of active research. In fact, the structural relationships to biological response varies not only with the exact chemical structure and geometry of the molecule but also with the tested species.

Table 2. Results of disc assays for cellulose modified through reaction with dibutyltin dichloride.

	Growth Inhibition		
%Sn	A. flavas	A. niger	A. fumagatus
44	3	3	2
41	4	4	3
37	4	4	3
Monomer	4	4	4

where: 4 = 100% Inhibition, 3 = 75% Inhibition, etc.

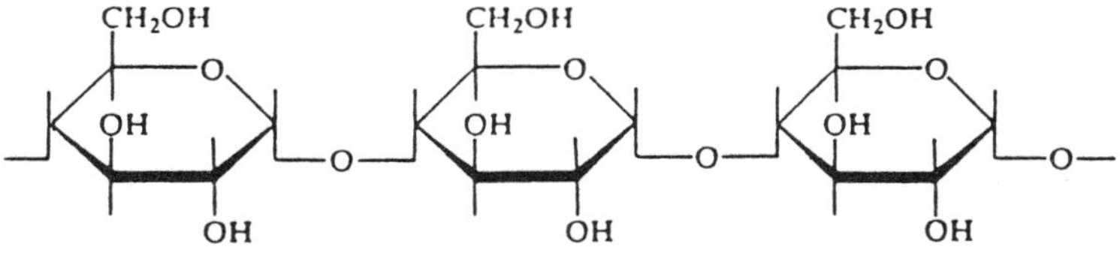

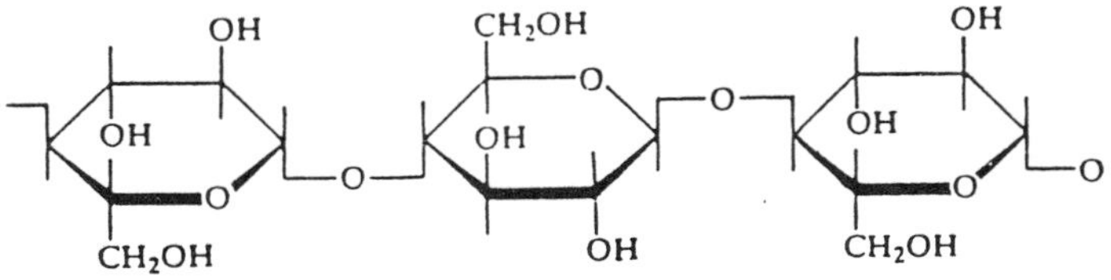

Figure 3. The structure of linear amylose and cellulose.

As noted above, polysaccharides are known to exhibit varying biological activities. Some are known to activate disease defense mechanisms such as interferon and immune systems. Some influence allergic responses and affect cell membrane permeability.

The most often cited toxicity relationships are that trialkyl>dialkyl>monoalkyl and methyl>ethyl>propyl>butyl>>>octyl. While this is the general trend with respect to mammals, it may be different depending on the particular organism tested. For instance, for fungi and bacteria, the order is butyl=propyl>ethyl>methyl (dependence of the biological activity of tri-n-alkyltin acetates;).[13] For a wide range of synthetic and natural modified polymers, the general toxicity order has been found to decrease with increase in alkyl chain length (see, for instance, Tables 4 and 5). While this is at odds with some of the reported toxicity trends, it must be remembered that the precise toxicity is dependent on such factors as hydrolysis rates and ease of going through a

Table 3. Growth on cellulose modified through reaction with dibutyltin dichloride.

Sample (% Dibutyltin Incorporation)	Growth of *T. reesei* (mg protein/ml)		Growth of *C. globosum* (mg protein/ml)	
	Dextrose	No Dextrose	Dextrose	No Dextrose
Cellulose	420	250	300	240
(44)	30	60	60	40
(41)	30	60	60	50
(37)	40	20	20	40

Portions taken from reference 9.

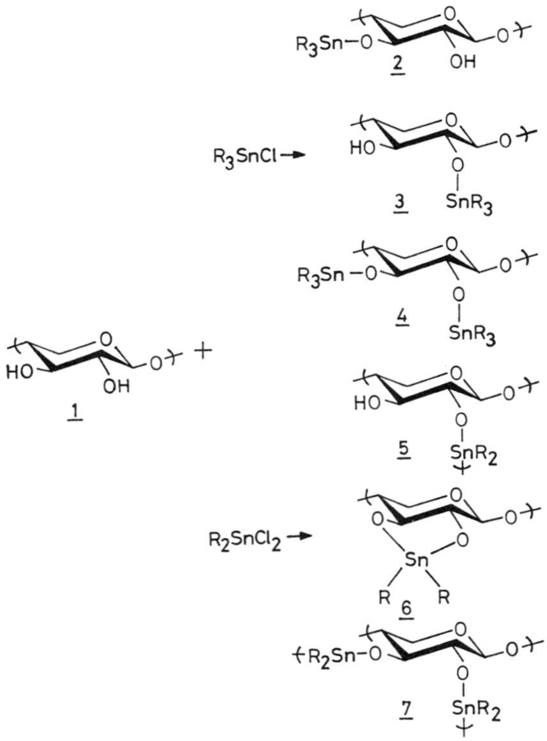

Figure 4. The structures of the organostannane-saccharide derivatives in this study.

membrane. Thus, while the overall trend with respect to the tested gram
negative and gram positive bacteria, fungi and other microorganisms is
consistent with that found for mammals and not that reported in some
studies with the tested microorganisms, the variabilities are such that
differing trends may be appropriate for these polymer-bound materials.
Further, while the trends shown in Tables 4 and 5 show an increase in
toxicity with decrease in alkyl chain length, other similarly bound
products may show other structural relationships with respect to
toxicity.

Table 4. Results as a function of alkylstannane.

Organostannane (%-Sn)	Polysaccharide	Inhibition Ccm			
		P. aueriginosa	E.Coli	S.aureus	C.albicans
Bu$_2$Sn(26)	Dextran	B	C	C	5(0.6)
Me$_2$Sn(12)	Dextran	0.1	0.3	0.4	0.6
Ly$_2$Sn(15)	Dextran	0	C	0	0
Bu$_2$Sn(41)	Cellulose	C	C	0	0
Me$_2$Sn	Cellulose	0.2	0.4	0.5	C

Table 5. Results as a function of alkylstannane for cellulose-
derived products.

Organostannane (%-Sn)	Inhibition (%)		
	A.flavus	A.Niger	A.fumigatus
Pr$_2$Sn(21)	100	100	75
Bu$_2$Sn(37)	100	100	75
Oc$_2$Sn(29)	50	0	0

SACCHARIDE/POLYSACCHARIDE

As noted before it is well established that the nature of the sac-
charide/polysaccharide can have a marked affect on their biological
properties. This is especially true for internal applications. This is
well established for simple external applications, particularly for
polysaccharides that are chemically modified. The data given in Table 6,
while not conclusive, indicates that the nature of the polysaccharide
does have an affect on the biological activity of the modified product.
For instance, for the products from dibutyltin, the dextran-modified
material inhibits *S. aureus* but does not inhibit *P. aueriginosa* whereas
the analogous product except derived from cellulose inhibits *P. aueriginosa*
nosa but does not inhibit *S. aureus*. Even so, more detailed studies need
to be done to better describe this dependency.

Table 6. Comparative inhibition of tin-containing materials.

Organostannane (%Sn)	Saccharide	Inhibition (cm)			
		P.aueriginosa	E.coli	S.aureus	C.albicans
Bu$_2$Sn(26)	Dextran	0	C	C	S (0.6)
Bu$_2$Sn(21)	Cellulose	C	C	0	0
Bu$_2$SnCl$_2$	Itself	C	0.1	C	S (0.4)
Ly$_2$Sn(15)	Dextran	0	C	0	0
Ly$_2$Sn(32)	Sucrose	0	0	0	0
Me$_2$Sn(12)	Dextran	0.1	0.3	0.4	0.6
Me$_2$Sn	Cellulose	0.2	0.4	0.5	C
Me$_2$Sn	Sucrose	0.4	0.3	0.9	0.1
Ph$_2$Sn(34)	Sucrose	0	0	0	0
Ph$_2$Sn(12)	Dextran	0	0	0	0
Ph$_2$SnCl$_2$	Itself	0.2	0.2	0.1	0.1
Ph$_3$Sn	Cellulose	0	0	0	S (1.5)
Ph$_3$Sn	Dextran	0	0	0	S (1.0)

REFERENCES

1. C. Schuerch, in: *"Bioactive Polymeric Systems,"* C. Gebelein and C. Carraher, Eds., Plenum, NY, 1985, Chapter 14.
2. *"The Pharmacological Base of Therapeutics,"* L. S. Goodman and A. Gilman, Eds., Macmillian, NY, 1980.
3. C. Carraher, Y. Naoshima, C. Butler, V. Foster, D. Gill, M. Williams, D. Giron and P. Mykytiuk, PMSE, **57**, 186 (1987).
4. C. Carraher, J. Schroeder, C. McNeely, D. Giron and J. Workman, Organic Coatings and Plastics Chemistry, **40**, 560 (1979).
5. Y. Naoshima, H. Shudo, M. Uenishi and C. Carraher, PMSE, **58**, 553 (1988).
6. C. Carraher, T. Gehrke, D. Giron, D. R. Cerutis and H. M. Molloy, J. Macrol. Sci-Chem., **A19**, 1121 (1983).
7. C. Carraher, D. Giron, J. Schroeder and C. McNeely, U. S. Patent 4,312,981 (Jan. 1982).
8. C. Carraher, P. Mykytiuk, H. Blaxall, D. R. Cerutis, R. Linville, D. Giron, T. Tiernan and S. Coldiron, Organic Coatings and Plastics Chemistry, **45**, 564 (1981).
9. C. Carraher, W. R. Burt, D. Giron, J. Schroeder, M. L. Taylor, H. M. Molloy and T. Tierman, J. Appl. Polym. Sci., **28**, 1919 (1983).
10. Y. Naoshima, H. Shudo, M. Uenishi and C. Carraher, J. Poly,. Mater., **8**, 51 (1991).
11. C. Carraher and C. Butler, U. S. Patent 5,043,463 (Aug. 1991).
12. Y. Naoshima, H. Shudo, M. Uenishi and C. Carraher, J. Polymeric Materials, **8**, 51 (1991).
13. M. J. Selwyn, in: *"Chemistry of Tin,"* P. Harrison, Ed., Chapman and Hall, NY, 1989.

ACETAN - A NEW BACTERIAL POLYSACCHARIDE

Victor J. Morris

AFRC Institute of Food Research
Norwich Laboratory, Norwich Research Park
Colney, Norwich, NR4 7UA, UK

The bacterial exopolysaccharide xanthan gum forms thixotropic aqueous dispersions. Xanthan will form thermoreversible gels when mixed with certain plant galactomannans or plant glucomannans. Such behavior is fairly well understood in terms of the chemical and stereochemical structure of xanthan. It would be useful to employ such information in order to predict the structural requirements which would be needed if a given polysaccharide was required to mimic the functional behavior of xanthan. In addition, one might wish to suggest structural modifications to such a polysaccharide which might be used to optimize or alter its functional behavior. The practicality of such an approach will be assessed by discussing the behavior of the bacterial polysaccharide acetan, which has a chemical structure similar to xanthan, and the behavior of 'modified acetans.' Various methods which have been used to modify the acetan structure will be discussed.

INTRODUCTION

New useful 'industrial polysaccharides' are normally discovered by the empirical screening of a range of extracellular polysaccharides produced by a variety of newly isolated microorganisms. If one completely understood the interrelationship between the chemical structure, the stereochemistry, and the functional properties of polysaccharides, then it would be possible to predict the structural features required in order to produce a given functionality. Further, one could suggest systematic modifications to given structures required in order to optimize a given functionality, or to produce a new type of functional behavior.

The feasibility of such an approach to polysaccharide design will be discussed in the light of experimental studies on the commercially available bacterial polysaccharide xanthan gum, the new bacterial polysaccharide acetan, and modified variants of the acetan structure.

Biotechnology and Bioactive Polymers, Edited by C. Gebelein
and C. Carraher, Plenum Press, New York, 1994

9

XANTHAN

Xanthan is the extracellular polysaccharide secreted by the bacterium *Xanthomonas campestris*. Aqueous dispersions of xanthan gum are thixotropic and xanthan is used commercially as a thickening and suspending agent. Such dispersions are 'weak gels.'[1] At low shear rates the dispersions show elastic behavior. However, at increased shear rate the structure breaks down and the system flows. This process is reversible and the structure reforms when the shear rate is reduced. When xanthan is mixed with certain glucomannans, or with certain galactomannans, then the resultant binary mixtures form permanent thermo-reversible gels.[2,3] These binary mixtures are used as commercial gelling agents. Such mixtures are true gels: upon deformation they remain elastic until fracture occurs.

The chemical structure[4,5] of xanthan (Figure 1a) consists of a cellulosic backbone (ß(1→4) linked D-glucose) regularly substituted on alternate glucose residues with trisaccharide sidechains. The attachment of this side-arm influences the stable solid-state conformation of the polysaccharide. Instead of adopting the normal two-fold ribbon-like conformation characteristic of cellulose, the xanthan backbone folds into a five-fold helical structure.[6,7] A wide variety of physico-chemical methods have been used to demonstrate that in solution xanthan exhibits a thermally-reversible order-disorder transition.[8] This order-disorder transition has been identified with a helix-coil transition.[8] The helix to coil transition can be induced by raising the temperature or by lowering the ionic strength of the solution.[8] The high viscosity and thixotropy of xanthan dispersions is observed under conditions which favor formation of the xanthan helix. The 'weak gel' network appears to be stabilized by adoption of the xanthan helix.[1]

ACETAN

Acetan is an anionic heteropolysaccharide produced extracellularly by the bacterium *Acetobacter xylinum* strain NRRL B42.[9] The partial chemical structure of acetan is shown in Figure 1b.[9] The polymer consists of a cellulosic backbone substituted upon alternate glucose residues with pentasaccharide sidechains. Both xanthan and acetan share a common backbone. The sidechains in both polymers are linked to alternate glucose residues. The nature of the sidechain-backbone linkage is common. Indeed the carbohydrate sequences for the first two sugars in the sidechains are identical for both polysaccharides. Whereas xanthan is partially substituted with both acetate and pyruvate, acetan contains only acetate as a substituent.[9]

The nature of the sidechain-backbone linkage of acetan suggests that acetan should adopt the five-fold helical conformation characteristic of xanthan.[10] Analysis of the X-ray diffraction patterns obtained from oriented fibers of acetan confirm that the solid-state conformation of acetan is a five-fold helix with a pitch (4.8 nm) almost indistinguishable from that of xanthan (4.7 nm).[9,11,12] Chiroptical studies on acetan solutions show clear evidence for an order-disorder transition (Figure 2).[11] The disordered conformation can be induced by raising the temperature or by lowering the ionic strength of the solution.[11] By analogy with the behavior of xanthan solutions it has been proposed that the order-disorder transition corresponds to a helix-coil transition. Under conditions favoring adoption of this ordered (helical) conformation aqueous dispersions of acetan exhibit viscosities and shear thinning behavior comparable to xanthan samples.[11] In general the mechanical spectra of

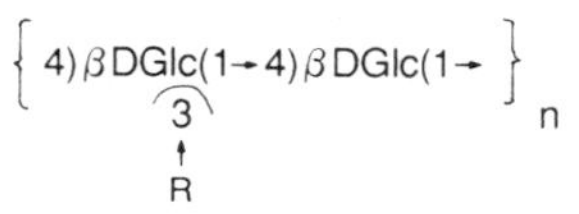

$$\left\{ 4)\beta\,\mathrm{DGlc}(1\rightarrow 4)\beta\,\mathrm{DGlc}(1\rightarrow \atop \underset{\uparrow}{3} \right\}_n$$
$$R$$

(a) XANTHAN

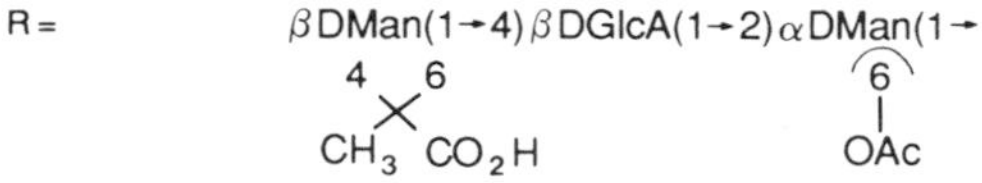

$$R = \quad \beta\,\mathrm{DMan}(1\rightarrow 4)\beta\,\mathrm{DGlcA}(1\rightarrow 2)\alpha\,\mathrm{DMan}(1\rightarrow$$

$$\underset{\mathrm{CH_3}\quad \mathrm{CO_2H}}{4 \diagdown \!\!\!\!\times\!\!\!\! \diagup 6} \qquad\qquad \underset{\mathrm{OAc}}{\overset{6}{|}}$$

(b) ACETAN

$$R = \mathrm{LRha}(1\rightarrow 6)\beta\,\mathrm{DGlc}(1\rightarrow 6)\alpha\,\mathrm{DGlc}(1\rightarrow 4)\beta\,\mathrm{DGlcA}(1\rightarrow 2)\alpha\,\mathrm{DMan}(1\rightarrow$$

+ 1-2 OAc per repeat unit

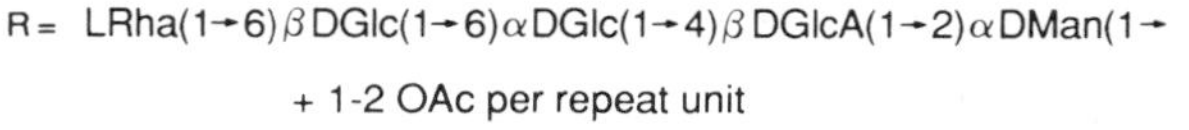

Figure 1. Chemical structures of the bacterial polysaccharides (a) xanthan, and (b) acetan. Note that acetyl and pyruvate substitution of xanthan is generally incomplete. Acetan contains no pyruvate. Some acetyl substitution occurs 6-linked to the α-D mannosyl residue. There appears to be another site of acetylation as yet undetermined.

acetan samples resembles that observed for entangled polymer solutions although, at the highest concentrations, evidence for 'weak gel' behavior can be observed (Table 1).[11] Heating acetan samples to temperatures above the order-disorder transition temperature results in a marked drop in viscosity and the loss of the dramatic shear thinning behavior.

SYNERGISTIC BINARY GELS

Xanthan forms thermo-reversible transparent gels when mixed with certain glucomannans, such as konjac mannan, or when mixed with certain galactomannans, such as tara gum or carob (locust bean) gum.[2,4] It has been proposed for some time that the junction zones of such gels were formed by intermolecular binding between xanthan and the other component of the binary gel.[2,3] Indeed the fact that, under the conditions for which the binary system gels, the isolated individual components do not gel, is consistent with the idea of intermolecular binding. The original model proposed an intermolecular binding of the glucomannan, or of an unsubstituted face of the galactomannan backbone, to the xanthan helix.[2,3]

Recently, X-ray diffraction studies of oriented fibers prepared from binary gels have provided direct evidence for intermolecular binding between the two polysaccharides.[13,17] Mixing experiments suggested that a prerequisite for gelation was a prior denaturation of the xanthan helix.[13,17] This observation, coupled with a qualitative analysis of the X-ray data obtained from oriented gel fibers, suggests an alternative model for intermolecular binding. Rather than binding to the xanthan helix, it is proposed that binding occurs between the denatured (cellulosic) xan-than backbone and the stereochemically similar backbones of the galacto-mannans or glucomannans.[13,15] The galactomannans consist of ß$(1\rightarrow 4)$ linked D-mannose backbones irregularity substituted with 6-linked α-D-galactose. The glucomannans are linear polymers containing

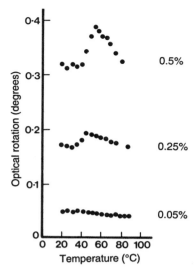

Figure 2. Optical rotation studies on acetan. Measurements made at
406 nm in a 10 cm cell for aqueous solutions of 0.5%,
0.25% and 0.5% acetan. The data show an 'order-disorder'
transition on heating.

both β(1→4) D-glucose and β(1→4) D-mannose in their backbones. D-glucose and D-mannose differ only in the orientation of the hydroxyl group at the 2-position in the ring. Thus, isomorphous replacement of D-glucose with D-mannose in a β(1→4) linked structure leads to a very similar two-fold ribbon-like structure.

It has been proposed that the junction zones in xanthan-galactomannan mixed gels may involve cocrystallization of the two poly-

Table 1. Mechanical spectrum for an aqueous 3% acetan sample.
Strain 0.25. G^1 and G^{11} are the storage and loss moduli
respectively.

Frequency (Hz)	G^1(Pa)	G^{11}(Pa)
0.10	15.2	9.93
0.14	17.3	10.7
0.19	19.5	11.6
0.27	21.9	12.6
0.37	24.5	13.6
0.52	27.2	14.6
0.72	30.2	15.8
1.00	33.4	16.9
1.39	36.8	18.3
1.93	40.4	19.6
2.68	44.5	21.1
3.73	48.7	22.7
5.18	53.6	24.4
7.20	58.9	26.4
10.0	65.1	28.6

saccharides.[16,17] In order to accommodate the longer sidechains of the
xanthan there would need to be appropriate vacancies in the galaetomannan
crystallites. Such vacancies could be provided by unsubstituted regions
of the galactomannan chain. On such a model the length of the xanthan
sidechain would dictate the galactose-mannose ratio of the galactomannan
for which gelation might occur. Similarly, increasing the length of the
sidechain might be expected to inhibit gelation with galactomannans such
as tara gum or carob gum. The previous models for gelation would suggest
that, provided the length of the sidechain did not alter the helical
structure of the polysaccharide, it should not inhibit gelation.

Acetan might be considered as a xanthan molecule with an extended
sidechain. Preliminary studies of acetan-tara and acetan-carob mixtures
suggest that the longer sidechain of acetan hinders and may completely
inhibit gelation. This appears to favor a model for gelation involving
intermolecular binding with the denatured xanthan backbone rather than
the xanthan helix. Preliminary studies on acetan-konjac mannan mixtures
have also, as yet, failed to reveal evidence of gelation.

Further studies require the preparation of a family of acetan-like
polysaccharide structures based on the acetan structure but containing
sidechains of reduced lengths.

MODIFIED ACETANS

In order to investigate the effect of the sidechain on the gelation
of acetan, and to modify the acetan in order to induce gelation, attempts
have been made to produce a family of acetan based polysaccharides containing reduced sidechains.

The most obvious approach is to modify the acetan structure chemically . The first stage would be to remove the terminal rhamnose residue
by selective acid hydrolysis. Acetan solutions were treated with sulfuric
acid of varying molarity. The effects of time and temperature were assessed by isolating the remaining polysaccharide and analyzing for the
presence of rhamnose. The optimal conditions for acid hydrolysis were
identified. Analysis of the sugar composition of the resultant polysaccharide confirmed the removal of terminal rhamnose. However, methylation
analysis of the polysaccharide revealed the presence of both terminal
glucose and terminal glucutonic acid. These studies suggested that the
acid hydrolysis was not specific and resulted in a mixture of
polysaccharide structures.

Enzymic hydrolysis of acetan should be more specific. Enzymes are
also available and have been used to cleave the glycosidic linkages
present in the acetan sidechain.[9] Preliminary studies on the removal of
terminal rhamnose have revealed that, although such hydrolysis is specific, it has not proved possible to achieve 100% conversion. Thus sequential enzymic hydrolysis would also produce a mixture of polysaccharide
structures.

An alternative strategy is to mutate the bacterium A. xylinum in
order to produce novel strains which synthesize novel variants of the
acetan structure. To this end A. xylinum has been subjected to chemical
mutagenesis using the chemical mutagen N-methyl-N[1]-nitro-nitrosoguani-
dine. The bacterium A. xylinum produces at least three extracellular
polysaccharides: cellulose,[18] a previously undetected branched neutral
mannan, and the anionic hetero-polysaccharide acetan.[9] Initially it was
decided to isolate stable cellulose minus mutants. Using chemical muta-

genesis it was possible to isolate several cellulose-minus, acetan-positive stable mutants. One of these stable cellulose-minus mutants, A. xylinum CR1, was subjected to further chemical mutagenesis. Inspection of the morphology of the colonies produced by surviving bacterial strains permitted the classification of the bacterial strains into two types: those showing normal levels of extracellular polysaccharide production and those showing reduced exopolysaccharide production. Sugar analysis of the exopolysaccharide extracts revealed that for those bacterial strains showing reduced exopolysaccharide production the exopolysaccharide extract did not contain rhamnose. Rhamnose was found to be present in the exopolysaccharide extracts from bacterial strains with normal colony morphology. On this basis reduced exopolysaccharide production was used as a potential selection method for isolating possible variants of the acetan structure. A number of such mutants have been isolated. At present only one of these mutants has been analyzed in detail.

The bacterial mutant A. xylinum CR1/4 has been selected for further study. Attempts to produce bulk (gram) quantities of exopolysaccharide for chemical and physical studies revealed that this mutant is unstable and eventually reverts to acetan production. However, by carefully monitoring the reversion it has been possible to isolate large quantities of the exopolysaccharide designated CR1/4.

The exopolysaccharide CR1/4 was isolated and purified by the following procedure. The bacterial broths were clarified by centrifugation to remove bacterial cells. Then alcohol precipitation was used to isolate a crude exopolysaccharide extract. Acidic and neutral exopolysaccharides were separated by CTAB precipitation. The neutral and acidic fractions were examined by methylation analysis. The analysis of the neutral fraction was found to be consistent with the presence of the previously identified neutral mannan.

The results of the methylation analysis on CR1/4 are shown in Table 2. Methylation of the unreduced polysaccharide revealed the presence of (1→4) glucose, (1→4,6) glucose, and (1→2) mannose. Analysis of the carboxy-reduced sample revealed, in addition, the presence of terminal glucuvonic acid. This analysis is consistent with a variant of the acetan structure in which the terminal rhamnose and both (1→6) glucose residues have been removed. Such a structure could also be obtained by removing the terminal mannose residue from the xanthan structure and has been called the xanthan tetramer. Further carbohydrate analysis is in progress to completely define the CR1/4 structure.

Preliminary physico-chemical and rheological studies are being conducted on the behavior of aqueous CR1/4 preparations. X-ray diffraction studies of oriented fibers prepared from oriented tetramer films

Table 2. Methylation analysis for CR1/4 before and after carboxy reduction. t-terminal, GLC glucose, Man-mannose and GlcA-glucuronic acid.

Linkage	Sample	
	CR1/4	CR1/4 carboxy-reduced
t-GlcA	–	19
(1→2) Man	33	27
(1→4) Glc	34	27
(1→3,4) Glc	33	27

Table 3. Mechanical spectrum for an aqueous 0.75% CR1/4 sample. Strain 0.25. G^1 and G^{11} are the storage and loss moduli respectively.

Frequency (Hz)	G^1(Pa)	G^{11}(Pa)
0.05	0.062	0.21
0.06	0.029	0.26
0.08	0.098	0.29
0.11	0.141	0.32
0.15	0.230	0.47
0.21	0.247	0.51
0.28	0.358	0.65
0.38	0.464	0.74
0.51	0.622	0.90
0.68	0.832	1.07
0.92	1.1	1.20
1.24	1.23	1.41
1.67	1.43	1.59
2.26	1.80	1.71
3.04	1.94	1.95
4.09	2.62	2.12
5.31	2.96	2.56
7.42	3.48	2.87
10.00	4.28	2.72

Table 4. Comparative shear thinning behavior of 0.5% aqueous samples of acetan and CR1/4. Measured using a cone and plate assembly (cone angle 2.4 deg).

0.5% Acetan		0.5% CR1/4	
Viscosity (cP)	Shear rate (s^{-1})	Viscosity (cP)	Shear rate (s^{-1})
27.5	2.50	333.0	0.063
26.1	3.15	210.0	0.10
26.0	3.96	56.8	0.16
27.8	5.00	77.6	0.25
26.4	6.28	175.0	0.40
26.2	7.91	159.0	0.63
24.6	9.95	150.0	1.00
26.4	12.5	140.0	1.58
26.0	15.8	128.0	2.50
25.2	19.9	113.0	3.96
24.2	25.0	96.9	6.28
23.2	31.5	81.6	9.95
22.2	39.6	68.5	15.8
21.1	49.9	55.4	25.0
19.5	62.8	44.2	39.6
18.5	79.1	34.4	62.8
17.2	99.5	26.9	99.5
15.8	125.0	20.6	158.0
14.7	158.0	15.9	250.0
13.7	198.0	12.1	396.0
12.7	250.0	9.46	628.0

have shown that the xanthan tetramer adopts the five-fold xanthan helical conformation.[10] Table 3 shows the mechanical spectra for an aqueous 1% sample of CR1/4. It can be seen that, like acetan and xanthan, CR1/4 yields aqueous preparations which exhibit high viscosity and reversible shear thinning behavior (Table 4). As expected, at a given polymer concentration, CR1/4 yields a higher viscosity yhan that of acetan under comparable conditions. We are presently conducting more detailed comparative studies of xanthan, acetan and CR1/4, and also investigating the possible gelation of CR1/4 mixtures with galactomannans and glucomannans.

ACKNOWLEDGEMENTS

The author is grateful to Jane Harris, Geoffrey Brownsey, Mike Ridout, Patrick Gunning and Caroline MacCormick for providing experimental data prior to publication.

REFERENCES

1. S. B. Ross-Murphy, V. J. Morris & E. R. Morris, Faraday Symp. Chem. Soc., **18**, 7 (1983).
2. I. C. M. Dea & A. Morrison, Adv. Carbohydr. Chem. Biochem., **31**, 241 (1975).
3. I. C. M. Dea & E. R. Morris, in: "*Extracellular Microbial Polysaccharides*," P. A. Sandford & A. Laskin, Eds., ACS Symp. Ser., **45**, 174 (1977).
4. P. E. Jansson, L. Kenne & B. Lindberg, Carbohydr. Res., **45**, 275 (1975).
5. L. D. Melton, L. Mindt, D. A. Rees & G. R. Sanderson, Carbohydr. Res., **46**, 245 (1976).
6. R. Moorhouse, M. D. Walkinshaw & S. Arnott, in: "*Extracellular Microbial Polysaccharide*," P. A. Sandford & A Laskin, Eds., ACS Symp. Ser., **45**, 90 (1977).
7. K. Okuyama, S. Arnott, R. Moorhouse, M. D. Walkinshaw, E. D. T. Atkins & C. H. Wolf-Ullish, in: "*Fiber Diffraction Methods*," D. A. French & K. H. Gardener, Eds., ACS Symp. Ser., **141**, 411 (1980).
8. E. R. Morris, D. A. Rees, G. Young, M. D. Walkinshaw & D. Darke, J. Mol. Biol., **110**, 1 (1977).
9. R. O. Couso, L. Iephi & A. Dankert, J. Gen. Microbial., **133**, 2123 (1987).
10. R. P. Millane & T. V. Narasaiah, Carbohydr. Polym., **12**, 315 (1990).
11. V. J. Morris, G. J. Brownsey, P. Cairns, G. R. Chilvers & M. J. Miles, Int. J. Biol. Macromolecules, **11**, 326 (1989).
12. M. J. Miles & V. J. Morris, in: "*Mucus and Related Topics*," E. Chantler & N. A. Ratcliffe, Eds., Company of Biologists Ltd, Cambridge, 1989, p. 403.
13. V. J. Morris, G. J. Brownsey, P. Cairns & M. J. Miles , Carbohydr. Res., **160**, 411 (1988).
14. P. Cairns, M. J. Miles, & V. J. Morris, Nature, **322**, 89 (1986).
15. G. J. Brownsey, P. Cairns, M. J. Miles & V. J. Morris, Carbohydr. Res., **176**, 329 (1988).
16. V. J. Morris, Food Biotechnology, **4**, 45 (1990).
17. V. J. Morris, in: "*Food Polymers, Gels and Colloids*," E. Dickinson, Ed., Royal Society of Chemistry, London, 1991.
18. L. Glaser, J. Biol. Chem., **232**, 627 (1958).

COAGULATION PROPERTIES OF CHEMICALLY MODIFIED HEPARINS

A.W. Stemberger, F. Bader, S. Haas, J.M. Walenga[1], and
G. Blümel

Institut für Experimentelle Chirurgie der Technischen
Universität München
Ismaninger Straße 22
8000 München 80
and
(1) Department of Pathology
    Loyola University
    Maywood, IL 60153

Divalent carbonic acid halides have been used to fix
heparin onto biomaterials of biological origin, however, the
effect of this process on the coagulation profile induced by
heparin has not yet been clearly determined. To investigate
this, commercially available unfractionated and low molecular
weight heparins were treated with the corresponding halides
of adipic and glutaric acid. Clotting tests and chromogenic
substrate assays were used to characterize the influence of
these derivatives on the clotting mechanism. In a thrombosis
model in rats, the anticoagulant and antithrombotic effects
of these substances were also analyzed. The *in vitro* data
indicate a loss of the anticoagulant profile. Heparin-
catalyzed acceleration of thrombin and factor Xa-inhibition
by AT III were diminished following derivatization, although
the heparin cofactor II activity was unaltered. However, in
an animal model the antithrombotic properties of these heparin derivatives were reduced but still significant. These
experiments were repeated with acetylchloride derivatization
revealing comparable results.

INTRODUCTION

In the literature, numerous methods have been described to prepare
non-coagulant surfaces by binding heparin onto biomaterials. Divalent
carbonic acids have been used to fix heparin onto biomaterials of biological origin. But the effect of these processes on the coagulation profile
induced by heparin has not yet been clearly determined. The treatment of
commercially available characterized heparins and the so-called low
molecular weight preparations was performed with the corresponding halides of dicarbonic acids. These chemically modified heparins were char-

acterized by various clotting assays *in vitro* and following administration in an animal model.

EXPERIMENTAL

Unfractionated and low molecular weight heparins of various pharmaceutical firms (Thrombophob[R] <T>, Fragmin[R] <F>, Fraxiparin[R] <Fr>) were dried, suspended in tetrahydrofuran (THF) and treated with the halides of glutaric and adipic acid as well as acetylchloride in a concentration of 0.01-1% (weight halides/weight THF) for 12 hours at room temperature.[6] The reagents were removed by repeated washing with THF and dried under vacuum. The same experiments were performed with sulfonyl- and thionyl chloride in THF, and hydrochloric acid/methanol preparations in THF. The molecular weight profile of heparin and its derivatized analogues was characterized by HPLC (Waters Chromatography Division, using TSK 3000 SW column). The effect on coagulation was investigated by clotting tests (APTT, Heptest[R]) using human plasma. Afterwards these coagulation assays were repeated after the addition of specific antibodies towards antithrombin III (gift of Behringwerke) and in addition, substrate assays (anti-Xa and anti-IIa activity, heparin cofactor II assay) were performed.

Derivatized heparins (except thionyl-, sulfonyl- and methanolic HCL-treated) were administered into rats using a thrombosis model according to Harbauer.[2] Isotonic saline served as control. Then the thrombus formation was investigated by the "Doppler"-technique and visual inspection. Heparin and heparin related effects were analyzed in blood samples.

RESULTS/DISCUSSION

The chemical modification of unfractionated and low molecular weight heparins with halides of the dicarbonic acids and acetic acid strongly reduced the anticoagulant action of heparin as demonstrated by *in vitro* clotting assays (APTT and Heptest[R]). Even at concentration rates of over 2 units per ml plasma (approximately 14 μg per mL plasma), these modified heparins showed only a minor prolongation of the clotting time as measured by APTT and Heptest[R] (Figure 1). Heparin-catalyzed acceleration of antithrombin III (AT III) induced inactivation of the serpins, thrombin (factor IIa) and factor Xa, was also reduced. Data for acetylchloride modified preparations are shown in Figures 2 and 3.

The influence of AT III on the APTT-assay in the presence of heparin and modified preparations was studied with and without antibodies towards this main coagulation inhibitor. As depicted in Figure 4, the prolongation of the clotting time induced by heparin was suppressed by the addition of anti-AT III antibodies. These results were comparable to the clotting time experiments where modified heparins were added to plasma without the addition of these antibodies. Thionyl- and sulfuryl chloride modifications revealed similar results.

The incubation of heparins with hydrochloric acid in methanol – corresponding to the theoretical concentration when the halides are hydrolytically cleaved – did not alter the pharmacological profile of heparin.[4] Modified heparin also did not alter the heparin cofactor II-induced inactivation of thrombin (Figure 5). Heparin shares this distinct pharmacological effect – the heparin cofactor II-induced inactivation of

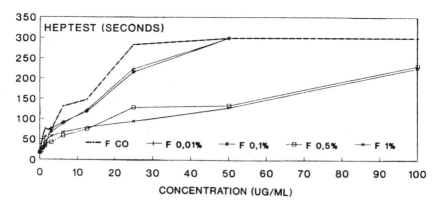

Figure 1. Derivatization of a low molecular weight heparin - dose
response effect following acetylchloride treatment (%).
Data analyzed by the Heptest. Control (CO) was done with
F, the native starting material.

thrombin - with a variety of other glycosaminoglycans and synthetic
heparinoids.

An analysis of the molecular weight profile of heparins treated with
the divalent halides of dicarbonic acids was performed by HPLC procedures
with respect to the corresponding acetylchloride and the untreated heparins (Figures 6A, 6B). No oligomers were demonstrated which means that
divalent halides of dicarbonic acids do not crosslink heparins.

These modified heparins were administered into rats in concentrations up to 280 units (i.e., according to label of the manufacturer for
unmodified heparins) per kg body weight. No plasma anticoagulation was
noted. Nevertheless, a statistically significant reduction of thrombus
formation could be proved. This result is still under investigation to
elucidate the underlying mechanism. *In vivo*, heparin and glycosaminoglycans induce the so-called lipoprotein-associated coagulation inhibitor

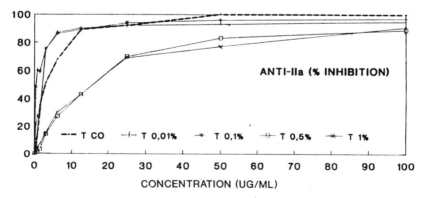

Figure 2. Derivatization of unfractionated heparin - dose response
effect following acetylchloride treatment (%). Data
analyzed by the anti IIa-activity. Control (CO) was done
with T, the native starting material.

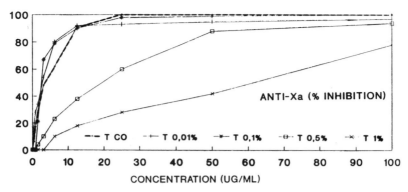

Figure 3. Derivatization of unfractionated heparin - dose response
effect following acetylchloride treatment (%). Data
analyzed by the anti Xa-activity. Control (CO) was done
with T, the native starting material.

(LACI) directed towards factor Xa but no LACI-related inactivation of
factor Xa was detectable in the presence of modified heparins.

SUMMARY

The treatment of heparin with halides of dicarbonic acids as glutaric and adipic acid results in a loss of the anticoagulant activity. The
AT III-induced accelerated inhibition of active coagulation factors
normally induced by heparin is strongly reduced following derivatization,
whereas the catalytic reaction towards heparin cofactor is unaltered.
Administration in an animal model revealed antithrombotic effects. Studies are in progress to characterize these chemical modifications on the
structure of heparin, elucidate the mode of action *in vivo*, in particular

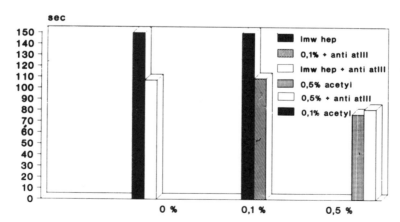

Figure 4. Dependence of the heparin-induced coagulation inactivation via AT III and the dose response effect of derivatization with acetylchloride (%).

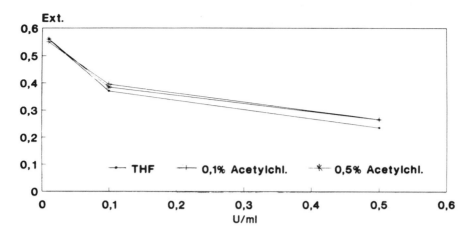

Figure 5. Heparin cofactor II activity of a low molecular weight
heparin and the corresponding acetyl chloride
derivatives (0.1% and 0.5%). Control experiments with F,
the starting material in THF only.

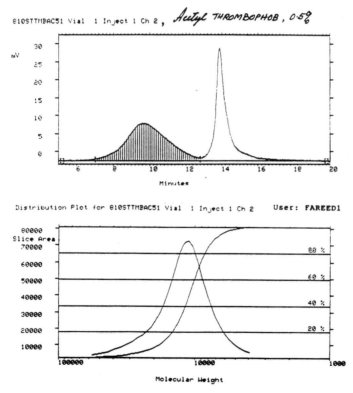

Figure 6. Molecular weight profile of an unfractionated heparin
analyzed by HPLC. Comparison of (A) (above) acetyl
chloride and (B) (opposite page) adipic dichloride
derivatization. (continued)

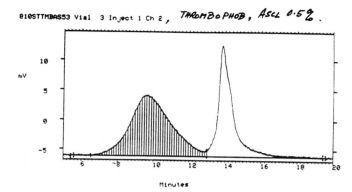

810STTMBAS53 Vial 3 Inject 1 Ch 2, *THROMBOPHOB, ASCL 0.5%*.

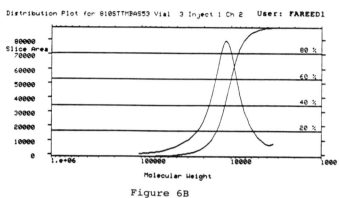

Distribution Plot for 810STTMBAS53 Vial 3 Inject 1 Ch 2   User: FAREED1

Figure 6B

their use as potential antithrombotic drugs. As stated in the literature, glycosaminoglycans and heparin derivatives depleted of anticoagulant activity by methods different from our modification procedures possess certain pharmacological activities such as antimetastatic and antiinflammatory effects, inhibition of angiogenesis and hypersensitivity reactions.[1,3,5,7] Therefore we intend to study the derivatives described herein for those activities.

ACKNOWLEDGEMENTS

The authors would like to thank Mrs. Annette Amper for the preparation of this manuscript.

REFERENCES

1. M. Bar-Ner, A. Eldor, L. Wasserman, Y. Matzner, J. R. Cohen, Z. Fuks and J. Vlodavsky, Blood, 70, 551-557 (1987). "Inhibition of Heparinase - Medicated Degradation of Extracellular Matrix Heparansulfate by Non-Anticoagulated Heparin Species."

2. G. Harbauer, W. Hiller, U. J. Uhl, and E. Wenzel, in:
   "Standardization of Animal Models of Thrombosis," K. Breddin and R.
   Zimmermann, Eds, Schattauer, Stuttgart, 1985, pp. 115-119. "A Venous
   Thrombosis Model in the Rabbit."
3. T. Irimura, M. Nakajima, and G. L. Nicolson, Biochem., **25**, 5322-5328
   (1986). "Chemically Modified Heparins as Inhibitors of Heparan
   Sulfate Specific Endo-ß-Glucuronidase (Heparinase) of Metastatic
   Melanoma Cells."
4. S. W. Levy and L. B. Jaques, Thromb. Res., **13**, 429-441 (1978).
   "Appearance of Heparin Antithrombin-Active Chains *in vivo* After
   Injection of Commercial Heparin and in Anaphylaxis."
5. Y. Matzner, G. Marx, R. Drexler and A. Eldor, Thromb. Haemostas, **52**,
   134-137 (1984). "The Inhibitory Effect of Heparin and Related
   Glycosaminoglycans on Neutrophil Chemotaxis."
6. A. Stemberger, A. Riedl, S. Haas, K. Breddin, J. Walenga and G.
   Blümel,, Sem. Thrombos. Hemos., **17**, Suppl 1, 80-84 (1991). "Action of
   Modified Heparins on Coagulation: *In vitro* and *in vivo* Studies."
7. M. S. Sy, E. Schneeberger, R. McCluskey, M. J. Greene, R. D.
   Rosenberg and B. Benacerraf, Cell. Immunol., **82**, 23-32 (1983).
   "Inhibition of Delayed-Type Hypersensitivity by Heparin Depleted of
   Anticoagulant Activity."

HYLAN GEL FOR SOFT TISSUE AUGMENTATION

Nancy E. Larsen, Cynthia T. Pollak, Karen Reiner, Edward
Leshchiner, and Endre A. Balazs

Departments of Biochemistry, Histology and Chemistry
Biomatrix, Inc.
Ridgefield, New Jersey

Hylan, a hyaluronan derivative, was chemically
crosslinked with divinyl sulfone to produce water insoluble,
elastoviscous hylan gel slurries. The physical and biological
properties of this gel slurry were examined by rheological
and particle size analysis and by *in vivo* studies. Hylan gel
slurries are made up of pseudoplastic, deformable gel
particles with greater elasticity (at all frequencies) and
greater viscosity (at low shear rate, e.g. 0.01 s^{-1}) than the
soluble hylan polymer.

Hylan gel slurry was injected intradermally and
subdermally in mice and was found to produce a minimal
reaction at 24 hours, thereafter (up to 7 weeks) there was no
significant tissue reaction. Intradermal injection of [^{3}H]-
hylan gel slurry in guinea pigs revealed a minimal tissue
reaction after 1 week, and measurement of radioactivity in
the tissue at 1, 2 and 4 weeks revealed only a slight
decrease in the total amount of injected radioactivity.
Immunogenic activity of hylan gel slurry was evaluated in
rabbits adjuvant; unmodified hylan gel slurry, degraded hylan
gel slurry and hylan gel slurry-ovalbumin conjugate were used
to immunize rabbits. No antibody production to any hylan gel
sample was detected, although control rabbits immunized with
ovalbumin developed titers >400 of anti-ovalbumin antibodies
by day 21, as measured by the passive cutaneous anaphylaxis
assay (PCA).

INTRODUCTION

A variety of materials have been used for the correction of soft
tissue defects including paraffin, silicone and the protein collagen.[1-3]
Paraffin and silicone cause intense foreign-body reactions and are known
to migrate from the site of injection. Collagen implants have been used
with great frequency and success for the correction of soft tissue
defects (dermal contour deformities). It has been widely documented,
however, that collagen implants are resorbed by the body necessitating

regular and repeated injections of the material.[4-6] Collagen therapy is also associated with a low incidence of hypersensitivity reaction and prospective patients must undergo skin testing before receiving collagen therapy.

Hylan gel slurry is a nonproteinaceous, insoluble, viscoelastic, hyaluronan (hyaluronic acid, HA) derivative; the polysaccharide chains are chemically crosslinked with divinyl sulfone to produce sulfonyl-bis-ethyl crosslinks between hydroxyl groups.[7,8] Chemical crosslinking alters the solubility and rheological properties of the HA molecule but does not change the biological compatibility of the native polymer.[7] The physical properties of hylan gel slurries are controlled in part by the degree of crosslinking and hence, a range of gel slurries may be produced from very soft and deformable gels to very firm, nondeformable gel slurries.

In this report we describe the results from animal studies in which hylan gel slurry was evaluated at intradermal and subdermal sites in mice and guinea pigs, and intramuscalarly (repeat injection) in an immunogenicity study in the rabbit. The results indicate that hylan gel slurry may provide a safe, effective alternative for soft tissue augmentation.

MATERIALS AND METHODS

1. PREPARATION OF HYLAN GEL SLURRY AND RADIOLABELED HYLAN GEL SLURRY

Hylan gel slurry was produced from soluble hylan polymer according to the procedure described in U.S. Patent #4,582,865 by a divinyl sulfone crosslinking procedure.[8] Hylan gel slurry used in this study was sterile, nonpyrogenic, noncytotoxic, and had a polymer content of 4.3 mg/mL. The shear viscosity was 497.6 and 2.096 Pa at 0.015 and 14.7 sec^{-1} respectively; G' was 50.4 Pa at 0.001 Hz, G" was 15.5 Pa at 0.001 Hz.

Tritiated hylan gel was produced by the addition of tritiated water during crosslinking, resulting in a [^{3}H]-label covalently attached to carbon in the crosslink and pendant groups. Specific activity was determined on a digested aliquot of gel and typically was between 250,000–350,000 dpm/mL (0.11–0.16 uCi/mL); polymer content ranged from 3.5–5.5 mg/mL. In the experiments described in this report, the [^{3}H]-hylan gel had a specific activity of 330,000 dpm/mL (0.15 uCi/mL) and a polymer content of 5.2 mg/mL. [^{3}H]-hylan gel was assayed for sterility (USP <71>),[9] pyrogen content (Limulus Amebocyte Lysate test, USP <85>),[9] and cytotoxicity (Agar Overly, USP <87>).[9]

2. *IN VIVO* ASSAYS

2A. Intradermal and Subdermal Administration of Hylan Gel in Mice

Nu Nu nude mice (Charles River Corporation) were housed in sterile microisolator cages (Lab Products, Inc.) equipped with HEPA filters, and were handled in a sterile environment. Anesthetized mice (sodium pentobarbital) were injected intradermally or subdermally with approximately 0.1 mL of hylan gel at three sites (back of neck = site #1, mid-back region = site #2, and tail region = site #3) and approximately 0.2 mL was injected intradermally or subdermally in the abdomen. Injection sites were marked with a Sharpie pen; mice were sacrificed at

24 hours, 1, 2, 3, 6, and 7 weeks. Mice were sacrificed by carbon dioxide overdose and weighed, photographed, and the injection site tissue excised and submerged in formalin before being processed for histologic evaluation.

2B. Intradermal Injection of [^{3}H]-Hylan Gel in Guinea Pigs

Guinea pigs (female, 300-350 grams each, Charles River Corporation) were anesthetized with ketamine HCl and acepromazine. The back of each animal was shaved and cleaned; each guinea pig received six intradermal (ID) injections; four of [^{3}H]-hylan gel (0.02 mL-0.25 mL) and two of 0.15 M NaCl. A 25-gauge needle was used for all of the injections. Each injection site was marked and monitored at 24 hours and then weekly, post-injection. Guinea pigs were sacrificed (3 per group) at the following time points: 30 minutes (immediately after injection), 7, 14, and 28 days by injection of an overdose of sodium pentobarbital into the peritoneal cavity. Each skin section was excised, weighed and either fixed in formalin (two of four [^{3}H]-hylan gel sites per animal) or digested with Unisol Tissue Solubilizer (Isolab, Inc.) for measurement of tritium (two [^{3}H]-hylan gel sites and two saline sites from each animal). Formalin fixed tissue was processed for histologic evaluation and tissue sections were stained with hematoxylin and eosin (H&E) or toluidine blue. Tissue digests were prepared by adding 5 mL of Unisol per gram tissue, followed by heating at 50°C for up to 24 hours. Total tritium per tissue sample was determined and percent present in the skin was estimated based on the amount injected.

2C. Rabbit Immunization

Male SPF (specific pathogen free) New Zealand white rabbits (2.5-3.0 kg, Hazleton Research Animals, Inc., seven per test group) were anesthetized with ketamine HCl/xylazine and 3 mL of blood was withdrawn from the auricular artery; pre-immune serum was prepared from each blood sample (blood was transferred to glass tubes and allowed to clot for 30 minutes, then serum was collected by centrifugation) serum was frozen at -90°C until use in the assay. The first immunization consisted of injection of 0.1 mL of sample (emulsified in Complete Freunds Adjuvant, CFA) into the footpad and injection of 0.9 mL of the same sample into the right thigh muscle. Subsequently, weekly immunizations consisted of injection of 1 mL of sample (emulsified in Incomplete Freund's Adjuvant, IFA) into the rear thigh muscle. At 3, 5, 12 and 20 weeks after the start of immunization, rabbits were anesthetized and blood was withdrawn from the auricular artery. Serum was prepared as described and frozen at -90°C until use.

Test samples of hylan gel (HY-G), degraded hylan gel (HY-G-D; gel liquefied by a sonication procedure) and hylan gel-ovalbumin conjugate (HY-G-OA, prepared as described by Richter 1974) were emulsified with equal volumes of CFA and IFA by repeated cycling through a syringe. Emulsification of the ovalbumin control (1 mg/mL) was achieved by the same procedure. The concentration of hylan polymer in the hylan gel used in this study was 2 mg/mL.

2D. Measurement of Serum Antibodies by Passive Cutaneous Anaphylaxis

Female SPF Hartley guinea pigs (250-350 grams, Hazelton Research Animals, Inc.) were anesthetized with ketamine HCl and acepromazine and their abdomens were shaved and cleaned. Each guinea pig received up to four intradermal (ID) injections of serum which had been diluted 5-fold or 10-fold; each guinea pig also received one 0.1 injection of 0.15 M sodium chloride as a negative control. Three hours after ID serum injections, each guinea pig was injected intravenously with 0.5 mL of Challenge solution (HY-G, HY-G-D, HY-G-OA, ovalbumin mixed with an equal volume of 1% Evans blue dye in 0.15 M sodium chloride. Thirty to sixty minutes after the IV injection guinea pigs were sacrificed by injection of an overdose of sodium pentobarbital (cardiac puncture) and the abdomen skin was excised. Skin lesions (injection sites) were evaluated using a scale of 0-4; 10 mm was used as the standard minimum lesion size, therefore, a blue lesion of 10 mm size rated a value of 1 when blueing lesions were obtained. The titer of PCA-reactive antibodies was determined from the reciprocal of the highest serum dilution which produced a blue lesion of 10 mm.[10]

RESULTS

Histopatholoy of excised skin samples from mice injected with hylan gel revealed that hylan gel was located in the dermis, deep dermis and subcutis. During the first 24 hours, injected hylan gel was associated with minimal to mild multifocal acute inflammation; at the remaining time points (1, 2, 3, 6 and 7 weeks) there was no tissue reaction to the hylan gel in 22 out of 24 injection sites. In the other 2 injection sites (2 weeks and 7 weeks) hylan gel was associated with a minimal inflammatory reaction (Table 1, Figure 1).

Intradermal injection of [3H]-hylan gel into the guinea pig resulted in a minimal inflammatory reaction. No inflammatory infiltration was noted on day 1, the infiltrate was apparent beginning on day 5. Most of the inflammatory cells were classic round cells and monocytes and macrophages, some were moderately vacuolated. The [3H]-hylan gel implant remains present through day 28 (Figure 2).

When injection sites were palpated, it was noted that at 24 and 48 hours post-injection 33 of 36 [3H]-hylan gel sites were palpable (36 of 36 [3H]-hylan gel injection sites were palpable immediately after

Table 1. Histopathology of injected hylan gel in mice.

Time Post-injection	Tissue Reaction	
24 hours	minimal to mild multifocal acute inflammation:	3/4 sites
	no tissue reaction	1/4 sites
1 week	no tissue reaction	8/8 sites
2 weeks	no tissue reaction	3/4 sites
	minimal focal reaction not associated with hylan gel	1/4 sites
3 weeks	no tissue reaction	4/4 sites
6 weeks	no tissue reaction	4/4 sites
7 weeks	no tissue reaction	3/4 sites
	minimal multifocal reaction	1/4 sites

Photomicrographs of the cross section of injection sites 1 and 6 weeks post-injection are shown in Figure 1.

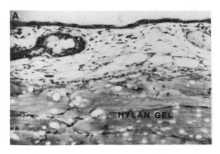
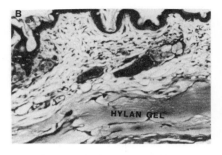

Figure 1. Photomicrographs of cross-sections of dermal and
subdermal tissue from Nu Nu nude mice injected
with hylan gel. A. Tissue take one week after
injection of hylan gel; no significant tissue
reaction associated with hylan gel. B. Tissue
taken 6 weeks after injection of hylan gel; no
significant tissue reaction. Toluidine blue stain,
x 25 original magnification.

injection), and at 1, 2, 3 and 4 weeks post-injection 8 of 36, 14 of 24,
8 of 12 and 5 of 12 sites were palpable, respectively.

Radioactivity was measured in Unisol-digested skin samples. In the
control (immediately after injection), an average of 89.1 ± 2.9% of the
injected radioactivity was recovered in the tissue. At time points of 5-

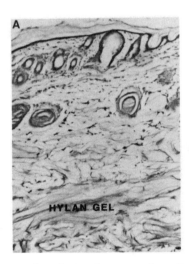
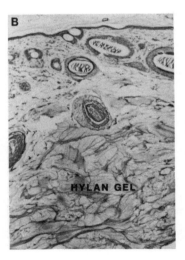

Figure 2. Photomicrographs of cross-sections of dermis from
guinea pigs containing [^{3}H]-hylan gel. A.
Intradermal [^{3}H]-hylan gel implant 14 days post-
injection. B. Intradermal [^{3}H]-hylan gel implant
28 days post-injection. Toluidine blue stain, x 25
original magnification.

Table 2. Guinea pig: intradermal injection of [^{3}H]-hylan gel.

Time point	Volume injected (average mL)	%Total DPM Recovered	%Control (zero time)
0	0.091 ± 0.04	89.1 ± 2.9	-
5-7 days	0.107 ± 0.08	93.1 ± 4.4	100
14 days	0.036 ± 0.02	74.5 ± 18.0	85
28 days	0.174 ± 0.05	77.1 ± 9.6	87

7, 14 and 28 days, the average percent recovery of tritium was 93.1 ± 4.4%, 74.5 ± 18 and 77.1 ± 9.6% respectively, or 100, 85 and 87% of control (zero time) (Table 2). The presence of the [^{3}H]-hylan gel was confirmed by histological evaluation as described in preceding paragraph.

Evaluation of serum from immunized rabbits by passive cutaneous anaphylaxis indicated that there was no specific antibody production to HY-G, HY-G-D, or HY-G-OA in any rabbit up to 20 weeks after immunization. Six of the seven rabbits immunized with HY-G-OA demonstrated slightly positive reactions to an OA challenge, suggesting the presence of low titers of anti-OA antibodies (Table 3A.) Evaluation of serum from the OA-immunized rabbits (positive control group) revealed the presence of significant titers of anti-OA antibodies (>400), indicating that the immunization procedure was effective and that the PCA assay was sensitive and functional (Table 3B.).

Table 3A. Evaluation of serum from hylan gel-immunized rabbits.

Immunizing Substance	PCA Titer in Serum				
	Day 0	Day 21	Day 35	Day 100	Day 150
Hylan Gel	0 (7)	0 (7)	0 (7)	0 (5)	0 (4)
					<5 (1)
Hylan Gel-Degraded	0 (7)	0 (2)	0 (3)	0 (3)	0 (1)
		<5 (5)	<5 (2)	<5 (2)	<5 (1)
Hylan Gel-OA	0 (7)	0 (7)	0 (5)	0 (5)	0 (5)
Hylan Gel-OA	0 (6)	0 (6)	0 (3)	0 (4)	0 (2)
(Hylan Gel Challenge)	<5 (1)	<5 (1)	<5 (2)	<5 (1)	<5 (3)
Hylan Gel-OA	0 (7)	0 (1)	0 (4)	0 (4)	0 (3)
(OA Challenge)		<5 (2)	<5 (1)	<5 (1)	<5 (1)
		>5 (4)			>5 (1)

Table 3B. Evaluation of Serum from Ovalbumin Immunized Rabbits

Immunizing Substance	PCA Titers				
	Day 0	Day 21	Day 35	Day 100	Day 150
Ovalbumin	0 (7)	>400 (5)	>400 (3)	≤800 (2)	>800 (1)
		100 (1)	200 (1)	400 (1)	>400 (4)
		0 (1)	0 (1)	200 (1)	
				>100 (1)	

DISCUSSION

Hylan gel slurries, prepared by crosslinking soluble hylan polymer were shown to be highly biocompatible (nonimmunogenic, noninflammatory, nontoxic), and apparently resistant to significant resorption and migration. The unique rheological properties of hylan gel allow injection through small diameter needles into relatively dense tissue (skin, muscle). Hylan gels are pseudoplastic and, therefore, shear 'thin' at high shear rates such as that which is present during passage through a needle, and regain their original viscosity in the absence of high shear rate (i.e., upon delivery to a tissue compartment). Hyaluronan (HA) is a ubiquitous component of all animal connective tissue and makes up a significant part of the intercellular matrix of the skin where one of its functions is to maintain tissue hydration.[11-15] Indeed, a correlation exists between water content and HA content of dermal tissue.[12,14,16] Alterations in HA content and HA metabolism are known to accompany maturation and aging of human skin;[17,18] aging of skin is also accompanied by a deterioration of mechanical properties which is a result of altered viscoelastic properties of the intercellular matrix.[19] These findings imply that there is a relationship between healthy young skin and the presence of a hydrated, viscoelastic HA network in the intercellular matrix.[20]

HA derivatives such as hylan gels are well suited for use in dermal augmentation because of their insolubility (and resistance to degradation), their high water content (and potential to mimic the natural hydrating functions of HA) and their viscoelasticity (which is similar to native HA). These properties, together with the fact that hylan gel implants do not elicit inflammatory, immunologic (cellular or humoral) or foreign reactions, contribute to their stability and compatibility in the tissue. Hylan gels may function by serving as scaffolding in directing and controlling tissue regeneration and as space-fillers in tissue.[21] Other implant materials, both natural and synthetic, do not have these properties. Collagen implants are known to resorb, and are known to elicit an inflammatory reaction.[6,15] Following injection, the collagen implant is recognized as foreign, is chemotactic to leukocytes and fibroblasts, and elicits a granulocytic response which is followed by infiltration of macrophages.[5] Macrophages produce collagenase, which may be responsible for digestion of the collagen implant.[4] Injected collagen apparently does not act as a scaffolding for tissue replacement by host connective tissue,[5] as suggested by Knapp.[3] Lack of persistence of injectable collagen in the tissue is well documented, and it has been shown that it is removed from the body without replacement of host connective tissue.[5,6] Repeated collagen injections are therefore needed to maintain the tissue augmentation. Other materials such as paraffin, silicone and teflon often elicit intense foreign body reactions, and other adverse reactions as well as tend to migrate from the site of injection,[22,23] and are no longer frequently used.

Aside from efficacy, the greatest advantage associated with the use of hylan gel for tissue augmentation is its non-immunogenic nature. It has been well documented that purified, noninflammatory HA does not elicit humoral or cell mediated immune reactions.[10,24] Hylan and hylan gels have been shown to behave identically to the native polymer in this regard.[20,26] The lack of antigenic or immunogenic activity may result from the fact that the chemical identity of the polysaccharide is maintained from species to species and that it is nonproteinaceous. This property in an implant is advantageous for several important reasons: little risk of hypersensitivity or rejection reactions, no prerequisite skin test for prospective patients, and little likelihood that the implant will be resorbed as a result of tissue reaction, and hence repeated injections may be reduced or eliminated.

In conclusion, hylan gel implant may provide a safe, efficacious, and convenient alternative to collagen implant therapy for soft tissue augmentation, thus avoiding undesirable side effects and eliminating the need for repeated injection to maintain the augmentation.

REFERENCES

1. J. L. Barton and W. J. Cunliff, in: "*Textbook of Dermatology*," A. Rook, D. B. Wilkonson, F. J. G. Ebling, R. H. Champion, & J. L. Burton, Eds., Oxford-Blackwell, p. 1870-1871 (1986).
2. V. J. Selmanowitz and N. Orentreich, J. Dermatol. Surg. Oncol., **3**, 597-611 (1977). "A Monographic Review."
3. T. R. Knapp, E. N. Kaplan and J. R. Daniels, Plast. Reconstruct. Surg., **60**, 898-405 (1977).
4. P. J. Donald, **95**, 607-614 (1986).
5. A. Arem, Clin. Plast. Surg., **12**, 209-220 (1985).
6. F. M. Kamer and M. M. Churukian, Arch. Otolaryngol., **110**, 93-98 (1984).
7. E. A. Balazs and A. Leshchiner, United States Patent #4,500,676 (1985).
8. E. A. Balazs and A. Leshchiner, United States Patent #4,582,865 (1986).
9. United States Pharmacopia. XXI, United States Pharmacopia Convention, Inc., 12601 Twinbrook Parkway, Rockville, MD 20852 <71>, <85>, <87> (1985).
10. A. W. Richter, Int. Arch. Allergy Appl. Immunol., **47**, 211-217 (1974).
11. J. R. Yates, in: "*Mechanism of Water Uptake by Skin*," N. R. Eden, Ed., Wiley Interscience, New York, p. 485 (1971).
12. J. A. Szirmai, in: "*The Amino Sugars*," Volume 11b, E. A. Balazs and R. W. Jeanloz, Eds., Academic Press, New York, p. 129 (1965).
13. R. H. Pearce and B. J. Grimmer, in: "*Advances in the Biology of Skin*," Volume II, W. Montagna, J. P. Bently and R. L. Dobson, Eds., Appleton, NJ, p. 89 (1970).
14. J. P. Bentley, H. Nakagawa and G. H. Davies, Biochim. Biolphys. Acta., **224**, 35 (1971).
15. E. A. Balazs and D. A. Gibbs, in: "*Chemistry and Molecular Biology of the Intercellular Matrix*," Volume 3, E. A. Balazs, Ed., Academic Press, New York, p. 1241 (1970).
16. M. Uzuka, K. Nakagima, S. Ohta and Y. Mozi, Biochim. Biophys. Acta., **627**, 199 (1980).
17. J. G. Smith, E. A. Davidson and R. W. Taylor, in: "*Advances in Biology of Skin*," Volume VI, W. Montagma, J. P. Bentley and R. L. Dobson, Eds., Appleton, New York, p. 211 (1966).
18. J. H. Poulsen and M. K. Cramers, in: "*Human Dermis, and a Material of Reference*," Scand. J. Clin. Lab. Invest., **42**, 545 (1982).
19. C. H. Daly and G. F. Odland, J. Invest. Dermatol., **73**, 84 (1979).
20. E. A. Balazs and P. Band, Cosmetics and Toiletries, p. 99 (1984).
21. E. A. Balazs and E. A. Leshchiner, in: "*Proc. Intern. Conf. on Cellulosics Utilization in the Near Future*," Elsevier, Applied Sci. Publ., Tokyo (1988).
22. G. Matton, A. Anseeuw and F. DeKeyser, Aesth. Plast. Surg., **9**, 133-140 (1985).
23. Editorial: Plast. Reconstr. Surg., **61**, 892 (1978).
24. A. W. Richter, E. M. Ryde and E. O. Zetterstrom, Int. Arch. Allergy Appl. Immunol., **59**, 45-48 (1979).

25. E. A. Balazs, J. L. Denlinger, E. A. Leshchiner, P. Band, N. Larsen,
A. Leshchiner and B. Morales, in: "Proc. *Fifth Ind. Conf. on
Biotech.*" (1988).
26. N. E. Larsen, M. B. Kling, E. A. Balazs, E. L. Leshchiner,
Transactions: Sixteenth Annual Meeting of The Society for
Biomaterials, Charleston, South Carolina, **302**, May 20-23, 1990.

ADSORPTION OF METAL IONS ON CHITOSAN AND CHEMICALLY MODIFIED CHITOSAN AND THEIR APPLICATION TO HYDROMETALLURGY

Katsutoshi Inoue, Kazuharu Yoshizuka and Yoshinari Baba

Department of Applied Chemistry
Saga University
Honjo-machi, Saga 840, Japan

Crosslinked copper(II)-complexed chitosan was prepared to avoid the dissolution in various aqueous acidic media since chitosan, as it is, dissolves in aqueous solutions of organic acids and inorganic acids in some concentration regions except for sulfuric acid solutions. Adsorption of various metal ions from aqueous ammonium nitrate solution on the crosslinked copper(II)-complexed chitosan was investigated in terms of pH dependence on the distribution coefficient of each metal ion. It was found to have much higher selectivity than a commercial iminodiacetic acid type of chelating resin for the mutual separation of metal ions; especially noticeable are : the high selectivities to gallium and indium over zinc, to aluminum and iron(III) over zinc and to nickel over cobalt. Based on these characteristic adsorption behaviors of chitosan, several practical hydrometallurgical processes were proposed for the purification of some metals.

Carboxymethylation of chitosan was carried out to prepare N-carboxymethyl- and N,N-dicarboxymethyl-chitosan as adsorbents for precious metals. These chemically modified chitosans exhibited excellent behaviors as well as chitosan itself for the adsorption of palladium(II) and platinum(IV) from hydrochloric acid. However, contrary to chitosan, as it is, or the crosslinked copper(II)-complexed chitosan, they exhibited a much superior elution behavior of these metals with certain concentrations of hydrochloric acid, which are considered to be promising for the practical application to the purification of these metals.

INTRODUCTION

As is well known, chitin, a natural marine polymer, and chitosan, its deacetylated derivative, can effectively adsorb not only various oganics including polychlorinated biphenyls, proteins and nucleic acids but also metal ions.[1,2] Because of its excellent adsorption behavior, chitosan has been employed as an effective coagulating agent in activated sludge plants and used for recovering proteins from food processing

Biotechnology and Bioactive Polymers, Edited by C. Gebelein
and C. Carraher, Plenum Press, New York, 1994

factories. In addition, many studies have been conducted on its application to chromographic separations of metal ions. Although its excellent adsorption behavior also suggests its feasible applications to separation, concentration and purification of metals for hydrometallurgy, only meager data have been available as the fundamental information necessary for its hydrometallurgical application.[3]

In addition, some attempts have been made to enhance or improve the adsorption of metal ions by chemically modifying chitosan, for example, by introducing Schiff's bases in order to enhance the adsorption capacity of uranium from sea water.[4]

In this work, we conducted fundamental investigations on the adsorption behaviors of metal ions on chitosan in order to obtain basic information necessary for the hydrometallurgical application of chitosan. Because chitosan as it is has some aqueous solubility in some acidic media, crosslinking between polymer chains of chitosan has been proposed in order to avoid the dissolution in aqueous solutions. However, the crosslinking resulted in significant lowering of the adsorption capacity. In this work, crosslinked chitosan prepared according to the method proposed by Ohga, et al., was employed in order to overcome these disadvantages.[5] In this method, chitosan was crosslinked by (chloromethyl)oxirane after complexation with copper(II) to protect the amino group of chitosan, effective for the adsorption, against the attack by the crosslinking agent and, at the same time, to preserve adsorption domains.

Further, we synthesized N-carboxymethylated and N,N-dicarboxymethylated chitosan as the chemically modified derivatives to improve the adsorption and elution behaviors of metals, palladium(II) and platinum(IV) in particular.

EXPERIMENTAL

1. PREPARATION OF THE ADSORBENTS

The crosslinked copper(II)-complexed chitosan and the carboxymethylated chitosan were prepared from completely deacetylated chitosan kindly supplied from Katokichi Co. Ltd., Kan-onji, Japan.

The former was prepared according to the method proposed by Ohga, et al, as mentioned earlier.[5] N-carboxymethylated chitosan was prepared according to the reaction expressed by Scheme 1 as follows: chitosan was stirred and refluxed together with monochloroacetic acid (two equivalents to a glucosamine unit of chitosan) using pyridine as a catalyst, in ethanol under nitrogen atmosphere for 3 days. The product gel was washed, sequentially, with 1 mol/dm^3 sodium hydroxide solution, 0.5 mol/dm^3 sulfuric acid solution and finally with deionized water after filtration and dried in vacuo to a constant weight before use.

N,N-Dicarboxymethylated chitosan was prepared via N,N-dinitrilated chitosan, the intermediate product, according to the reaction expressed by Scheme 2 as follows: the latter was prepared by stirring chitosan in a great excess amount of aqueous glycoronitril solution at about 350K for 54h. The product gel was filtered followed by hydrolysis by stirring together with 30 wt% sodium hydroxide solution also at about 350 K for 72 h to obtain the final product gel, which was washed sequentially with deionized water, 0.5 mol/dm^3 sulfuric acid and again deionized water after filtration, and dried in vacuo to a constant weight before use. The

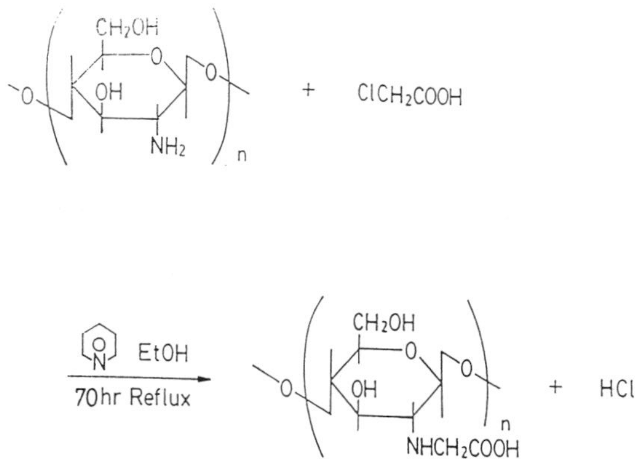

Scheme 1. Route of the synthesis of N-carboxymethylated chitosan.

fractional conversions of the chitosan into the N-carboxymethylated and N,N-dicarboxymethylated chitosan were about 80-90% and 50-60% respectively.

2. PROCEDURE

All of the experimental works of adsorption and elution of metal ions was carried out batch wise as follows: a weighed amount of dried gel (0.05-0.1 g) was shaken together with a constant volume of an aqueous solution, containing a known concentration of metal ion, in a flask immersed in a thermostated water bath maintained at 303°K for about 24 h, within which all kinds of metal ions had been confirmed to be attained to equilibrium of adsorption or elution. All metal ions, except for palladium(II) and platinum(IV), were adsorbed from 1 mol/dm^3 ammonium nitrate solutions while palladium and platinum were adsorbed from various concentrations of hydrochloric acid solutions. The concentration of metal ions

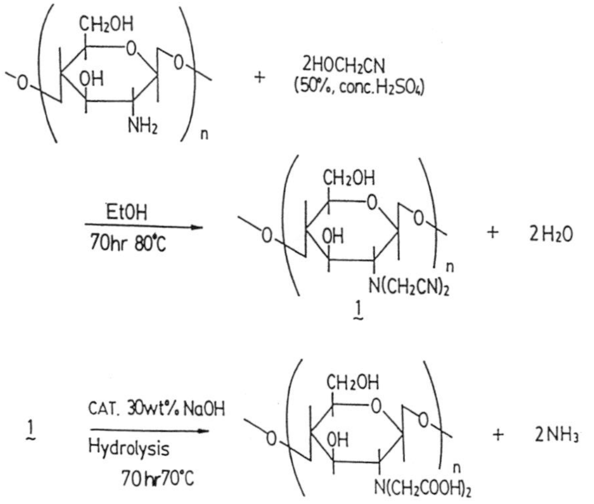

Scheme 2. Route of the synthesis of N,N-dicarboxymethylated chitosan.

in aqueous solutions were measured by titration with EDTA or by using a Nippon Jarrell-Ash model AA-782 atomic absorption spectrophotometer. The amount of adsorption were calculated from the concentration change before and after the adsorption and the weight of the dried gel.

RESULTS AND DISCUSSION

1. ADSORPTION OF METAL IONS ON CROSSLINKED COPPER(II)-COMPLEXED CHITOSAN

Figure 1 shows the pH dependence on the distribution coefficient of various metals in the adsorption from 1 mol/dm^3 aqueous ammonium nitrate solutions on the crosslinked copper(II)-complexed chitosan. This figure apparently suggests that the plots for n-valent metal ions are lying on the straight lines with the slope of n except for cobalt(II). From this result, and other experimental results suggesting that there is no concentration dependence of nitrate ion on the distribution ratio of copper(II), it can be concluded that n-valent metal ions are adsorbed on chitosan as the metal-chelate with the composition of 1:n metal:glucosamine unit releasing n ions of hydrogen ion per one metal ion, as expressed by Scheme 3 as an example for the adsorption reaction of copper(II), though some authors have concluded that metal ions are adsorbed only by their coordination to nitrogen or/and oxygen atoms.[6,7] The plots for cobalt(II) are lying on a straight line with the slope of 2, which might suggest that divalent cobalt is oxidized to trivalent state during chelation, as observed in the solvent extraction with chelating extractions from aqueous ammoniacal solutions.[8] Compared with commercial chelating resins, especially with iminodiacetic acid type of chelating resins, the similar tendencies are observed in the sequence of the adsorption of these metal ions;[9] however, it should be noteworthy that chitosan has much higher selectivity than these chelating resins, for example, to copper(II) over nickel(II), to nickel(II) over cobalt(II), to iron(III) over copper(II), and so on. Specifically, some aspects in the adsorption behaviors on chitosan are noticeable in terms of practical application to hydrometallurgy as follows;

(1) Gallium(III) and indium(III) are adsorbed at much lower pH than zinc(II), which enables selective adsorption of small amount of gallium(III) and indium(III) in the presence of great amount of zinc(II) and is, therefore, practically applicable to the recovery of these rare metals from zinc refinery by-products or zinc leach residue.

(2) Iron(III) and aluminum(III) are adsorbed also at much lower pH than zinc(II), which enables effective separations of these two metals from zinc(II) and is applicable to the selective removal of iron and aluminum impurities in zinc plating bath of steel.

(3) The difference between the adsorption lines of nickel(II) and cobalt(II) is as great as about 1 pH unit, which enables mutual separation of these metals with great efficiency and can be applied to the production of high purity cobalt by selective removal of nickel impurity by chitosan.

From the loading test with copper(II) from 1 mol/dm^3 aqueous ammonium nitrite solution, the loading capacity of the original chitosan, the crosslinked copper(II)-complexed chitosan and a commercial iminodiacetic acid type chelating resin was found to be 2.31, 1.37 and 1.94 mol/kg, respectively; apparently, chitosan, as it is, has higher

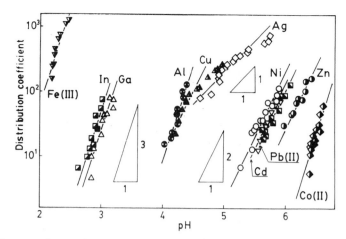

Figure 1. pH dependencies of the distribution coefficients of
various metal ions in the adsorption on crosslinked
copper(II)-complexed chitosan from 1.0 mol/dm^3 ammonium
nitrate solutions.

adsorption capacity of copper(II) than commercial chelating resin, and
crosslinking greatly decreases the adsorption capacity in spite of
protecting the adsorption site on chitosan by complexation with
copper(II) in advance.

2. ADSORPTION OF PALLADIUM(II) AND PLATINUM(IV) ON CHEMICALLY MODIFIED CHITOSAN

It is well known that palladium and platinum are strongly extracted
by nitrogen-containing extractants such as long-chain alkylamines or

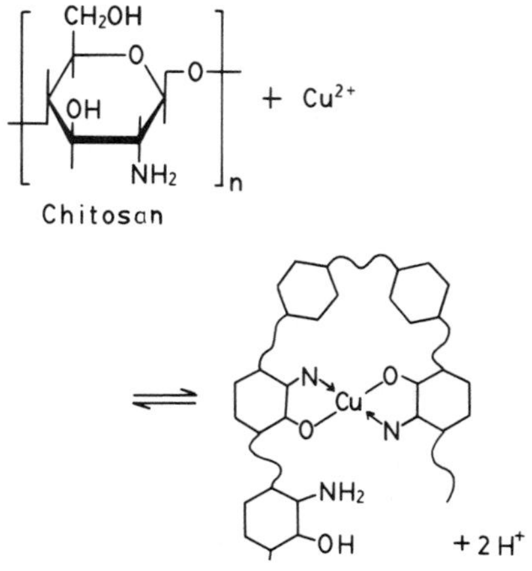

Scheme 3. Complex-formation of copper(II) with chitosan.

39

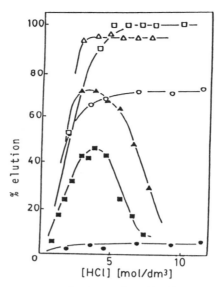

Figure 2. The percentage elution of palladium(II) and platinum(IV) from three kinds of adsorbents by varying concentrations of hydrochloric acid.

LEGEND:

Pd	Pt	Adsorbent
o	●	Crosslinked Cu(II) complexed chitosan
□	■	N-carboxymethylated chitosan
△	▲	N,N-dicarboxymethylated chitosan

strongly adsorbed on weak base anion exchangers. However, on the other hand, it is very difficult to strip or elute the loaded palladium or platinum from these extractants or adsorbents even with very high concentrations of hydrochloric acid. Consequently, the development of the solvent extraction reagents or adsorbents with excellent stripping or elution behaviors of these metals has been strongly demanded in metallurgical industry of precious metals. In the adsorption of palladium(II) and platinum(IV) from hydrochloric acid solution on the two types of carboxymethylated chitosan, the distribution coefficients of these metals are slightly lower than chitosan as it is and the crosslinked copper(II)-complexed chitosan though no appreciable differences were observed between the latter two adsorbents in this adsorption system. However, the maximum loading capacity of palladium(II) was considerably enhanced by the carboxymethylation while that of the platinum(IV) was not affected or only slightly decreased.

Figure 2 shows the percentage of elution from these adsorbents loaded with palladium or platinum with various concentrations of hydrochloric acid. It is noteworthy that the elution behavior of these metals is remarkably improved or enhanced by carboxymethylation. That is, complete elution of palladium(II) can be achieved from N-carboxymethylated chitosan though only 70% of the loaded palladium is eluted with high concentrations (more than 5 mol/dm^3) of hydrochloric acid from the original chitosan. The percentage of elution of platinum(IV) was greatly enhanced up to about 50% from mono-carboxymethylation and as high as about 70% from di-carboxymethylation though only about 5% of the loaded platinum can be eluted from the original chitosan. Also by taking account of the experimental result that any of aluminum and iron were not adsorbed on these carboxymethylated chitosan from hydrochloric acid solutions, these excellent adsorption and elution behaviors can be expected to be practically applied to the hydrometallurgy of precious metals, for example, in the recovery of palladium and platinum from waste catalysts for effluent gas from automobiles.

REFERENCES

1. K. Takemoto, Gendai Kagaku, (6), 57 (1981).
2. S. Hirano, Kagaku, **43**, 155 (1988).
3. I. Blazquez, F. Vicente & B. Gallo, J. Appl. Poly. Sci., **33**, 2107 (1987).
4. P. L. Lopez-de-Alba, B. Urbina, J. C. Alvarado, G. A. Andreu & J. A. Lopez, J. Radioanal. Nucl. Chem., Lett., **118**, 99 (1987).
5. K. Ohga, Y. Kurauchi & H. Yanase, Bull. Chem. Soc. Jpn., **60**, 444 (1987).
6. K. Ogawa, K. Oka, T. Miyanishi & S. Hirano, in: *"Chitin, Chitosan and Related Enzymes,"* J. P. Zikakis, Ed., Academic Press, Orlando, 1984, p. 327.
7. S. Schlick, Macromolecules, **19**, 192 (1986).
8. P. Guesnet, J. L. Sabot & D. Bauer, J. Inorg. Nucl. Chem., **42**, 1459 (1980).
9. S. Tomoshige, M. Hirai & T. Shibata, Fusen, **29**, 210 (1982).

CHITIN AND CHITOSAN: ECOLOGICALLY BIOACTIVE POLYMERS

Shigehiro Hirano, Hiroshi Inui, Hideto Kosaki, Yoshitaka
Uno, and Tsuyoshi Toda

Department of Agricultural Biochemistry and Biotechnology
Tottori University
Tottori, Japan 680

Chitinase and lysozyme activities were stimulated by
treating with chitin, chitosan, or derivatives in a culture
of plant and animal cells. (1) Chitinase activity was
stimulated about 1.5 times by treating with chitin or
chitosan in the calluses of rice, cabbage leaves and soybean.
(2) Two chitinase isozymes were detected in untreated rice
callus, and an additional isozyme was detected in rice callus
treated with O-carboxymethylchitin. (3) Radish seed chitinase
was stimulated 1.3 times by coating with low-molecular-
chitosan (d.p. 20) or chitosan-oligosaccharides (d.p. 2-9) at
the growing stage of germination. (4) Extracellular lysozyme
activity was stimulated by chitin or chitin-oligosaccharides
(d.p. 7,8) in the culture of chicken embryo fibroblast cells.
(5) Serum lysozyme activity was enhanced by intravenous
injection of chitosan-oligosaccharides (d.p. 2-9). A
mechanism for the induction of chitinase isozyme in plant
cells was proposed, its ecological significance was
discussed, and some applications in the field of
biotechnology were described.

INTRODUCTION

Chitin is (1→4)-linked 2-acetamido-2-deoxy-ß-D-glucan, and chitosan
isN-deacetylated chitin (Figure 1). These biopolymers are main structural
components in cuticles of crustacean, insects and mollusks, and in cell
walls of pathogens.[1] Their hydrolyzing enzymes [lysozyme (EC 3.2.1.17),
chitinase (EC 3.2.1.14), and chitosanase] are widely distributed in the
tissue and body fluids of animals and plants, and also in soil and
hydrospheres (Figure 2).[2] These hydrolyzing enzymes play a role for
protecting plants and animals against infection with insects and
pathogens.[3] Therefore, the stimulation and induction of these enzymes are
of importance in the agroindustry using insecticides and pesticides at
low levels.

Chitin and Chitosan are biologically synthesized at an estimated
amount of one billion tons per year, and are biodegradable without accu-

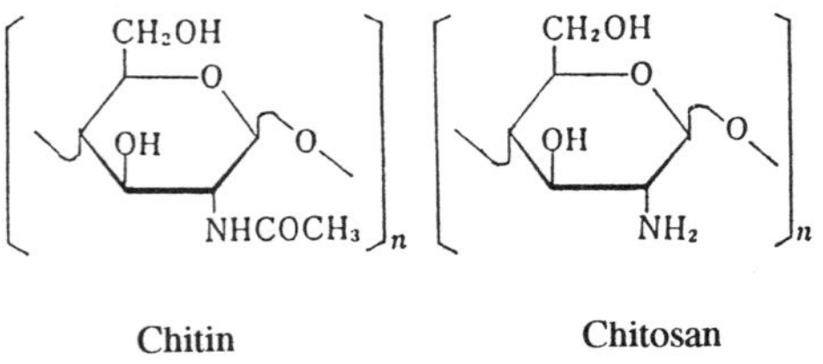

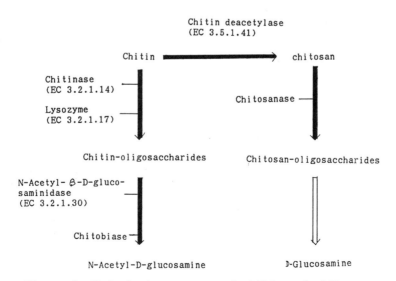

Chitin   Chitosan

Figure 1. Repeating units for the chemical structures of
chitin and chitosan.

mulation on the earth. This is called as "chitin cycle" and probably
conserves our environment and ecology on the earth.[4] Chitin and chitosan
are almost non-toxic, antigenically inactive, and biocompatible in both
animal and plant tissue.[5] It is of significance to use these biodegradable biopolymers as a novel material in the field of biotechnology.

We now wish to report that chitin and chitosan are ecologically
bioactive, and are potentially usable as a biocompatible material in the
field of biotechnology.

EXPERIMENTAL

1. MATERIALS

The following compounds were prepared by conventional methods, and
are used in the present work: crab shell chitosan, N-acetylchitosan (d.s.

Figure 2. Hydrolyzing enzymes of chitin and chitosan.

1.0 for Ac), O-carboxymethyl (CM) chitin (Na salt, $[\alpha]^{30}_D$ -6° (c 1.0, water), d.s. 0.58 for CM, d.s. 0.9 for Ac), O-glycolchitin ($[\alpha]^{30}_D$ -13° (c 1.0, water), d.s. 0.8 for glycol, 0.9 for Ac), O-glycolchitosan ($[\alpha]^{20}_D$ -15° (c 0.5, water), d.s. 0.8 for glycol), two mixtures of chitin-oligosaccharides (d.p. 2-6 and 7-8), a mixture of chitosan-oligo-saccharides (lactate salt, d.s. 2-8), a mixture of (1→4)-α-D-galactosam-inan-oligosaccharides (d.p. 2-7), and low-molecular-weight (LMW)-chitosan (d.p. about 20).

2. METHODS

2A. Rice Callus and Chitinase Assay

The callus of rice (Oryzae sativa, L. var. Koshihikari) was cultured on Murashige-Skoog's (MS) agar (0.9%) medium at 25°C in the dark. The stimulation of chitinase activity was tested by two ways: (1) a steri-lized suspension or solution (0.2 or 0.5 mL) of chitosan (2.5 mg), a chitin derivative (2.5 mg), or microbial cell walls (3.0-3.5 mg) in 20 mM sodium phosphate buffer (0.5 mL, pH 6.0) was dropped on the surface of the callus (about 1.5 g by fresh weight). Only the buffer solution was dropped over the callus as a control. (2) The callus was cultured on MS agar medium containing 0.01, 0.05, or 0.1% of a compound. After culturing appropriate period (see Tables 1-3), the callus was suspended at 4°C with Polyclar AT (10% of callus weight) in 50 mM sodium phosphate buffer (pH 6.0) containing 1 mM EDTA and 1 mM phenylmethanesulfonyl fluoride, dis-rupted with a Physcotron homogenizer, and centrifuged at 20,000g for 10 min. The supernatant solution was dialyzed against 20 mM sodium phosphate buffer solution (pH 6.0) containing 1 mM EDTA to give a crude enzyme extract. Protein content was quantitatively determined according to Lowry et al[6] using bovine serum albumin as a standard.

Chitinase activity was assayed with shaking at 37°C for 1 hr in a suspension (total 2 mL) of N-acetylchitosan (20 mg) and a crude enzyme extract (1 mL) in 0.1 M citric acid-0.2 M Na_2HPO_4 buffer solution (1 mL, pH 3.0). The enzyme reaction was stopped by addition of 1 mL of 6.7% Na_2WO_4 in 1/3 N H_2SO_4. After centrifuging at 1500g for 10 min, the value of reducing sugar in the supernatant solution was measured by a modified Schales' method.[7] One unit (U) for enzyme activity is defined as the amount of chitinase activity which produces 1 µmol of reducing sugars as N-acetyl-D-glucosamine per min under the present condition.

2B. Chromatography on a DEAE-cellulofine Column

For the analysis of chitinase isozymes, the crude enzyme extract was adjusted to 80% saturation by addition of solid $(NH_4)_2SO_4$ at 4°C. The resultant precipitate was dissolved in 25 mM imidazole-HCl buffer (20 mL, pH 6.8) containing 1 mM EDTA, and dialyzed against the same buffer, and chromatographed on a DEAE-cellulofine column (1.5 x 14 cm) which was previously equilibrated with the same buffer solution. The isozymes were eluted with 50 mL of the same buffer solution, and then with 400 mL of the same buffer solution at a linear gradient of 0 to 0.3 M of NaCl.

2C. Chromatography on a Sephadex G-75 Column

Gel filtration was performed on a Sephadex G-75 column (1.5 x 76 cm) which was previously equilibrated with 0.2 M sodium phosphate buffer (pH 6.0). Bovine serum albumin (66 kDa), ovalbumin (45 kDa), α-chymotrypsino-gen A (bovine pancreas, 26 kDa), and lysozyme (chicken egg-white, 14 kDa) were used as reference compounds for molecular weights.

2D. Electrophoresis on Polyacrylamide Gel

Electrophoresis was conducted on a polyacrylamide gel at pH 4.0 and 8.0 according to Reisfeld, et al[8] and to Williams and Reisfeld.[9] Chitinase activity in the gel was detected with O-glycolchitin as a substrate in 0.2 M sodium acetate buffer solution (pH 3.0) according to Trudel and Asselin.[10]

2E. Chicken Embryo Fibroblast Cells, and Lysozyme Assay

Chicken embryo fibroblast cells, which were prepared from the chicken embryo of eggs fertilized for 12-13 days, were suspended at about 1.0 x 10^6 cells/mL in a growth medium [Eagle's minimum essential medium (90 mL), 5% calf bovine serum (5.0 mL), 5% tryptose phosphate broth (5.0 mL), 2.92% L-glutamic acid (1.5 mL), 0.1 mL of a mixture of penicillin (100 units/mL) and streptomycin (100 μg/mL) and 2.9 mL of 7.5% NaHCO$_3$ solution]. Ten mL of the cell suspension was placed in a sterilized glass-petri dish (diameter 9.0 cm). After incubating in 5.0% CO$_2$ at 37°C for 4-6 days, the lysozyme activity was determined in the presence of 0.1, 0.5, 1.0, or 10 mg/dish for each compound in a concentration of 10 mg/mL. An appropriate amount (see footnote in Table 5) was intravenously injected daily for 5 days through a 0.2 μm filter into the vein of male rabbit conchae. For the analysis of the serum lysozyme activity, the blood (2 mL) was drawn from rabbits, and kept at room temperature for 30 min in a glass-test tube, and centrifuged at 1,500g at 0°C for 20 min to give a serum fraction. The serum (0.2 mL) was mixed with 2.8 mL of a suspension of the lyophilized cell walls (0.25 mg/mL) of *M. lysodeikticus* (M-3770, Sigma) in 0.1 M phosphate buffer solution (pH 7.4). The suspension was incubated at 37°C with mechanical shaking, and the turbidity was monitored at 600 nm by a Shimadzu spectrophotometer (UV-2200). The turbidity decreased proportionally with the elapse of reaction-time for up to 60 min. One unit (U) for the enzyme activity is defined as a decrease in the absorption of 0.001 at 600 nm per min.

RESULTS AND DISCUSSION

1. THE STIMULATION OF CHITINASE ACTIVITY IN CALLUSES

Rice callus, which was cultured for 35 days on MS agar medium, was transplanted on the fresh medium. As shown in Table 1, the dry weight of the callus was decreased slightly by treatment with 0.5 mL of 0.05 or 0.1% solution of CM-chitin, but decreased little by treatment with its 0.01% solution. Treating the callus with 0.1% solution of CM-chitin stimulated the callus chitinase activity, but decreased the dry weight with a lightly brown color. The similar results were found by treatment with O-glycolchitin or with the cell walls of *M. lysodeikticus* in rice and soybean calluses.[11] Therefore, the callus was treated with 0.01% solution (0.2 or 0.5mL) of a test-compound, and the stimulation of chitinase activity in the callus was analyzed at the same culture-age. The rice callus chitinase was stimulated in the following sequence: CM-chitin > cell walls of *M. lysodeikticus* > chitin oligosaccharides > chitosan oligosaccharides > O-glycolchitin > cell walls of *Helminthosporium oryzae* > chitosan chitin (Table 2). In cabbage leaf callus, the stimulation rankings were in the sequence: LMW chitosan > chitosan oligosaccharides,[2] and in soybean callus the sequence was: LMW-chitosan > CM-chitosan > O-glycolchitosan.[11] The chitinase activity in radish seeds, which were coated with chitosan or chitosan derivatives, was stimulated at the growing period of germination (Table 3). Its stimulation sequence was: LMW chitosan ≥ chitosan-oligosaccharides > chitosan, but D-glucosamine was inactive.[2] Although the time of initial response and stimulating extent varied with compounds and with their concentration, the induction of chitinase activity appeared within a few hours to 3 days after treatment (Figure 3). Their chitinase activity increased about 1.5-fold and was kept for at least 25 days.

2. THE INDUCTION OF CHITINASE IN RICE CALLUS

As shown in Figure 4, the crude enzyme extract of the rice callus untreated was chromatographed on a DEAE-cellulofine column. Two chiti-

Table 1. The dry weight and chitinase activity of rice callus treated with CM-chitin.[a]

Concentration (%)	Relative dry weight	Chitinase activity (U/g of dry weight)
0 (control)	1.0	4.95 ± 0.20
0.01	0.9	5.64 ± 0.16
0.05	0.7	6.40 ± 0.25
0.1	0.6	7.28 ± 0.54

(a) 0.5 mL each of 0.01, 0.05, and 0.1% solutions of CM-chitin in 20 mM sodium phosphate buffer solution (0.5 mL, pH6.0) was dropped onto the surface of rice callus (about 1.5 g by fresh weight), and the callus was cultured for 15 days. Each value is the mean of three experiments.

Table 2. The stimulation of chitinase in rice callus by treating with chitin or chitosan derivatives.

Days of culture after treatment	Chitinase activity (U/g of dry weight)[a]				
	0	1	2	3	7
Untreated	18.9	21.6	21.7	19.7	20.3
O-Carboxymethyl Chitin	n.d.	n.d.	n.d.	36.1	40.5
Cell walls of M. lysodeikticus	n.d.	32.1	39.0	32.9	n.d.
Chitin-oligosaccharides	n.d.	31.8	30.5	29.9	n.d.
Chitosan-oligosaccharides	n.d.	n.d.	28.5	n.d.	n.d.
O-Glycolchitin	n.d.	n.d.	n.d.	28.0	28.5
Cell walls of H. oryzae	n.d.	26.9	26.6	24.0	n.d.
Chitosan	n.d.	22.4	22.2	21.3	n.d.

(a) Each value represents the mean of three experiments.

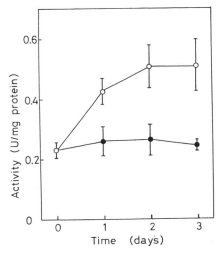

Figure 3. Cellular response to chitin-oligosaccharides in
rice callus. Onto the surface of callus, 0.5 mL of
20 mM sodium phosphate buffer solution (pH 6.0)
with (o) or without (●) chitin-oligosaccharides
(2.5 mg) was dropped, and the callus was cultured
for additional 3 days. Each experiment point is
the mean of five experiments.

nase-active fractions (isozymes I and II) appeared, and an additional
chitinase-active fraction (isozyme III) appeared in the extract of the
callus treated with CM-chitin. The isozyme I was significantly stimulated
by treatment with CM-chitin, but the isozyme II activity was only
slightly stimulated. When chromatographed on a Sephadex G-75 column, the
isozymes I, II, and III were eluted at the molecular mass of 28, 32, and
32 kDa, respectively. After purification by re-chromatography on a CM-
cellulose or a DEAE-cellulofine column, and on a Sephadex G-75 columns,
each of these isozymes gave a single band, as analyzed by electrophoresis
on a polyacrylamide gel at pH 4.0 (isozyme I) and at pH 8.0 (isozymes II

Table 3. The stimulation of the extracellular activity of
lysozyme in the culture of chicken embryo fibroblast cells by treating with chitin or chitosan
derivatives.[a]

Compound	The extracellular activity of lysozyme
Untreated (control)	inactive
Chitin	0.0394 ± 0.0033
Chitin-oligosaccharides (d.p. 7-8)	0.0020 ± 0.0006
Chitin-oligosaccharide (d.p. 2-6)	almost inactive
Chitosan	almost inactive

(a) The test was done by incubating at 10 mg each of the compounds per
dish at 37°C for 5 days. Each value is the mean of three
experiments.

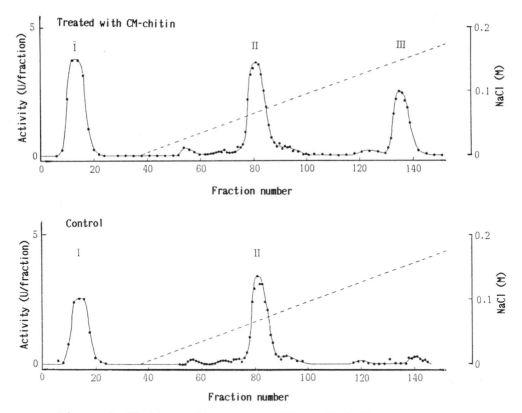

Figure 4. Elution patterns on a DEAE-cellulofine column for isozymes of rice chitinase from the calluses treated and untreated with CM-chitin. Chitinase activity was determined in each fraction (2 mL).

and III). These isozymes did not hydrolyze cell walls of *M. lysodeikticus* (Sigma), indicating no lysozyme activity. All the isozymes showed the maximum activity at pH 2.3-2.7, and stable even after 15 min incubation at pH 2-4 and 50°C. The Km values of the isozymes I, II, and III using N-acetylchitosan were 1.7-2.2, 4.5 and 17 mg/mL, respectively. Tri- and tetra-saccharides as main products, with hydrolysate by saccharides as minor products, were detected in a hydrolysate by each of the chitinase isozymes, as examined by thin-layer chromatography with a solvent system of n-propanol:ethyl acetate:water (5:1:3, v/v). The result indicates that the isozymes are endo-hydrolases.

3. THE STIMULATION OF EXTRACELLULAR LYSOZYME ACTIVITY IN THE CULTURE OF CHICKEN EMBRYO FIBROBLAST CELLS

At time-intervals of the culture of chicken embryo fibroblast cells, the decrease of turbidity at 600 nm (lysozyme activity) was 0.0124 ± 0.0032 for 24 h, 0.630 ± 0.057 for 72 h, and 0.0532 ± 0.0012 for 120 h in the presence of chitin (10 mg/dish), but was little in the absence of chitin as a control (Figure 5). Almost no intracellular activity was detected in both the presence and absence of chitin. A mixture of chitin-

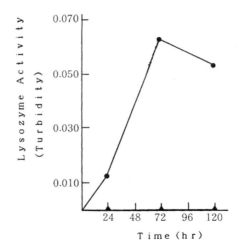

Figure 5. The time course of the extracellular activity of
lysozyme in chicken embryo fibroblast cells
treated with chitin, Lysozyme activity with (●) or
without (▲) of chitin (10 mg/dish). Each
experiment point is the mean of three experiments.

oligosaccharides (d.p. 7-8), was active, but that of chitin-oligosaccharides (d.p. 2-6) was almost inactive (Table 4).

Lysozyme activity in rabbit serum was stimulated by the intravenous
injection of a mixture of chitosan-oligosaccharides, but it was only
little stimulated by the injection of a mixture of chitin-
oligosaccharides or of galactosaminan-oligosaccharides at the same dosage

Table 4. The stimulation of chitinase activity in seedlings
of radish seeds coated with chitosan derivatives.[a]

Compound used for coating	Chitinase activity	
	mU/g dry seedlings	mU/mg protein
Uncoated (control)	590 ± 15	1.7 ± 0.3
Chitosan	640 ± 20	1.8 ± 0.3
Chitosan-oligosaccharides	740 ± 1	2.2 ± 0.5
O-Glycolchitosan	590 ± 29	1.6 ± 0.5
LMW-chitosan		
(0.01g/60g seeds)	660 ± 29	2.1 ± 0.4
(0.1g/60g seeds)	790 ± 31	2.3 ± 0.3
(1.0g/60g seeds)	600 ± 30	1.7 ± 0.3
D-Glucosamine	590 ± 15	not determined

(a) Analyzed on the 4th day after germination. Each value is the mean of
3 to 5 experiments.

(Table 5). A mixture of chitin-oligosaccharides (d.p. 7,8) was active in the cell culture, but almost inactive in intravenous injection, indicating different cellular responses to chitin or chitosan. The conflict is under investigation. Orally administrated chitosan is digested by chitinase and chitosanase, which are secreted from intestinal microorganisms in animal digestive organs.[12] LMW-chitosan and chitosan-oligosaccharides are absorbed in their intestinal organs, and transported into the blood, where the activity of blood lysozyme is stimulated. This is in agreement with an inhibitory activity of chitin-oligosaccharides against infection with *Listeria momocytogens* in mice.[13]

4. A PROPOSED MECHANISM FOR THE INDUCTION OF CHITINASE ISOZYMES IN PLANTS

The chitinase isozymes are classified into class I-chitinase for basic isozymes with cysteine-rich domain at their N-terminal end, class II-chitinase for acidic isozymes similar to class I-chitinase but lacking the cysteine-rich domain, and class III-chitinase for the isozymes with a conserved sequence different from classes I and II-chitinases.[14] Acidic isozymes of chitinase are induced by infection with fungal phytopathogens in tomatoes,[15] and basic isozymes by treatment with ethylene in beans[16,17] and tobaccos.[18] However, both acidic and basic isozymes were induced in tobaccos by infection with tobacco mosaic virus.[19] The basic isozymes (class I-chitinase) are generally found inside cells such as vacuole,[20-22] and the class II and III-chitinases are found outside cells.[15,23] Their enzyme inductions are essentially similar in insects,[24] plants, and animals. The stimulation of enzyme activity is a signal for a cellular response to chitin or chitosan. The cellular response is accompanying with the activation of macrophage system in animals,[25] with the stimulation of phenylalanine ammonia-lyase for lignin and phytoalexin syntheses in plants,[26] or with the production of antibacterial proteins in insects.[27]

As shown in Figure 6, chitin or a derivative is a trigger for the cellular response and informational delivery in plant cells, and the signals of the cellular response are the stimulation and induction of the activity of chitinase and phenylalanine ammonia-lyase. Chitin or deriva-

Table 5. The stimulation of serum lysozyme activity in rabbits by the intravenous injection of chitosan or derivatives.

Compound[a] injected	Blood Lysozyme Activity (U/mL serum) Day after the last injection			
	1st	3rd	5th	60th
Saline (control)	4.4 ± 1.2	4.4 ± 2.0	n.d.[b]	4.3 ± 1.2
Chitosan-oligosaccharides	9.2 ± 2.2	7.7 ± 2.2	6.9 ± 2.4	4.7 ± 2.0
Chitin-oligosaccharides	4.4 ± 1.5	3.7 ± 1.6	n.d.	n.d.
Galactosaminan-oligosaccharides	4.5 ± 0.9	4.4 ± 0.8	n.d.	n.d.

(a) Daily injected for 5 days at a dosage of 7.1-8.6 mg/(kg per body weight) per day into rabbits weighing 3.5-4.2 kg per rabbit. Each value is the mean of three experiments in three to six rabbits.
(b) Not determined.

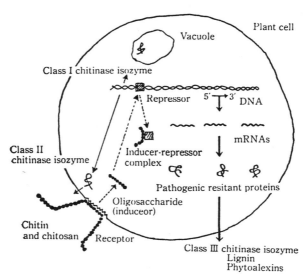

Figure 6. A proposed mechanism for the induction of chitinase
isozymes in plants (see the text).

tive binds as an elicitor to a receptor site over the plant cell wall,
and is hydrolyzed by class II-chitinase to give chitin or chitinase
oligosaccharides in a trace amount. The oligosaccharides enter into
cytoplasm through the cell wall, translocate in a receptor site of a
repressor on DNA and remove a repressor located on the DNA. The transcription of novel mRNAs begins to produce novel proteins involving class
III-chitinase, phenylalanine ammonia-lyase (EC 4.3.1.5), and other proteins, which are "ecologically bioactive proteins." This system is one of
the cellular responses of plants for protecting plant cells against
infection with pathogens.

5. SOME APPLICATIONS IN THE FIELD OF BIOTECHNOLOGY

Chitin and chitosan have a potential value in biotechnological
applications as an ecologically bioactive material, which enhances the
self-defensive function of plant and animal cells against pathogenic
infection. Some of the possible applications are:

(1) Adding chitin or chitosan to the culture media of plant and
animal cells and tissues in order to enhance their growth with
preventing infection with pathogens or germs.[28]

(2) Coating plant seeds with chitin or chitosan in order to stimulate seed chitinase activity (Table 3),[29] and to increase plant
production.[2,30]

(3) Adding chitin and chitosan into soils in order to stimulate the
growth of ray fungi involving *Actinomycetes* and to inhibit the
growth of mold fungi involving *Fusarium*.[31]

(4) Sprinkling chitin or chitosan over plant leaves and soils in
order to activate plant cells and tissues.[31]

(5) Covering or smearing plant and animal wounds with chitosan, or dipping the wounds in a chitin or chitosan solution in order to accelerate wound-healing with the prevention of pathogenic infection in plants and animals.[32]

(6) Adding chitosan to animal feeds as an ingredient in order to stimulate the lysozyme activity of bloods and tissues in animals (Table 5), and to improve the distribution of intestinal microorganisms.[12,33]

(7) Adding chitin or chitosan to food-manufacturing processes in order to prevent the growth of molds in foods even at low levels of NaCl,[34] and to produce biofunctional foods decreasing blood cholesterol level in humans.[12]

REFERENCES

1. D. Knorr, Food Technol., **85**, (1984).
2. S. Hirano, M. Hayashi, K. Murae, H. Tsuchida, & T. Nishida, in: "*Applied Bioactive Polymeric Materials*," C. G. Gebelein, C. E. Carraher, Jr., & V. R. Foster, Eds., Plenum Publ. Corp., New York, 1988, p. 45.
3. A. Schlumbaum, F. Mauch, U. Vogeli, & T. Boiler, Nature, **324**, 365 (1986).
4. S. Hirano, Y. Koishibara, S. Kitaura, T. Tanaka, H. Tsuchida, K. Murae & T. Yamamoto, Biochem. System. Ecol., **19**, 379 (1991).
5. S. Hirano, H. Seino, Y. Akiyama & I. Nonaka, in: "*Progress in Biomedical Polymers*," C. G. Gebelein & R. L. Dunn, Eds., Plenum Publ. Corp., New York, 1990, p. 283.
6. O. H. Lowry, N. J. Rosebrough, A. L. Farr & R. J. Randall, J. Biol. Chem., **193**, 265 (1951).
7. T. Imoto & K. Yagishita, Agric. Biol. Chem., **35**, 1154 (1971).
8. R. A. Reisfeld, U. J. Lewis & D. E. Williams, Nature, **195**, 281 (1962).
9. D. E. Williams & R. A. Reisfeld, Ann. New York Acad. Sci., **121**, 373 (1964).
10. J. Trudel & A. Asselin, Anal. Biochem., **178**, 362 (1989).
11. S. Hirano & S. Okuno, to be published.
12. S. Hirano, C. Itakura, H. Seino, N. Kanbara, Y. Akiyama, I. Nonaka & T. Kawakami, J. Agr. Food Chem., **38**, 1214 (1990).
13. A. Tokoro, M. Kobayshi, N. Takewaki, K. Suzuki, Y. Okawa, T. Mikami & M. Suzuki, Microbiol. Immunol., **33**, 357 (1989).
14. H. Shinnshi, J. M. Neuhaus, J. Ryals & F. Meins, Plant Mol. Biol., **14**, 357 (1990).
15. M. H. A. J. Joosteen & P. J. G. M. DeWitt, Plant Physiol., **89**, 945 (1989).
16. K. E. Broglie, J. J. Gaynor & R. M. Broglie, Proc. Natl. Acad. Sci. USA, **83**, 6820 (1986).
17. K. E. Broglie, P. Biddle, R. Cressman & R. Broglie, Plant Cell, **1**, 599 (1989).
18. D. Keefe, U. Hinz & F. Meins, Plant, **182**, 43 (1990).
19. M. Legrand, S. Kauffmann, P. Geoffroy & B. Fritig, Proc. Natl. Acad. Sci. USA, **84**, 6750 (1987).
20. J. Malamy, J. P. Carr, D. F. Kleisig & I. Raskin, Science, **250**, 1002 (1990).
21. F. Mauch & L. A. Staehelin, Plant Cell. **1**, 447 (1989).
22. T. Boiler & U. Vogeli, Plant Physiol., **74**, 442 (1984).
23. N. Benhamou, M. H. A. J. Joosteen & P. J. G. M. DeWitt, Plant Physiol., **92**, 1108 (1990).

24. R. S. Anderson & M. L. Cook, J. Invertebr. Pathol., **33**, 197 (1979).
25. K. Nishimura, S. Nishimura, N. Nishi, F. Numata, Y. Tone, S. Tokura & I. Azuma, Vaccine, **2**, 93 (1984).
26. L. A. Hadwiger & J. M. Beckman, Plant Physiol., **66**, 205 (1980).
27. G. P. Kaaya, C. Flyg & C. H. Boman, Insect Biochem., **17**, 309 (1987).
28. S. Hirano & N. Nagoa, Agric. Biol. Chem., **53**, 3065 (1989).
29. S. Hirano, T. Yamamoto, M. Hayashi, T. Nishida & H. Inui, Agric. Biol. Chem., **54**, 2719 (1990).
30. L. A. Hadwiger, R. Fristensky & R. C. Riggleman, in: "*Chitin, Chitosan, and Related Enzymes*," J. P. Zikakis, Ed., Academic Press, Orlando, 1984, p. 291.
31. S. Hirano, "Nogyogijutsutaikei (Japanese)", **7**, 156·18 (1992).
32. S. Hirano, in: "*A Hand-book for Novel Medical Materials*," M. Seno & O. Otsubo, Eds., R & D Planning, Tokyo, 1986, p. 235.
33. P. R. Austin, C. J. Brine, J. E. Castle & J. P. Zikakis, Science, **212**, 749 (1981).
34. Y. Uchida, Food Chemicals (Japanese), February, p. 22 (1988).

RECOVERY OF PROTEINS FROM WHEY USING CHITOSAN AS A COAGULANT

John F. Kennedy, Marion Paterson, David W. Taylor and
Maria P. C. Silva

Research Laboratory for the Chemistry of Bioactive
Carbohydrates and Proteins,
Department of Chemistry
The University of Birmingham, P O Box 363
Birmingham B15 2TT. England

In recent years the attitude towards whey proteins has
changed and they are no longer considered as waste products,
but rather as valuable nutrients that can be used in the food
industry. A study of the effectiveness of chitosan as a
coagulant for removing proteins from whey elucidated the
conditions at which chitosan operates most efficiently. Such
factors as optimum dosage, pH of reaction, time of mixing and
sedimentation rates were studied. The highest recovery of
proteins from whey was *ca.* 40%, estimated by amino acid
analysis.

INTRODUCTION

Whereas so much of current lifestyle is linked to what is frequently
called the "disposable age", the principle of items being readily
disposable is at variance with the terms "recovery" and "recycling" which
are also currently given precedence.

Proteins have a variety of function and specifities, and therefore
can have considerable values, such values make recovery and recycling
worthwhile. The problems of this recovery of proteins, because of the
sensitivity of their structures needs to be carried out using mild
procedures and mild reagents.

Separations in downstream processing have become very important but
large scale effective chromatography has yet to be developed to an
economic level. Furthermore, chromatographic separations not only yield
material of high value quality, but must also yield material of highly
safe quality suitable for use, e.g., in foodstuff and pharmaceutical
preparations. Many polysaccharides, particularly those with carboxyl
groups, have a variety of uses in complexation with other molecules, and
such uses include a phase change. Chitosan is different in that it takes
a positive charge, is basic, offers other complexing abilities, and is
readily available. Its application is, however, underdeveloped.

Biotechnology and Bioactive Polymers, Edited by C. Gebelein
and C. Carraher, Plenum Press, New York, 1994

55

This paper deals with the application of chitosan to protein isolation and purification from whey. Whey is a great source of materials such as carbohydrates, amino acids, and proteins, including valuable fractions such as ß-lactogloblin, and α-lactalbumin and immunoglobulins. The whey by-product from cheese manufacture is already crudely separated into high- and low-molecular weight fractions by ultrafiltration into protein going for animal feed and carbohydrate going for fermentation to alcohol.

Whey may be defined in a very general sense as the liquid remaining after removal of casein from milk. Whey composition varies depending on milk composition and method of casein removal, but it typically contains about 50% of the total solids of the original milk and includes fat, lactose, minerals and vitamins.

The major proteins remaining in whey precipitation of caseins consist of ß-lactoglobulin (2.0-4.0 g/L), α-lactalbumin (1.0-1.7 g/L), serum albumin (0.1-0.4 g/L, immunoglobulins (0.6-1.0 g/L) and components of the proteose-peptone fraction (0.6-1.8 g/L), which include post-translational proteolysis products of caseins and several minor whey proteins.[1,2]

In recent years, the attitude towards whey products has changed and they are no longer considered as waste products, but rather as valuable nutrients that can be used in the food industry. However, this use is dependent on, among other factors, the possibility of tailoring whey proteins for specific end-applications and recovering them with intact functional properties.

The use of chitosan ((1→4)-2-amino-2-deoxy-ß-D-glucan) (Figure 1) as a coagulant for the removal of proteinaceous solids from whey is known.[3] The glucosamine backbone of chitosan gives it a polycationic character which places it in a similar class with other polyelectrolytes being used as coagulating and flocculating agents. This paper demonstrates the effectiveness of chitosan as a coagulant for selectively removing proteins and elucidates the conditions at which chitosan operates most effectively. Such factors as optimum dosage, pH of reaction, time of mixing and sedimentation rates have been studied.

EXPERIMENTAL

1. EFFECT OF THE CONCENTRATION OF CHITOSAN ON THE EXTRACTION OF PROTEINS FROM WHEY

Chitosan from shrimp, supplied by Proteins Laboratories, Inc. (Washington), was solublized in acetic acid (1% v/v, 10 mg/mL). The solubilized chitosan was added to whey (pH 3.5, 5 mL), the concentration of which ranged from 0 to 800 mg/L, and was mixed using a magnetic stirrer, at room temperature, for 1 h. The pH was increased to 9.0 with sodium hydroxide solutions and the suspension centrifuged at 3000 rpm for 10 min. As a control, acetic acid solution (1%, v/v) without chitosan was added to whey under the same conditions described above. The absorbances of the supernatants (dilution 1:20 in distilled water, of whey with and without chitosan and of the solubilized chitosan were measured using a UV/vis spectrophotometer at 280 and 450 nm (Figure 2). The concentration of protein (mg/mL) in solution was considered as the difference between the absorbances at 280 and 450 nm.

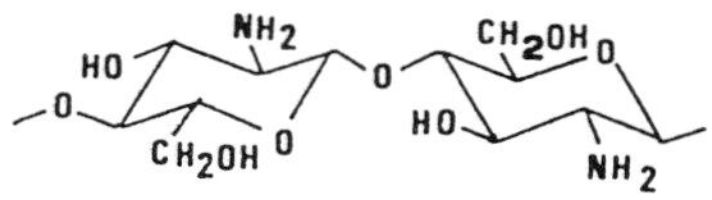

Figure 1. Chemical structure of chitosan.

2. EFFECT OF THE TIME OF MIXING ON THE EXTRACTION OF PROTEINS FROM WHEY BY CHITOSAN

Chitosan from shrimp was solubilized in acetic acid (1% v/v, 10 mg/mL). The solubilized chitosan (400 μL) was added to whey (pH 7.0, 5 mL) and mixed using a magnetic stirrer at reaction times ranging from 1 to 15 h, at room temperature. The same procedure was carried out as described in Section 1.

3. EFFECT OF THE PH OF REACTION ON THE EXTRACTION OF PROTEINS FROM WHEY USING CHITOSAN

Chitosan from shrimp was solubilized in aqueous acetic acid (1% v/v, 10 mg/mL). Treatment A: The solubilized chitosan (400μL) was added to whey (pH ranging from 3.5 to 7.0, 5 mL) and mixed using a magnetic stirrer, at room temperature, for 2 h, at the desired pH. As controls, acetic acid solutions (1% v/v, 400 μL) without chitosan were added to whey (pH ranging from 3.5 to 7.0, 5 mL) and were treated under the same conditions described above. The same procedure was carried out as described in Section 1. Treatment B: solubilized chitosan (160 or 320 μL) was added to the supernatant (pH adjusted to 7.0 for all samples, 2 mL) obtained from treatment A and using a magnetic stirrer, at room temperature, at pH 7.0, for 2 h. The same procedure was carried out as described in Section 1. The concentration of protein (mg/mL) in the original whey was calculated taking into account the dilutions during the treatments.

4. EFFECT OF THE PH OF REACTION AND CONCENTRATION OF CHITOSAN ON THE EXTRACTION OF PROTEINS FROM WHEY

Chitosan from shrimp was solubilized in acetic acid (1% v/v, 10 mg/mL). Treatment A: Solubilized chitosan (400 or 800 μL) was added to whey (pH 3.5 or 7.0, 5 mL) and mixed using a magnetic stirrer, at room temperature for 2 h. As controls, acetic acid solutions (1% v/v, 400 μL) without chitosan were added to whey (pH 3.5 or 7.0, 5 mL) and were treated under the same conditions described above. The same procedure was carried out as described in Section 1. Treatment B: Solubilized chitosan (160 or 320 μL) was added to the supernatant (pH 3.5 or 7.0, 2 mL) obtained from treatment A and mixed using a magnetic stirrer, at room temperature, at pH 3.5 or 7.0, for 2 h. The same procedure was carried out as in Section 1. The concentration of protein (mg/mL) in the original whey was calculated taking into account the dilutions during the treatments.

5. SEDIMENTATION RATES (FREE FALL) OF PRECIPITATES OBTAINED FROM WHEY TREATED WITH CHITOSAN

The sedimentation rates (free fall) of the precipitates obtained

57

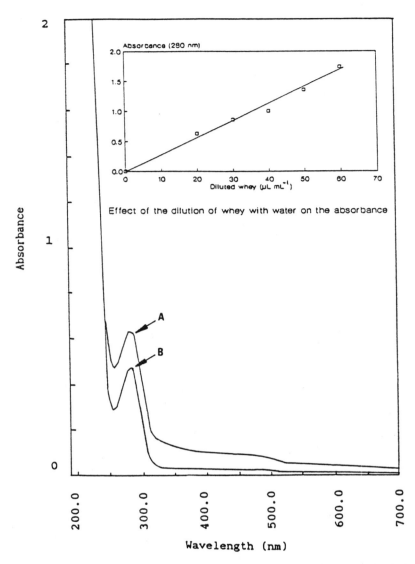

Figure 2. UV/vis spectra of whey (A) and whey treated with
chitosan (B).

from whey treated with chitosan were followed by measuring the volume of
precipitates immediately after increasing the pH to 9.0 (before
centrifugation) for 1 h (Treatment A) and for 30 min (Treatment B).
Treatment A: The concentration of chitosan added to whey was 800 mg/L and
the reaction was carried out at pH 3.5. Treatment B: The concentration of
chitosan added was 86.4 mg/L of the original whey and the reaction was
carried out at pH 7.0.

6. DRY MATTER OF WHEY AND OF WHEY TREATED WITH CHITOSAN (SUPERNATANT   AND
   PRECIPITATE)

   Whey (500 μL), supernatant (500 μL) and precipitate obtained from treatment A were lyophilized until dry and weighed. The contribution of dry matter (mg) of each component in the original whey (mL) was calculated taking into account the dilutions during the treatment.

7. AMINO ACID COMPOSITIONS OF WHEY AND OF WHEY TREATED WITH CHITOSAN (SUPERNATANT AND PRECIPITATE)

Protein samples were hydrolyzed in hydrochloric acid (5.8 mol/L) at 110°C, for 24 h. An automatic amino acid analyzer was used for the measurement of amino acids in hydrolyzed proteins.

8. EXTRACTION OF CHITIN FROM CRABS

8A. Method A[4,5]

Crab claws were cleaned by washing and scraping under running water, and dried in an oven at 100°C. The claws were ground to powder using a Glen Creson ball mill (50 mesh). Ground claws (100 g) were soaked in sodium hydroxide (100 g/L, 500 mL), at room temperature, for 3 days. Fresh sodium hydroxide solution was prepared every day. The deproteinized material was washed with water until free of alkali and with ethanol (95% v/v) to remove pigments. The product was dried, treated with hydrochloric acid (32%) at -20°C for 4 h, centrifuged for 30 min at 3000 rpm and the supernatant discarded. This procedure was repeated twice. The product was then washed with distilled water, the pH neutralized using a solution of sodium hydroxide (5.5 mol/L), centrifuged for 30 min at 3000 rpm, and the supernatant discarded. This washing procedure was repeated at least five times. The chitinous material was dried and ground again.

8B. Method B[6,5]

Crab claws were cleaned by washing and scraping under running water, and dried in an oven at 100°C. The claws (100 g) were digested for 5 h with hydrochloric acid (2 mol/L, 1 L), at room temperature, washed, dried and ground to a powder using a Glen Creson ball mill (50 mesh). The ground material was extracted for 2 days with hydrochloric acid (2 mol/L) at 0°C, the content of the flask being vigorously agitated from time to time. The collected material was washed and extracted for 12 h with sodium hydroxide (1 mol/L, 250 mL) at 100°C under occasional stirring. The alkali treatment was repeated four times. The chitinous material was dried and ground again.

9. DEACETYLATION OF CHITINOUS MATERIAL EXTRACTED FROM CRABS[7,8,5]

The chitinous materials (ca. 85 mg) extracted from methods A and B (Section 8) were treated with sodium hydroxide (40 mg/mL, 4.1 mL) at 108°C, using a 2-methylpropan-1-ol dryer, for 24 h. Deacetylated samples were then thoroughly washed with distilled water, and pH neutralized using a solution of hydrochloric acid, filtered and dried in oven at 60°C.

10. COMPARISON OF CHITOSAN FROM CRAB WITH CHITOSAN FROM SHRIMP AS COAGULANTS FOR THE EXTRACTION OF PROTEIN FROM WHEY

Chitosan preparations from crab obtained by methods A and B (Section 8) followed by deacetylation (Section 8) and from shrimp (Protan) were solubilized in aqueous acetic acid (1% v/v, 10 mg/mL). The treatment carried out were as described in Section 4 at room temperature.

RESULTS AND DISCUSSION

Polyelectrolytes are special classes of polymers containing certain functional groups along the polymer backbone which may be ionizable and have the ability to destabilize or enhance the flocculation of the constituents in the aqueous medium.[9] Several polysaccharides have been used as polyelectrolytes, e.g., chitosan, alginate, carboxymethyl cellulose, carrageenans, etc. The destabilization mechanism operative with polyelectrolytes is complex and cannot be collectively ascribed to one particular phenomenon. However, it is possible to set down two principal mechanisms which are based on: (1) a bridging model, where polyelectrolytes segments are adsorbed on the surfaces of adjacent colloids thereby binding them together, and (2) a model whereby ionic polyelectrolytes bearing a charge of opposite sign to the suspended material, are adsorbed and thereby reduce the potential energy of repulsion between adjacent colloids. In some instances the two mechanism are operative conjointly, whereas in others there is a predominance of one over the other. The several stages in the binding mechanisms is as follows:[10]

1. Dispersion of polyelectrolytes in the suspension;
2. Adsorption at the solid-liquid interface;
3. Compression or settling down of the adsorbed polyelectrolyte and;
4. Collision of adjacent polyelectrolyte coated particles to form bridges and thereby increasingly larger flocs.

Each of these stages are schematically represented in Figure 3. At excess flocculent concentrations, surfaces become saturated with adsorbed polymer and the particles are re-stabilized. Not only is bridging then prevented but the particles may also be sterically stabilized (Figure 4).[11] For the case of non-ionic and anionic polyelectrolytes applied to a negatively charged colloidal dispersion, the bridging model destabilization mechanism accounts for the phenomenon taking place.

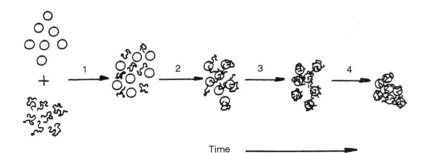

Time

Figure 3. Stages in the bridging mechanism of destabilization with polyelectrolytes. (1) Dispersion; (2) adsorption; (3) compression or settling down and (4) collision.

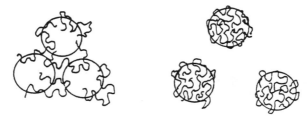

Figure 4. Flocculation and restabilization by adsorbed polymers.

However, for systems where strong electrostatic attraction between
polyelectrolyte and particle surface exists, an electrostatic patch
mechanism is proposed.[11] Rather than adsorption of polyelectrolyte at
only a few sites, with the remainder of the chain extending into solution
in the form of closed loops, virtually complete adsorption of added
polyelectrolyte onto the particle surface takes place with such systems.
The adsorbed polyelectrolyte chains thus form a charged mosaic with
alternating regions of positive and negative charge as shown in Figure 5.
Destabilization occurs when the charge mosaics of adjacent particles
align to provide strong electrostatic attraction.

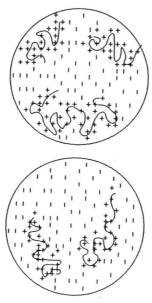

Figure 5. Electrostatic patch model for the interaction of
negatively charged particles with adsorbed cationic polymer.

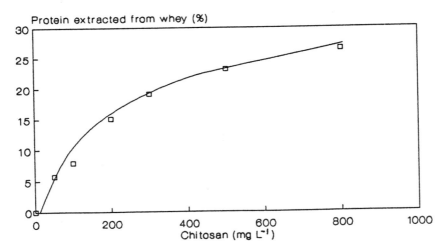

Figure 6. Effect of the concentration of chitosan on the
extraction of proteins from whey.

1. EFFECT OF THE CONCENTRATION OF CHITOSAN ON THE EXTRACTION OF PROTEINS
FROM WHEY

The effect of the concentration of chitosan on the extraction of
proteins from whey is shown in Figure 6. An increase of turbidity of the
supernatant was observed at a concentration of 1 g/L. The optimum dosage
of chitosan required for extraction of proteins from whey was *ca.* eight
times higher than that required for extraction of single cell protein
from fermentation broth.[12]

For a given particle mass concentration the optimum dosage of
polyelectrolyte depends on the type of material present, i.e., smaller
particle suspensions require higher dosage at optimum destabilization
(because of the higher total surface area). Furthermore, the larger the
range of particle sizes of a given suspension the wider is the range of
polyelectrolyte dosage giving a good performance.

The effect of the coagulant dosage on destabilization depends very
much on the destabilization mechanism operative. The increase of
turbidity of the treated whey observed with concentrations of chitosan of
ca. 1g/L is a common phenomenon which occurs with adsorption of
coagulants to colloidal particles.[9]

The availability of adsorption sites on particles to accommodate
polyelectrolyte loops from neighboring particles is an important factor
during bridging. If an excess of polyelectrolyte is added, too many
adsorption sites per particle will be occupied and bridge formation is
prevented: the particles effectively become restabilized (Figure 4).

2. EFFECT OF THE TIME OF MIXING ON THE EXTRACTION OF PROTEINS FROM WHEY
USING CHITOSAN

There are two stages in the flocculation process: the first, which
is perikinetic flocculation, arises from thermal agitation (Brownian
movement) and is a naturally random process. Flocculation during this
stage commences immediately after destabilization and is complete within

seconds since there is a limiting floc size beyond which Brownian motion has no or little effect.

The second stage is orthokinetic flocculation and arises from induced velocity gradients. The degree or extent of flocculation is governed by both applied velocity gradients and time of flocculation.

The rapid mixing stage is possibly the most important operation in the process of coagulation/flocculation process since at this stage destabilization reactions occur and primary flocs are formed. The effect of time of mixing on the extraction of proteins from whey by chitosan is shown in Figure 7. It is apparent that for this particular application, the rate of flocculation was maximized at a mixing period of 2 h. Extending the period of mixing disrupts the polyelectrolyte bridges and gives rise to desorption and/or rearrangement of looped chains on the particle. A decrease of 17% in the recovery of protein from whey was observed at mixing period of 15 h compared to 2 h. The effect of time of mixing and concentration of chitosan is shown in Table 1. In treatment A, an increase of concentration produced restabilization of the particles and decrease of protein extracted. On the other hand, an increase of mixing time to 4 h resulted in an increase of protein extracted. In treatment B (using the supernatants from treatment A), with the increase of both concentration and time of mixing, an increase of protein extracted was observed.

3. EFFECT OF THE PH OF REACTION ON THE EXTRACTION OF PROTEINS FROM WHEY USING CHITOSAN

The pH is another important factor in destabilization of colloids. Table 2 shows the effect of the pH on the extraction of proteins from whey using chitosan. The maximum removal of proteins occurred at pH 3.5, 5.5 and 7.0. Variations of the isoelectric points of the biopolymers present in the whey is a point to be considered. Changes in the extraction of proteins from whey in response to different concentrations of chitosan and different pHs are shown in Table 3. In treatment A, an increase in the chitosan concentration at pH 3.5, produced a

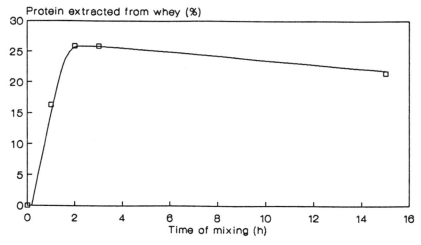

Figure 7. Effect of the time of mixing on the extraction of proteins from whey by chitosan.

Table 1. Extraction of protein from whey using different
concentrations of chitosan and different mixing
times.

Control Protein in whey mg/mL	Treatment A[a]				Treatment B[b]		
	Chitosan in whey mg/mL	Time h	Protein in whey mg/mL	Protein extracted %	Chitosan in whey mg/mL	Protein in whey mg/mL	Protein extracted %
12.6	800	2	9.7	23.0	864	6.4	49.2
10.4	1600	2	9.6	7.7	1728	5.1	51.0
11.0	1600	4	9.2	16.4	1728	5.2	52.7

(a) Reaction carried out at pH 3.5.
(b) Reaction carried out at pH 7.0.

restabilization of particles and a consequent decrease in protein extracted. In contrast, at pH 7.0 increasing the chitosan concentration produced a very slight increase in protein extracted. In treatment B (supernatant from treatment A), samples treated at pH 7.0 produced the highest extraction of proteins from whey. The best conditions for the extraction of proteins from whey would be pH 7.0, for treatment A and B, with a chitosan concentration of 1600 mg/L. However, the most economical treatment would possibly be pH 3.5 and pH 7.0 for treatment A and B, respectively with a chitosan concentration of 800 mg/L.

4. COMPARISON OF CHITOSAN PREPARATIONS FROM CRAB AND FROM SHRIMP AS COAGULANTS IN THE EXTRACTION OF PROTEINS FROM WHEY

Charge density and molecular weight of a particular polyelectrolyte both influence the destabilization mechanism and resulting floc

Table 2. Extraction of proteins from whey at various pHs of
reaction using chitosan as coagulant.

Control Protein in whey mg/mL	Treatment A[a]			Treatment B[b]		
	pH	Protein in whey mg/mL	Protein extracted %	pH	Protein in whey mg/mL	Protein extracted %
13.4	3.5	10.2	23.9	7.0	6.8	49.3
11.6	4.0	10.0	13.8	7.0	6.5	44.0
11.2	4.5	9.8	12.5	7.0	7.1	36.6
11.3	5.0	10.0	11.5	7.0	5.9	47.8
11.8	5.5	8.7	26.3	7.0	5.7	51.7
11.3	6.0	9.1	19.5	7.0	6.2	45.1
11.5	6.5	9.0	21.7	7.0	6.2	46.1
12.0	7.0	8.9	25.8	7.0	5.8	51.7

(a) Concentration of chitosan in whey: 800 mg/L.
(b) Concentration of chitosan in whey: 864 mg/L.

Table 3. Extraction of proteins from whey using different concentrations of chitosan and pH of reaction.

Control Protein in whey mg/mL	Treatment A[a]				Treatment B[b]			
	Chitosan in whey mg/mL	pH	Protein in whey mg/mL	Protein extracted %	Chitosan in whey mg/mL	pH	Protein in whey mg/mL	Protein extracted %
11.8	800	3.5	10.0	15.3	864	3.5	6.4	45.8
11.7	800	3.5	9.4	19.6	864	7.0	5.5	53.0
10.3	1600	3.5	9.6	6.8	1728	3.5	7.5	27.2
10.6	1600	3.5	9.6	9.4	1728	7.0	5.6	47.2
12.5	800	7.0	9.4	24.8	864	7.0	6.2	50.4
12.5	800	7.0	9.3	25.6	864	3.5	8.7	30.4
13.0	1600	7.0	9.4	27.7	1728	7.0	6.0	53.8
13.0	1600	7.0	9.6	26.2	1728	3.5	6.8	47.7

(a) Concentration of chitosan in whey: 800 mg/L.
(b) Concentration of chitosan in whey: 864 mg/L.

formation. The results of the extraction of proteins by crab chitosan, produced by methods A and B (Experimental, Section 9) and shrimp chitosan shown in Table 4. Variations in the degrees of acetylation and molecular weights of the tested chitosans are possibly the reasons for the variations in the concentrations of protein extracted from whey, at the different pHs. Compared to shrimp chitosan, the extractions of proteins from whey using crab chitosan was ca. 14% higher.

5. SEDIMENTATION RATES OF WHEY TREATED WITH CHITOSAN

For industrial purposes, the knowledge of the sedimentation rate in the process of recovery of proteins extracted from whey is necessary. In batch sedimentation the suspension is allowed to settle for the required

Table 4. Extraction of proteins from whey using chitosan from shrimp and crab.

Chitosan	Control	Treatment A[a]		Treatment B[b]	
	Protein in whey mg/mL	Protein in whey mg/mL	Protein extracted %	Protein in whey mg/mL	Protein extracted %
crab method A	11.9	8.4	29.4	4.8	59.7
crab method B	11.9	8.6	27.7	6.8	42.9
shrimp	11.9	10.1	15.1	5.7	52.1

(a) Concentration of chitosan in whey: 800 mg/L; reaction carried out at pH 3.5.
(b) Concentration of chitosan in whey: 864 mg/L; reaction carried out at pH 7.0.

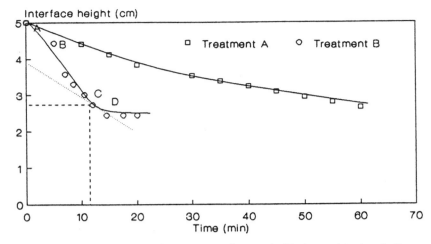

Figure 8. Sedimentation rate of precipitates obtained from
whey treated with chitosan.

time, after which sludge and supernatant are drawn off separately.
Retention time is then a major design parameter.

The suspension settling properties of whey treated with chitosan are
shown in Figure 8. The characteristics of a batch settling system are the
induction period zone (a-B), the contrast period zone (B-C) and the first
falling rate period zone (C-D). D is the compression point, a

Table 5. Amino acid compositions of whey and whey treated
with chitosan from shrimp

Amino Acid	MW	Whey			Supernatant B		
		nmol/mL	µg/mL	Ratio AA:Ala	nmol/mL	µg/mL	Ratio AA:Ala
Asp	133.11	6100	812.0	14.5	2800	372.7	14.7
Thr	119.12	4500	536.0	10.7	2100	250.1	11.
Ser	105.10	4100	430.9	9.8	1900	199.7	10.0
Glu	147.14	8800	1294.8	21.0	4100	603.3	21.6
Pro	115.14	3900	449.0	9.3	1800	207.3	9.5
Gly	75.07	2100	157.6	5.0	950	71.3	5.0
Ala	89.10	4200	374.2	10.0	1900	169.3	10.0
Cys	121.16	360	43.6	0.9	200	24.2	1.1
Val	117.15	3300	386.6	7.9	1400	164.0	7.4
Met	149.22	1100	164.1	2.6	500	74.6	2.6
Ile	131.18	3200	419.8	7.6	1300	170.5	6.8
Leu	131.18	6000	787.1	14.3	2700	354.2	14.2
Tyr	181.20	1200	217.4	2.9	500	99.7	2.9
Phe	165.20	1600	264.3	3.8	740	122.2	3.9
His	155.16	1200	186.2	2.9	560	86.9	2.9
Lys	146.19	4700	687.1	11.2	2100	307.0	11.1
Arg	174.21	1200	209.1	2.9	460	80.1	2.4
Total		57560	7419.8	−	26060	3357.1	−

Table 6. Amino acid composition of precipitation obtained
after treatment of whey with chitosan from shrimp.

Amino acid	MW	Precipitate A			Precipitate B		
		nmol/mg	µg/mg	Ratio AA:Ala	nmol/mg	µg/mg	Ratio AA:Ala
Asp	133.11	98	13.0	14.6	88	11.7	14.4
Thr	119.12	69	8.2	10.3	67	8.0	11.0
Ser	105.10	85	8.9	12.7	67	7.0	11.0
Glu	147.14	150	22.1	22.4	130	19.1	21.3
Pro	115.14	63	7.3	9.4	54	6.2	8.9
Gly	75.07	45	3.4	6.7	36	2.7	5.9
Ala	89.10	67	6.0	10.0	61	5.4	10.0
Cys	121.16	5.4	0.7	0.8	5.2	0.6	0.9
Val	117.15	56	6.6	8.4	49	5.7	8.0
Met	149.22	17	2.5	2.5	16	2.4	2.6
Ile	131.18	48	6.3	7.2	42	5.5	6.9
Leu	131.18	86	11.3	12.8	84	11.0	13.8
Tyr	181.20	18	3.3	2.7	18	3.3	3.0
Phe	165.20	25	4.1	3.7	23	3.8	3.8
His	155.16	39	6.1	5.8	39	6.0	6.4
Lys	146.19	71	10.4	10.6	68	9.9	11.1
Arg	174.21	20	3.5	3.0	9.7	1.7	1.6
Total		962.4	123.7	–	856.9	110.0	–

(hypothetical) point where free sedimentation ceases and the suspension is regarded as a continuous network in compression 8 (Figure 8).

At any time during the batch test, the interface settling velocities (u) can be evaluated by drawing tangents to the height-time curve shown below, where H is height of interface at time t. It is apparent that the settling velocities for the first treatment (A) of whey are much lower than for the second treatment (B).

$$u = dH/dt$$

6. AMINO ACID COMPOSITION OF PROTEIN EXTRACTED FROM WHEY USING CHITOSAN

The amino acid compositions of the original whey, the whey treated with chitosan (supernatant) and of the precipitates (chitosan plus proteins) are shown in Tables 5 and 6. The alanine: amino acid rates of the whey are similar to the treated whey (supernatant) and to the precipitates, suggesting that the extraction of proteins from whey by chitosan was not in a selective way.

The results of dry matter of the original whey, the whey treated with chitosan (supernatant), and of the precipitate (chitosan plus proteins) are shown in Table 7. These results show a difference of ca. 4% between the total initial matter (chitosan and whey) and the total final matter (treated whey (supernatant) and precipitates).

The mass balance of proteins, by amino analysis and UV spectrophotometry, during the process of extraction of proteins from whey by chitosan are given in Table 8. There is a difference of 30% between the total proteins extracted from the whey and the total proteins present in the precipitate. Loss of proteins in the material which adhered to the

Table 7. Dry matter of original whey and of supernatant and precipitate obtained from whey after treatment with chitosan.

Component	Dry matter in whey mg/mL
Whey (1)	66.2
Chitosan (2)	0.8
Total (1 & 2)	67.0
Supernatant (after treatment A)(3)	57.6
Precipitate (after treatment A)(4)	6.8
Total (3 & 4)	64.4

sides of the glass tube during the process of extraction of proteins from whey have contributed to this difference. The total protein content extracted from whey analyzed by amino acid analysis was ca. 25% lower than that determined by UV spectrophotometry.

Table 8. Mass balance of proteins during the process of extraction of proteins from whey using chitosan preparation from shrimp as coagulant. The protein contents were determined by UV and amino acid analysis.

Treatment Component	Amino acid analysis		UV analysis	
	Protein in whey mg/mL	Protein extracted %	Protein in whey mg/mL	Protein extracted[b] %
Control whey (1)	7.4	–	11.7	–
A Whey & chitosan (2)	7.4	–	11.7	–
Supernatant (3)	6.5[a]	–	9.5	–
Precipitate (4)	0.9	12.2	2.2[b]	18.8
B (3) & chitosan (5)	6.5[a]	–	9.5	–
Supernatant (6)	4.5	–	5.5	–
Precipitate (7)	1.2	16.2	4.0[c]	34.2
Total Precipitates (4+7)	2.1	28.4	6.2	53.0
Whey (1)- Supernatant (6)	2.9	39.2	6.2	53.0
Total (6+8)	6.6	–	11.7	–
Total (6+9)	7.4	–	11.7	–

(a) Estimated by the difference of (1) and (4).
(b) Estimated by the difference of (1) and (3).
(c) Estimated by the difference of (5) and (6).

REFERENCES

1. R. McL. Whitney, J. R. Brunner, K. E. Ebner, H. M. Farrell, Jr., R.

V. Josephson, C. V. Moor, and H. E. Swaisgood, J. Dairy Sci., **59**, 795-815 (1976).

2. D. M. Mulvihill and J. E. Kinsella, Food Technol., Sep. 102-111 (1987).

3. Bough and Landes, J. Dairy Sci., **59**, 1874-1880 (1976).

4. R. S. Whistler, and J. N. BeMiller, J. Org. Chem., **27**, 1161-1163 (1962).

5. R. A. A. Muzzarelli, Chitin, p 87-154, Pergamon Press Ltd., Oxford (1977).

6. R. H. Hackman, Austr. J. Biol. Sci., **7**, 168-178 (1954).

7. M. L. Wolfrom, G. G. Maher, and A. Chaney, J. Org. Chem., **23**, 1990-1991 (1958).

8. M. L. Wolfrom, and T. M. Shen-han, J. Am. Chem. Soc., **81**, 1764-1766 (1959).

9. J. Bratby, Coagulation and Flocculation, Uplands Press Ltd. Croydon.

10. R. J. Akers, in: The Scientific Basis of Flocculation, K. J. Ives, ed., p 131-163, Sijthoff and Noordhoff, Alpen aan den Rijn, (1978).

11. J. Gregory, in: The Scientific Basis of Flocculation, K. J. Ives, ed., p 101-130, Sijthoff, and Noordhoff International Publishers, B. V., Alpen van der Rijn, (1978).

12. C. R. Holland, in: Chitin and Chitosan - Sources, Chemistry, Biochemistry, Physical Properties and Applications, G. Skjak-Braek, T. Anthonsen and P. Samford, eds., p 559-566, Elsevier Applied Science, New York, (1989).

USE OF LIGNIN-BASED MATERIALS IN THE SELECTIVE INHIBITION OF

MICROORGANISMS

Charles E. Carraher, Jr., Dorothy C. Sterling, Cynthia
Butler and Thomas Ridgway

Florida Atlantic University
Departments of Chemistry and
Biological Sciences
Boca Raton, FL 33431
              and
University of Cincinnati
Department of Chemistry
Cincinatti, Ohio 45221

Lignin, modified through reaction with organostannane
halides, and SINS, formed through the co-reaction of lignin
and poly(ethylene glycols), and block polymers of poly(ethylene glycol-co-dimetylsiloxane-co-ethylene glycol) with organostannane halides exhibit selective inhibition of *C. albicans*, the microorganism responsible for yeast infections.

INTRODUCTION

Lignin is the second most abundant natural, renewable material. It
is produced at an annual rate of about 2×10^{10} tons and is present in
the biosphere at a level of 3×10^{11} tons.[1] Based on pulp and paper
production, lignin is produced worldwide from woody plants in mills at an
annual rate of 5×10^7 tons.[1-3] USA production of lignin from mills is
estimated to be about one half of this or about 200 lbs. per each person
in the USA. For isolated lignin, the major use is fuel.

Lignin serves as a basis for the industrial surfactant industry
being the most utilized feedstock for the compounding of surfactants. It
has been used as a filler and reinforcing agent in styrene-butadiene
rubber.[4-7] Its use as an additive to phenol-formaldehyde and urea-
formaldehyde resins has also been described.[8-24] Even so, industrial use
of lignin as a filler and additive has been limited due to the tendency
of lignin to internally hydrogen-bond.

The use of lignin as a component in copolymers has had wide study
although there have been only a few instances where industrial
application has occurred. As already noted, lignin-based materials serve
as a foundation to the surfactant industry. These lignin materials are

Biotechnology and Bioactive Polymers, Edited by C. Gebelein
and C. Carraher, Plenum Press, New York, 1994

sulfonated lignins.[1,11,25] These sulfonated lignins can be further modified through additional crosslinking, oxidation and carboxylation to give products with varying degrees of solubility and stability.[26-35] The water solubility has also been modified through use of grafting of hydrophilic groupings onto the units composing the lignin.[1,36-40]

As a structural material, lignin has been incorporated into a number of polymer matrixes and along with a number of polymers.[1,12,21,41-47] These products include reaction with isocyanates,[48-56] hydrocarbon polyethers such as poly(ethylene glycols),[56-60] lignin epoxidation,[61-63] and reaction with amines and polymeric amines.[1]

Even with the vast number of potential uses, lignin continues to be a major underused feedstock. It is the purpose of our studies to illustrate the use of lignin-based products as microorganism inhibitors suitable for use as fillers, additives or as bulk materials in applications ranging from biomedical applications to additives to specialty coatings. Here we describe the biological inhibition profile of products derived from the reaction of organostannane dihalides and lignin.

EXPERIMENTAL

A one quart Kimex emulsifying jar placed on a Waring Blendor (Model 1120) with a no-load stirring rate of about 1,000 rpm was employed as the reaction system. The jar lid has a hole through which a large mouthed glass funnel is placed and through which the second phase is introduced into the reaction jar. The aqueous solution containing the lignin and added base is placed in the reaction jar. The blender is turned on and the organic phase containing the organostannane halide is introduced into the reaction jar. After a designated time stirring is stopped and the blender jar contents poured into a separatory funnel. The product is collected as a precipitate, washed repeatedly with water and the organic liquid, transferred to a preweighed glass petri dish, and allowed to air dry.

The chemicals were used as received without further purification. Indulin AT and C were obtained as gifts from Westva Company, Charleston, SC. Indulin C is the sodium salt of Kraft Pine lignin. Indulin AT is a purified form of lignin from the Krafting process.

Infrared spectra were obtained using potassium bromide pellets employing an Mattson Instruments Galaxy 4020 FTIR. Mass spectroscopy (DIP) was carried out employing a Kratos MS-50 Mass Spectrometer operating in the EI mode, 8 KV acceleration and a ten second per decade scan rate with variable probe temperature (Midwest Center for Mass Spectrometry Operating, Lincoln, Nebraska) and a DuPont 21-491 Mass Spectrometer at 1.8kv. Elemental analysis was carried out by Galbraith Labs. (Knoxville, TN).

Microorganism studies were conducted in the usual manner. That is, culture plates containing a suitable growth medium were seeded with suspensions of inocula that would produce an acceptable lawn of a test organism after 24 hours of incubation at 37°C. Shortly after the plates were seeded, the test compounds were introduced as solids. Following incubation, observations on inhibition were made.

RESULTS AND DISCUSSION

1. STRUCTURE

Lignin has a somewhat two-dimensional structure that can be viewed as being composed of aromatic and aliphatic alcohols. Its structure varies from source to source, etc., but can be considered as having a general "idealized" structure that can be treated in a quantitative manner with respect to the number and type of hydroxyl groups present. Commonly, lignin is treated as consisting of a C_9 repeat formula unit where the superstructure contains aromatic and aliphatic alcohols and ethers, aliphatic aldehydes and vinyl units.

The precise structure of lignin varies with respect to plant variety, location in the plant, time of year and climatic conditions, recovery procedures, degradation techniques employed, analysis procedures utilized, etc. For the present material, analysis procedures and results are described in detail elsewhere.[64] Following is a brief treatment of the structural results.

The lignin-containing products can be considered to be one of two general types designated as organostannane modified lignins (Sn-L) and simultaneous interpenetrating networks (SINs) derived from lignin and hydroxyl-capped polymers containing ethylene oxide units connected by organostannane moieties. The lignin and hydroxyl-capped reactants are connected through Sn-O-R bonding. The evidence for the formation of this ether bridge is derived from a number of physical results. From infrared spectroscopy, the products exhibit a reduction in the OH band (ca. 3400 cm^{-1}) consistent with the reaction of the tin moiety with the hydroxyl on lignin. Mossbauer results are also consistent with the formation of $R_2Sn(OR)_2$ moieties giving a quad. splitting value of about 2.97 (mm/sec) and an isomer shift of about 1.23 (mm/sec).[64]

Finally, mass spectral results show ion fragments that contain portions of the lignin and organostannane and in the case of SINs, fragments containing portions derived from lignin, the hydroxyl-caped reactants and the organostannane. For instance, for the product derived from lignin, Bu_2SnCl_2 and PEG bands characteristic of products of the general form <u>1</u> (Figure 1) where x is OCH_2 and OCH_2CH_2 are present.

Again, for the analogous product, except derived from dilauryltin dichloride, ion fragments of the general <u>1</u> form are present for X = OCH_2 and OCH_2CH_2.

The proportions of substitution or hydroxyl-groups converted to organostannane ethers varies according to the nature of the organostannane. For instance, for one set of reaction conditions products derived from reaction of lignin with dibutyltin dichloride had two of every three hydroxyl groups converted to organostannane ethers whereas the products derived from reaction with dilauryltin dichloride showed only one ether formed for every five hydroxyls. This is consistent with the greater space requirement of the lauryl group in comparison to the requirements of the butyl group. In both cases, the aliphatic moieties are linear.

1. BIOLOGICAL ACTIVITIES

Of importance to the present study, prior results indicate that the

Figure 1. General Lignin Structural form 1.

percentage incorporation of a specific organostannane affects the extent of inhibition but not the microorganisms inhibited.[65]

The most often cited toxicity structural relationships are for alkylorganostannanes and include toxicity orders such that toxicity decreases as the length of the alkyl chain increases, toxicity increases as the number of alkyl groups on the tin increases from one through three and that the toxicity for organostannanes attached through carbon is less that if attached through more highly negative atoms such as nitrogen and oxygen. The latter two generalities are typically followed, but the former one is only true for select species such as mammals. For bacteria, the relationship with respect to alkyl length is complex and even varies as to whether the microorganism is gram negative or gram positive. The trend with respect to tin attachment is probably a result of the tendency of either physically or biologically induced hydrolysis of the tin-containing moiety.

The precise structural-biological property relationship for tin-containing materials are among the best known for organometallic-containing materials because of their widespread use as inhibitors of microorganisms, yet the relationships are still not well known nor are they well understood.

The biological activity was measured by the size of the zone of inhibition on test plates. Those products which exhibit the largest zone of inhibition are considered the most effective towards that particular microorganism. Four microorganisms were used in the testing: *Escherichia coli*, *Pseudomonas aeruginosa*, *Staphlococcus aureaus*, and *Candida albicans* (Table 1). The first two microorganisms are gram positive bacteria while the third is gram negative. *Candida albicans* is a fungus commonly known as the organism responsible for yeast infections in women.

More recently, selective toxicities have been reported for some polymer-bound organostannanes such that common microorganisms are not susceptible to the products, but that pathogenic microorganisms are killed on contact with the organostannane-bound polymeric materials. The most notable of these materials is poly(vinyl alcohol) modified through reaction with organostannane halides. These products showed little or no toxicity towards *E. coli*, *P. Aeruginosa* and *S. aureus* but were highly toxic towards the yeast-causing *C. albicans*. Here toxicity followed the trend of methyl>ethyl>propyl>, with butyl-containing products exhibiting little toxicity towards the *C. albicans*.[66]

For the present products, only butyl and lauryl-containing organostannanes were tested. Toxicity was of the general order with butyl>lauryl. Of interest was the somewhat selective nature of the toxicity such that the butyl-containing products were generally more toxic against *C. albicans* relative to the other microorganisms tested. Also of interest is the generally good toxicity of the butyl-products themselves against *C. albicans* since, as noted before, similar products,

Table 1. Inhibition as a function of modified lignin.

Reactants	Inhibition (cm)			
	P. aueriginosa	*E. Coli*	*S. Aureus*	*C. albicans*
Bu_2Sn/Lignin/PEG (2000)	0	0	0	0
Bu_2Sn/ignin	0	0	0	0.3
Bu_2Sn/Lignin/Silol (2400)	0	0	0	0.5
Bu_2Sn/Lignin/Silol (1200)	0	0	0.3	0.5
Bu_3Sn/Lignin	0	Contact	Cont.	0.5
La_2Sn/Lignin	0	0	0	0
Bu_2Sn/Hydroquinone	0	0	0.4	0.3
La_2Sn/Lignin/PEG (2000)	0	0	0	0
Bu_2SnCl_2	Contact	0.1	Cont.	0.1

except derived from poly (vinyl alcohol), showed little toxicity towards *C. albicans* for the butyl-derived materials. Further, as noted before, the toxicity decreases with increase in alkyl chain length for mammals where butyl-containing organostannanes are much less toxic than the corresponding methyl, ethyl and propyl organostannanes. Thus, the butyl-containing lignin products may offer a distinct advantage over other products since these products exhibit good inhibition to *C. albicans*.

Second, inhibition is dependent on both the nature of the organo portion of the organostannane moieties and on the nature of the modified reactants, i.e., lignin, PEG or Silol.

Third, the inhibition pattern for the modified products differs from that of the organostannane halide itself. Toxicity of the dibutyltin dichloride is derived from two sources - the organotin moiety and the acid chloride nature of the tin halide.

The results are such that further work is necessary to define the toxicity limits better and also the potential uses of these materials as anti-yeast agents.

REFERENCES

1. W. Glasser and S. Kelley, in: "*Encyclopedia of Polymer Science and Engineering*," Vol. **8**, 795, 2nd Ed., John Wiley, N.Y., 1988.
2. L. Sperling and C. Carraher, in: "*Encyclopedia of Polymer Science and Engineering*," Vol. **12**, 658, 2nd Ed., John Wiley, N.Y. 1988.
3. L. Sperling and C. Carraher, "*Renewable-Resource Materials*," Plenum, N.Y., 1986.
4. S. I. Falkehag, Appl. Polym. Symp. **28**, 247 (1975).
5. U. S. Pat. 3,991,022 (Nov. 9, 1976), M. S. Dimitri (to Westvaco Corp.).
6. A. F. Sirianni and J. E. Puddington, Rubber World **165**(6), 40 (1972).
7. A. F. Sirianni, C. M. Barker, G. R. Barker and I. E. Puddington, Pulp Pap. Mag. Can., **73** (11), 61 (1972).

8. U. S. Pat. 3,984,362 (Oct. 5, 1976), A. F. Sirianni and I. E. Puddington (to Canadian Patents and Development, Ltd.).
9. M. A. DePaoli and L. T. Furlan, Polym. Degrad. Stability, **11**, 327 (1985).
10. H. H. Nimz, in: "*Wood Adhesives: Chemistry and Technology*," A. Pizzi, Ed., Marcel Dekker, Inc., New York, 1983, Chapt. 5, pp. 247-288.
11. S. Y. Lin, Prog. Biomass Conversion **4**, 31 (1983).
12. U.S. Pat. 3,864,291 (Feb. 4, 1975) T. U. E. Enkvist.
13. U.S. Pat. 4,105606 (Aug. 8, 1978) K. G. Forss and A. G. M. Fuhrmann (to Keskuslaboratorio-Central-laboratorium Ab, Finland).
14. U.S. Pat. 4,113,675 (Sept.12,1978), M. R. Clarke and A. J. Dolenko (to Canadian Patents and Development Ltd., Ottawa, Canada).
15. U.S. Pat. 4,113,542 (Sept. 12, 1978).
16. R. C. Gupta and V. K. Sehgal, Holzforsch, Holzverwert, **30**, 4-5, 85 (1978).
17. U.S. Pat. 4,306,999 (Dec.22,1981) J. W. Adams and M. W. Schoenherr (to American Can Co.).
18. C. Ayla, Holzforschung, **36** (2), 93 (1982).
19. W. Lange, O. Faix, C. Ayla, and H. Georg, Adhesion, **11**, 16 (1983).
20. P. C. Muller and W. G. Glaser, J. Adhes., **17**, 157(1984).
21. P. C. Muller, S. S. Kelley, and W. G. Glasser, J. Adhes. **17**, 185 (1984).
22. D. Gardner and T. Sellers, PhD Dissertation, Mississippi State University, Mississippi State, Miss., 1984.
23. F. Troeger and R. Diebold, Holz Roh Werkst., **43**, 152 (1985).
24. Ger. Pat. 2,758,572 (Apr. 26, 1979), K. G. Forss and A. Fuhrmann (to Keskuslaboratorio-Central-laboratorium Ab, Finland).
25. G. G. Allan, in: "*Renewable-Resource Materials*," L. Sperling and C. Carraher, Eds., Plenum, NY, 1986, Chapt. 13.
26. Chem. Eng. News **62**, 19 (Sept. 24, 1984).
27. H. L. Chum, S. K. Parker, D. A. Feinberg, J. D. Wright, P. A. Rice, S. A. Sinclair, and W. G. Glasser, "*The Economic Contribution of Lignins to Ethanol Production from Biomass*," SERI/TR-231-2488, Solar Energy Research Institute, Golden, Colo. 1985.
28. Swed. Pat. 160,770 (1957), E. Adler and E. Gagglund.
29. Ger. Pat. 2,322,928 (Apr. 11, 1974), J. Benko and G. Daneault.
30. Product Bulletin 500E, BORREGAARD Co., Sarpsborg, Noway, 1979.
31. U. S. Pat 4,332,589 (June 1, 1982), S. Y. Lin (to American Can Co.).
32. H. Kaneko, S. Hosoyo, and J. Nakano, Mokuzai Gakkaishi, **26**, 752 (1980).
33. W. G. Glasser and W. Sandermann, Sven. Paperstidn., **73**, 246 (1970).
34. U.S. Pat 3.956,261 (May 11, 1976), S. Y.Lin (to Westvaco Corp).
35. W. Lange, Wood Sci. Technol., **14**, 1 (1980).
36. J. J. Meister, "*Renewable Resource Materials: New Polymer Sources*," Vol. 33, C. E. Carraher and L. H. Sperling, Eds., Polymer Science and Technology, Plenum Press, New York, 1985, pp. 305-322.
37. J. J. Meister, D. R. Patil, L. R. Field, and J. C. Nicholson, J. Polym. Sci., Polym. Chem. Ed., **22**, 1963 (1984).
38. J. J. Meister, D. R. Patil, and H. Channell, Ind. Eng. Chem. Prod. Res. Dev., **24**, 306 (1985).
39. J. J. Meister, D. R. Patil, Macromolecules **18**, 1559 (1985).
40. U.S. Pat 4,454,066 (June 12, 1984), P. Dilling and P. T. Sarjeant (to Westvaco Corp.).
41. H. H.Nimz, I. Gurang, and I. Mogharab, Liebigs Ann. Chem., **1976**, 1421 (1976).
42. C. Ayla and H. H. Nimz, Holz Roh Werkst., **42**, 415 (1984).
43. K. C. Shen and L. Calve, Adhes. Age, **23** (8), 25 (1980).
44. A. J. Dolenko and M. R. Clarke, for. Prod. J., **28** (8), 41 (1978).
45. K.Forss and A. Fuhrmann, Pap. Puu, 58,817 (1976).
46. K. Forss and A. Fuhrmann, For. Prod. J., **29**, 39 (1979)
47. T. G. Rials and W. G. Glasser, Holzforschung, **40** (6), 353 (1986).

48. K. Kratzl, K. Buchtela, J. Gratzl, J. Zauner, and O. Ettingshausen, Tappi, **45** (2), 117 (1962).
49. O. H-H. Hsu and W. G. Glasser, Appl. Polym. Symp., **28**, 297 (1975).
50. O. H-H. Hsu and W. G. Glasser, Wood Sci., **9** (2), 97 (1976).
51. W. G. Glasser, O. H-H. Hsu, D. L. Reed, R. C. Forte and L. C.-F. Wu, ACS Symp. Ser., **172**, 311 (1981).
52. V. P. Saraf and W. G. Glasser, J. Appl. Polym. Sci., **29**, 1831 (1984).
53. U.S. Pat. 3,546,199 (Dec. 8, 1970), D. T. Christian, M. Look, A. Nobell and T. S. Armstrong.
54. U.S. Pat. 3,476,795 (Nov. 4, 1969), G. G. Allan (to Weyerhaeuser Co., Tacoma, Wash).
55. L. C.-F. Wu and W. G. Glasser, J. Appl. Polym. Sci., **29**, 1111 (1984).
56. W. G. Glasser, C. A. Barnett, T. G. Rials, and V. P. Saraf, J. Appl. Polym. Sci., **29**, 1815 (1984).
57. T. G. Rials and W. G. Glasser, Holzforschung, **38**, 263 (1984).
58. V. P. Saraf, W. G. Glasser, G. L. Wilkes, and J. E. McGrath, J. Appl. Polym. Sci., **30**, 2207 (1958a)
59. V. P. Saraf, W. G. Glasser, and G. L. Wilkes, J.Appl. Polym. Sci., **30**, 3809 (1985b).
60. S. S. Kelley, Ph.D. Dissertation, Virginia Polytechnic Institute and State University, Blacksburg. Va., March, 1976.

LIVING POLYMERIZATION OF PROTEINS: ACTIN AND TUBULIN. A REVIEW

Stoil Dirlikov

Coatings Research Institute
Eastern Michigan University
122 Sill Hall
Ypsilanti, MI 48197

Living polymerization of the two major proteins in eucaryotic cells: actin and tubulin is reviewed. Actin is a large globular protein built of 375 alpha-amino acid residues and has a molecular weight of 42,000. It polymerizes *in vivo* and *in vitro* with the formation of long linear filaments (up to 50 microns) with two distinguished ("plus" and "minus") ends. Actin filaments are in equilibrium with the surrounding monomeric actin and undergo treadmilling, i.e., continuous polymerization at the plus end and depolymerization at the minus end which allows their spacial movement without changing the filament length. Tubulin is a larger globular protein (dimer) with a molecular weight of 100,000 which exhibits similar polymerization behavior. Microtubulin filaments, however, characterize with dynamic instability, i.e., they polymerize slower but undergo very fast depolymerization. Cell utilization and regulation of actin and tubulin polymerization is briefly reviewed as well.

INTRODUCTION

A recent review by Webster,[1] has summarized the general features of anionic, cationic, covalent, and free radical living polymerizations. In addition to these synthetic methods, living polymerization proceeds in living organisms as well. The present paper reviews the peculiar living polymerization of two high molecular weight proteins: actin and tubulin which proceeds both *in vitro* and *in vivo*.

ACTIN POLYMERIZATION[2,3]

1. ACTIN MONOMER

Actin monomer is a globular protein known as G-actin. It consists of 375 alpha-amino acid residues and has molecular weight of about 42,000.

Biotechnology and Bioactive Polymers, Edited by C. Gebelein
and C. Carraher, Plenum Press, New York, 1994

It is the most abundant protein in eucaryotic cells and constitutes more than 5 percent of their total dry protein weight.

Cells utilize actin homopolymerization and copolymerization with other cell proteins in many cellular activities and most actin mutations, therefore, have lethal effect. Actin, therefore, is a highly conserved protein. Its genome, which produce actin, has not undergone much change in the last several million years. There is only a slight difference in the amino acid order and content of actins in chicken and yeast. As a matter of fact, actins of different species copolymerize *in vitro*.

2. ACTIN HOMOPOLYMER

Actin homopolymers or the so-called actin filaments, or F-actin, are long linear double-stranded helical polymers with a diameter of 8 nm and length of usually 5-10 microns which could reach 50 microns in the long cells of the nervous system (Figure 1).

Actin filaments consist of uniformly oriented actin monomers. They are polar assembles with two structurally different plus and minus ends. In a way, their polarity resembles the head and tail ends of the synthetic polymers.

3. ACTIN POLYMERIZATION[4]

Actin polymerization, as any other polymerization, characterizes with three stages: initiation, propagation, and termination.

3A. Initiation

In vitro initiation is induced by an increase of the ionic strength of an aqueous solution of actin monomers to that in cells. In contrast, the depolymerization of actin filaments proceeds at lower salt concentration which allows isolation of actin monomers by muscle extraction (Figure 1).

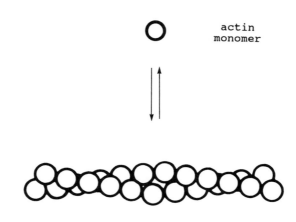

actin
monomer

filament

Figure 1. Polymerization of actin and schematic representation of actin filament structure.

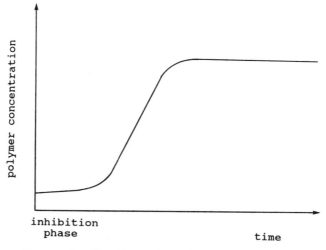

Figure 2. Kinetics of actin polymerization.

In *vivo* initiation or the so-called nucleation (in biology) follows similar patterns. The initiation step requires simultaneous interactions of three actin monomers in specific configuration which is observed as a lag (or inhibition) of actin polymerization. This is illustrated in Figure 2. After a while, the aqueous solution of actin rapidly increases its viscosity due to the formation of actin filaments.

3B. Propagation

Actin polymerization obeys peculiar propagation kinetics. Actin monomers and filaments are in equilibrium and propagation proceeds by the addition of actin monomers to both ends of the filament (Figure 3).

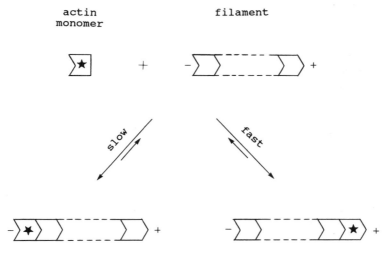

Figure 3. Schematic representation of actin directional polymerization.

Actin polymerization, however, is directional. It proceeds 5 to 10 times faster on the so-called plus end, than on the other minus end. This is observed microscopically if a small actin filament, labeled with myosin heads, is exposed to free actin monomers. After a while, the newly formed actin filament at the plus end is much longer than the corresponding filament at the minus end.

The directional actin polymerization is due to conformational changes which actin monomers undergo upon polymerization. As a matter of fact, both reactions, actin polymerization and conformational change, are coupled. A change of actin monomer into a less stable conformation is required for its addition (polymerization) to the minus filament end. This pathway is kinetically unfavorable. In contrast, actin addition to the plus end precedes its conformational change. This favors actin addition to plus filament end as schematically shown in Figure 3.

The equilibrium actin polymerization at both ends of the filament is schematically shown in Equation 1 with the corresponding polymerization (k_{on}) and depolymerization rate constants (k_{off}):

(Equation 1)

The equilibrium constant, K for each end is equal to:

$$K^+ = \frac{k_{on}^+}{k_{off}^+} \quad \text{and} \quad K^- = \frac{k_{on}^-}{k_{off}^-}$$

Due to the conformational changes of actin monomer, the polymerization rate constant at the plus filament end, k_{on}^+ is much larger than that at the minus end, k_{on}^-:

$$K_{on}^+ > K_{on}^-$$

Due to the same reasons, the depolymerization rate constant at the plus filament end, k_{off}^+ is as many times larger than that at the minus end, k_{off}^-:

$$k_{off}^+ > k_{off}^-$$

As a result, for true equilibrium polymerization the equilibrium constants at both ends are equal:

$$K^+ = K^-$$

The critical concentration of free actin, C_c at which the number of actin monomers, which polymerize at one of the filament ends, is equal to the number of actin monomer units which depolymerize from the same filament end, is equal to:

$$C_c = 1/K = k_{off}/k_{on}$$

For true equilibrium polymerization, the critical concentration of the plus filament end, c_c^+ is equal to that of the minus end, c_c^-:

$$c_c^+ = c_c^-$$

Actin polymerization, however, is coupled with ATP (adenosine triphosphate) hydrolysis.[5] ATP is an energy molecule of the cells. Each actin monomer has one tightly noncovalent bound ATP molecule. ATP hydrolysis to ADP (adenosine diphosphate) proceeds after actin monomer becomes incorporated onto the filament, the so-called process of delayed hydrolysis, as shown in Equation 2, where P_i is an abbreviation for pyrophosphate (and energy release):

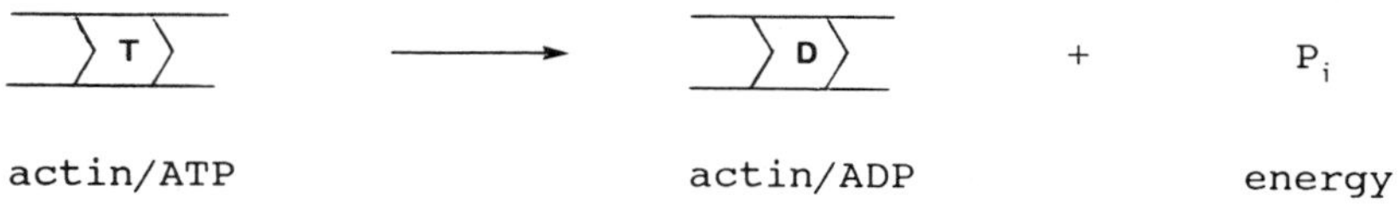

actin/ATP   actin/ADP   energy

(Equation 2)

Actin polymerization does not require ATP hydrolysis (and energy). ATP hydrolysis, however, makes actin polymerization not a true equilibrium polymerization.

The actin/ATP monomer, abbreviated T, has a much higher polymerization rate constant (k_{on}^T) than the actin/ADP monomer which is abbreviated D (k_{on}^D). In addition, the filament actin/ATP monomer units (T units) have much higher affinity to the filament (i.e. smaller depolymerization rate constant, k_{off}^T) than its actin/ADP units (D units) which have larger k_{off}^D. As a result, the "T-cap" decreases the depolymerization rate at the plus filament end, as shown in Equation 3. The longer the T-cap is, the faster the polymerization at the plus end proceeds.

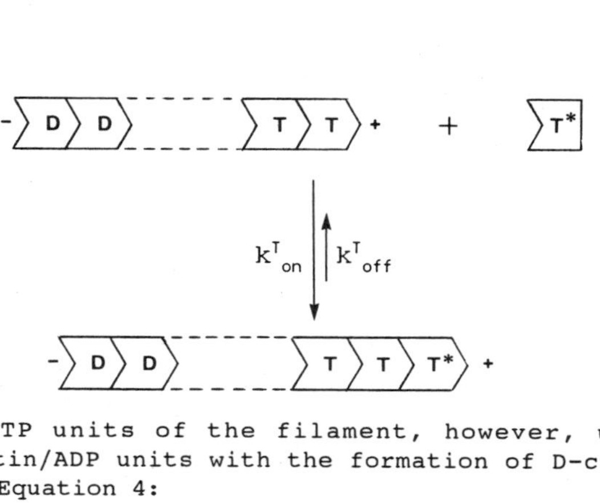

(Equation 3)

The actin/ATP units of the filament, however, undergo a slow hydrolysis to actin/ADP units with the formation of D-cap at the minus end, as shown in Equation 4:

$$-\boxed{T \rangle T \rangle T} \cdots \boxed{T \rangle T \rangle T} + \qquad \longrightarrow$$

$$-\boxed{D \rangle D \rangle D} \cdots \boxed{T \rangle T \rangle T} + \qquad \overset{k^D_{off}}{\underset{k^D_{on}}{\rightleftharpoons}}$$

$$\boxed{D} \; + \; -\boxed{D \rangle D} \cdots \boxed{T \rangle T \rangle T} +$$

(Equation 4)

The actin/ADP monomer units have lower affinity to the filaments, i.e., much higher depolymerization rate constant, k^D_{off} than the actin/ATP monomer units (k^T_{off}). As a result, the D-cap at the minus filament end increases its depolymerization rate and filaments rapidly depolymerize therefrom.

As mentioned above, the actin polymerization is not a true equilibrium polymerization as a result of ATP hydrolysis. Actin polymerizes as a T-monomer and depolymerizes as a D-monomer, as shown in Equation 5:

$$-\boxed{D^* \rangle T} \cdots \boxed{T} + \quad + \quad \boxed{T^*} \qquad \longrightarrow$$

$$-\boxed{D^* \rangle T} \cdots \boxed{T \rangle T^*} + \quad \longrightarrow \quad \boxed{D^* \rangle D} \cdots \boxed{T \rangle T^*} +$$

$$\longrightarrow \quad \boxed{D^*} \; + \; -\boxed{D} \cdots \boxed{T \rangle T^*} +$$

(Equation 5)

Cells need to replenish the energy for actin/ATP monomer formation:

$$actin/ADP \longrightarrow actin/ATP$$

As a result, the equilibrium constant at the plus end is higher than that of the minus end, and the critical monomer concentration at the plus end is smaller than that of the minus end:

$$K^+ > K^-$$

$$c^+_c > c^-_c$$

Thus, actin/ATP $\longrightarrow$ actin/ADP hydrolysis creates lower critical concentration for the plus filament end and higher critical concentration for its minus end. This is illustrated in Figure 4 where the dependence of filament growth rate at both filament ends on actin monomer concentration is given.[6] Positive growth rates correspond to

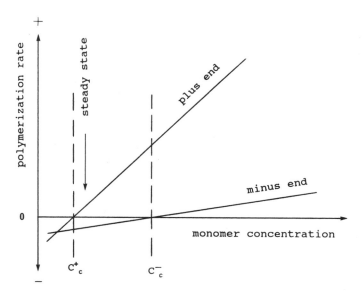

Figure 4. Dependence of the rate of the directional actin
polymerization on actin monomer concentration.

polymerization, whereas, negative growth rates - to depolymerization.
Obviously, the plus and minus filament ends characterize with different
growth rates. Both ends characterize with different critical
concentration as well. The region between the critical concentrations of
the plus (c_c^+) and minus ends (c_c^-) corresponds to simultaneous
polymerization at the plus end and depolymerization at the minus end.

$$c_c^+ < c_c^-$$

At steady state, shown with an arrow in Figure 4, the rate of
depolymerization of the minus end is equal to the polymerization rate at
the plus end, and actin filaments undergo the so-called treadmilling
without change in their length, as in Equation 6:

$$- \rangle\ a\ \rangle\ b\ \rangle\ c\ \rangle\ -\ -\ -\ -\ \rangle\ w\ \rangle\ x\ \rangle\ y\ \rangle\ +\ \ +\ \rangle\ z\ ]\ \longrightarrow$$

$$\rangle a\ ]\ \ +\ \ -\rangle\ b\ \rangle\ c\ \rangle\ \ \rangle\ -\ -\ -\ \rangle\ x\ \rangle\ y\ \rangle\ z\ \rangle\ +\ \ \longrightarrow$$

$$\rangle b\ ]\ \ +\ \ -\rangle\ c\ \rangle\ \rangle\ \rangle\ -\ -\ -\ \rangle\ y\ \rangle\ z\ \rangle\ a\ \rangle\ +\ \ \longrightarrow$$

(Equation 6)

Actin monomer z polymerizes at the plus filament end; simultaneously, monomer unit a depolymerizes at the minus filament end. Then, monomer
a (or any other free actin monomer in the surroundings of the filament)
polymerizes at the plus end; monomer unit b depolymerizes simultaneously
from the minus filament end, and so on. Thus, treadmilling allows "spacial movement" of actin filaments at a speed of about 1 micron per minute
at the expense of the ATP $\longrightarrow$ ADP hydrolysis energy.

3C. Termination

Cells accomplish the termination of *in vivo* actin polymerization by different "capping" proteins which cap both ends of the actin filaments. Capped actin filaments do not undergo either depolymerization or treadmilling.

4. CELL UTILIZATION OF ACTIN POLYMERIZATION

Cells utilize actin polymerization in many cellular activities. Actin filaments play an important role in muscle contraction, together with myosin.[7,8] Filamentous actin is a major component of cellular cytoskeleton.[9,10] It is located just beneath the cell membrane and provides its mechanical support.

Cells extend microspikes and lamellipodia (different types of cell protrusions), which are parallel actin filaments covered with cellular membrane, by actin polymerization at the plus end of the actin filaments which is located at the spike tip.[11,12] It allows cell crawling over substrates.

Phagocytosis, a process in which white blood cells engulf and disintegrate bacterial and other cells, is based on actin polymerization and depolymerization as well. Actin depolymerization allows the cytoskeleton (and the cell membrane) of the white cell to bend inward at the initial contact point between the white and bacterial cells. Then, actin polymerization pushes the membrane of the white cell outward around the bacterial cell and allows it to completely engulf the latter (Figure 5). The scanning electron micrograph in Figure 6 shows a macrocytic white blood cell engulfing a yeast cell.[13] The letters L point to the edges where actin polymerization proceeds at the plus filament ends.

Rapid actin polymerization is utilized by some invertebrate species during fertilization.[14] Their sperm cells contain a large amount of actin in the form of 1:1 complex with another much smaller protein which prevents the initiation of actin polymerization but does not interfere later with actin propagation. When a sperm cell approaches the surroundings of the corresponding egg, its intracellular pH changes and leads to decomposition of the actin complex. The light micrographs in Figure 7, taken in 0.75 seconds interval, show a sperm cell activated artificially by changing pH. Actin explosive polymerization follows with the formation of a harpoon-like spike which penetrates the egg membrane and both cells fuse (not shown).

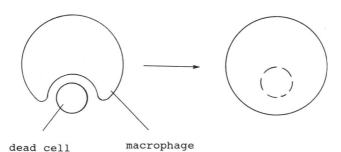

dead cell   macrophage

Figure 5. Schematic representation of phagocytosis. A macrophage engulfing a dead cell.

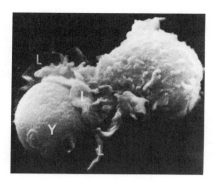

Figure 6. Scanning electron micrograph of phagocytil white blood cell (L) engulfing a yeast cell (Y). Magnification 2800X. (Reproduced from CELL, 1981, vol. 24, p 907, by copyright permission of the Cell Press.)

TUBULIN POLYMERIZATION[15,16]

1. TUBULIN MONOMER[17]

Tubulin monomer is another globular protein. It is a heterodimer and consists of two tightly linked related globular proteins: alpha- and

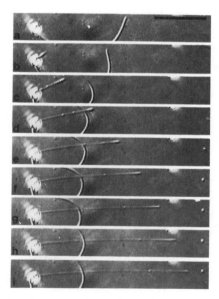

Figure 7. Light micrographs of actin polymerization in a sperm cell with the formation of harpoon-like spike. Scale bar, 20 microns. (Reproduced from the Journal of Cell Biology, 1982, vol. 93, p 822, by copyright permission of the Rockerfeller University Press).

beta-tubulin. Each tubulin is built of 450 alpha-amino acid residues and has molecular weight of about 50,000.

Tubulin is also very abundant in all eucaryotic cells. It is highly conserved cellular protein produced by closely related genes.[18] Tubulin from different species copolymerize with each other and with other cellular proteins for the same reasons as those discussed above for actin.

Tubulin is isolated from vertebrate brain where it constitutes about twenty percent of the soluble proteins. This reflects the high density of tubulin in the elongated axons and dendrites of the nerve cells.

2. TUBULIN HOMOPOLYMERS

Tubulin homopolymers or the so-called microtubules, consist of 13 parallel protofilaments, which form a long cylindrical hollow filament with an outer diameter of 25 nm and length in the range from 10 to 200 microns.[19,20] The microtubule structure is schematically illustrated in Figure 8 with only one protofilament. Each protofilament is a linear alternating copolymer of alpha- and beta-tubulins.

Protofilaments are directional, i.e., they have two distinguished plus (fast-growing) and minus (slow-growing) ends. All thirteen protofilaments have their plus ends on one and the same end of the microtubule; their thirteen minus ends are located on the other microtubule end. Thus, microtubules are directional as well with plus (fast-growing) and minus (slow-growing) ends.

3. TUBULIN POLYMERIZATION

3A. Initiation

Tubulin polymerization has its own peculiarities.[21] The initiation

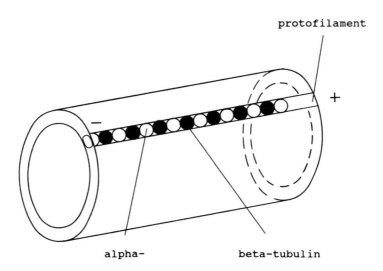

protofilament

alpha-   beta-tubulin

Figure 8. Schematic representation of microtubule structure.

of tubulin polymerization *in vivo* starts from a centriole (or its surroundings).[22-24] A centriole is a small cylindrical cell organelle which consists of microtubules and related proteins. It is usually located close to the cell nucleus or the cilia base. The initiation process is not well understood. Each centriole initiates only a certain predetermined number of microtubules. It is therefore believed that an unknown protein initiates tubulin polymerization. The centriole caps (holds) the minus ends of the microtubules and *in vivo* polymerization proceeds radically only at their plus ends. In contrast, actin polymerization proceeds at both ends.

3B. Propagation

In general, microtubule propagation follows the patterns of actin polymerization. It is not a true equilibrium polymerization. Conformational changes and nucleotide hydrolysis favor plus end polymerization. Tubulin, however, has tightly bound guanosine triphosphate, (GTP) instead of ATP, which is another cellular energy molecule.[25] GTP hydrolysis to guanosine diphosphate (GDP) is accompanied with energy release. It proceeds after tubulin monomer is incorporated onto the microtubule (delayed hydrolysis), as shown in Equation 7.

| tubulin/GTP | tubulin/GDP | energy |

(Equation 7)

Using a representation similar to that used above for actin, tubulin polymerization and depolymerization at the plus end is illustrated schematically here, in Equation 8. Remember, the microtubule minus end is capped and does not participate in *in vivo* polymerization.

(Equation 8)

The tubulin/GTP monomer (T) has a much higher rate of polymerization than the tubulin/GDP monomer (D). The T-cap of the plus end stabilizes the filaments and accelerates the polymerization. Microtubule growth continues until the last tubulin/GTP unit at the plus end undergoes random hydrolysis into a tubulin/GDP unit. Microtubules with D-cap plus end have practically no chance to add new tubulin monomers and the T-cap is never regained. The tubulin/GDP monomer units (D) of the microtubules have much higher depolymerization rate than tubulin/GTP monomer units (T). As a matter of fact, the D-cap end depolymerization rate is about a hundred times higher than that of a T-cap end. This leads to rapid catastrophic and complete depolymerization from the D-cap plus end. The polymerization starts again from the same tubulin initiation center.

In contrast to actin filaments, microtubules do not have a constant length and a "steady" state. They characterize with the so-called process of dynamic instability.[26,27] Microtubules either grow slowly *in vivo* at their plus ends, until the last T-monomer unit undergoes hydrolysis into D-monomer unit, or depropagate rapidly. Each of these two dynamic states lasts 15 seconds to several minutes.

In vitro tubulin polymerization proceeds spontaneously by similar mechanism, but both ends are active independently. Figure 9 shows slowly growing and rapidly shrinking microtubulin (ends), photographed sequentially in dark-field microscope at 1 minute intervals.[28] The plus end undergoes three times faster polymerization and three times faster depolymerization than the minus end.

3C. Termination

Cells accomplish the termination of *in vivo* tubulin polymerization by different "capping" proteins which cap the plus microtubule end. Capping proteins are usually located in the cell membrane or cytoskeleton. Microtubules with capped minus end (in a centriole) and capped plus end with a capping (membrane) protein do not exhibit dynamic instability and they are able to retain their length (Figure 10).

4. CELL UTILIZATION OF TUBULIN POLYMERIZATION

Cells utilize tubulin polymerization in many cellular activities;[29,30] two of them: locomotion,[31] and cell division,[32] are very important.

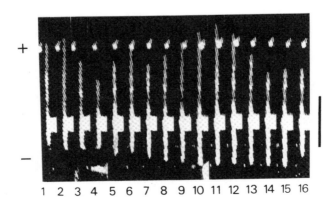

Figure 9. Dynamic instability of plus and minus ends of a single microtubule recorded sequentially in dark-field microscope at 1 minute intervals. Scale bar, 5 microns. (Reproduced with permission from Nature, 1986, vol, 321, p 606, copyright (1986) Macmillan Magazines Limited).

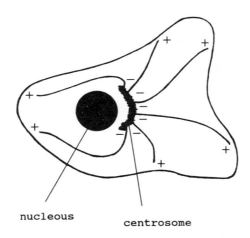

nucleous   centrosome

Figure 10. Schematic representation of a cell with capped
        microtubules.

Cells use cilia and flagella, which are hair-like cell protrusions,
to move themselves or to move fluid and food over their surface
(locomotion).[33] The cone of both the cilia and flagella are built of
microtubules. Their minus ends are embedded in the cilia/flagella base,
whereas, their plus ends are again located at the tip. Thus, tubulin
polymerization (at the plus end) extends cilia and flagella length.

Figure 11 shows a schematic representation of cilia and flagella of
different cells. Cilia of epithelial cells sweep layers of mucus produced
in the respiratory tract. Cilia located along the oviduct sweep the ova
(egg). Flagella proper (human, etc.) sperm cells, and so on. As mentioned
above, microtubules form the cone of axons and dendrites of nerve cells,
as well.

Microtubules provide cells also with a machinery for cell
division.[34-36] The microtubules are initiated radially in all directions
from the two centrosomes (a type of centriole),[37] which are located near
to the two newly formed nuclei. Some of these microtubules do not find
capping proteins in the cytoskeleton and they undergo rapid and complete
depolymerization (Figure 12). Then, they start again and again from the
same centrosome until they capture "matching" membrane proteins, which
cap their plus ends and prevent their depolymerization. These
microtubules pull and separate the two nuclei and later divide the cell
(Figure 13). It is believed that the dynamic instability of microtubules
allows them to find certain capping proteins in the cell membrane or
cytoskeleton and to establish the correct plane for cell division.

CELL REGULATION

Little is known of cell regulation of tubulin and actin,[38]
polymerization. A large variety of different proteins, however, are able
to copolymerize with actin,[39] and tubulin,[40] monomers or to form
complexes with their polymers: actin filaments and microtubules.
Different drugs,[41] block their polymerization or prevent their filaments
from depolymerization; they arrest cell movement and all other tubulin
and actin associated cell activities. Colchicine, an alkaloid, tightly

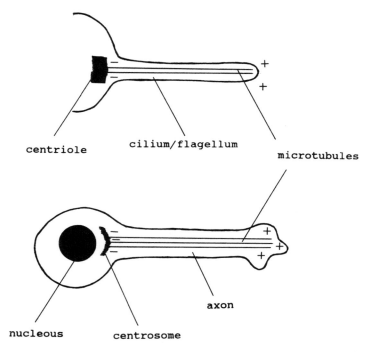

centriole   cilium/flagellum

microtubules

axon

nucleous   centrosome

Figure 11. Schematic representation of different cells with a
cilium, flagellum, or axon.

binds to tubulin monomer and prevents its polymerization.[42] It blocks
cell division in mitosis. Its effect is reversible; cell division
continues upon removal of colchicine. Taxol, another drug, tightly binds
to microtubules.[43] It prevents their depolymerization and cell division
as well. It is used effectively for treatment of cancer patients by
stopping cell division of the rapidly dividing cancer cells.

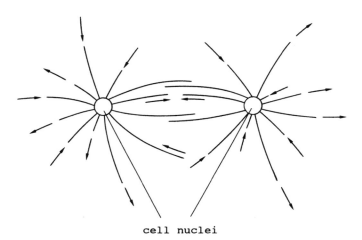

cell nuclei

Figure 12. Microtubule dynamic instability provides machinery
for cell division.

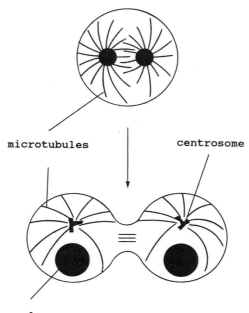

microtubules   centrosome

nucleous

Figure 13. Schematic representation of cell division.

CONCLUSION

At the present moment, the polymerization of actin and tubulin appears as a curiosity in biological science with its peculiar features: treadmilling, dynamic instability, etc. There is no doubt, however, that in the near future further studies on such types of protein polymerization will create a new exciting chapter in polymer chemistry.

REFERENCES

1. O. W. Webster, Science, **251**, 887 (1991).
2. T. D. Pollard & J. A. Cooper, Ann. Rev. Biochem., **55**, 987 (1986).
3. E. D. Korn, Physiol. Rev., **62**, 672 (1982).
4. T. D. Pollard & S. W. Craig, Trends Biochem. Sci., **7**, 55 (1982).
5. E. D. Korn, M.-F. Carlier & D. Pantaloni, Science, **238**, 638 (1987).
6. E. M. Bonder, D. J. Fishkind & M. S. Mooseker, Cell, **34**, 491 (1983).
7. R. Cooke, CRC Crit. Rev. Biochem., **21**, 53 (1986).
8. H. E. Huxley, Science, **164**, 1356 (1969).
9. M. Schliwa, *"The Cytoskeleton – An Introductory Survey,"* Springer Verlag, New York, 1986.
10. T. P. Stossel, C. Chaponnier, R. M. Ezzell, J. H. Hartwig, P. A. Janmey, D. J. Kwiatkowski, S. E. Lind, D. B. Smith, F. S. Southwick, H. L. Yin & K. S. Zaner, Ann. Rev. Cell Biol., **1**, 353 (1985).
11. Y. Wang, J. Cell Biol., **101**, 597 (1985).
12. L. G. Tilney, E. M. Bonder & D. J. DeRosier, J. Cell Biol., **90**, 485 (1981).
13. J. Boyles & D. F. Bainton, Cell, **24**, 905 (1981).
14. L. G. Tilney & S. Inoue, J. Cell Biol., **93**, 820 (1982).

15. P. Dustin, "*Microtubules*," 2nd. Ed., Springer Verlag, Berlin, (1984).
16. P. Dustin, Scient. Amer., **243**(8), 66 (1980).
17. D. W. Cleveland & K. F. Sullivan, Ann. Rev. Biochem., **54**, 331 (1985).
18. E. C. Raff, J. Cell Biol., **99**, 1 (1984).
19. E.-M. Mandelkow, R. Schultheiss, R. Rapp, M. Muller & E. Mandelkow, J. Cell Biol., **102**, 1067 (1986).
20. L. A. Amos & T. S. Baker, Nature, **279**, 607 (1979).
21. M. Kirschner & T. Mitchison, Cell, **45**, 329 (1986).
22. D. N. Wheatley, "*The Centriole: A Central Enigma of Cell Biology*," Elsevier, New York, 1982.
23. B. R. Brinkley, Ann. Rev. Cell Biol., **1**, 145 (1985).
24. J. R. McIntosh, Mod. Cell Biol., **2**, 115 (1983).
25. M.-F. Carlier, Cell Biophys., **12**, 105 (1988).
26. T. Mitchison & M. Kirschner, Nature, **312**, 237 (1984).
27. K. W. Farrell, M. A. Jordan, H. P. Miller & L. Wilson, J. Cell Biol., **104**, 1035 (1987).
28. T. Horio & H. Hotani, Nature, **321**, 605 (1986).
29. M. de Brabander, Endeavor, **6**, 124 (1982).
30. R. D. Allen, Scient. Amer., **256**(2), 42 (1987).
31. I. R. Gibbons, J. Cell Biol., **91**, 107s (1981).
32. S. Inoue, J. Cell Biol., **91**, 131s (1981).
33. L. T. Haimo & J. L. Rosenbaum, J. Cell Biol., **91**, 125s (1981).
34. R. D. Sloboda, Am. Sci., **68**, 290 (1980).
35. E. Schulze & M. Kirschner, J. Cell Biol., **102**, 1010 (1986).
36. J. S. Hyams & B. R. Brinkley, "*Mitosis Molecules and Mechanism*," Academic Press, London, 1989.
37. T. Mitchison & M. Kirschner, Nature, **312**, 232 (1984).
38. J. A. Cooper, J. Cell Biol., **105**, 1473 (1987).
39. A. Weeds, Nature, **296**, 811 (1982).
40. J. B. Olmstead, Ann. Rev. Cell Biol., **2**, 421 (1986).
41. L. Wilson, Life Sci., **17**, 303 (1975).
42. E. D. Salmon, M. McKeel & T. Hays, J. Cell Biol., **99**, 1066 (1984).
43. M. de Brabander, et al., Int. Rev. Cytol., **101**, 215 (1986).

CELL ADHESIVE PROPERTIES OF BIOELASTIC MATERIALS CONTAINING CELL ATTACHMENT SEQUENCES

Alastair Nicol, D. Channe Gowda, Timothy M. Parker and Dan W. Urry

Laboratory of Molecular Biophysics
School of Medicine
The University of Alabama at Birmingham, VH300
Birmingham, Alabama 35294-0019

The biocompatibility, conformational and inverse temperature transition properties of poly(Val^1-Pro^2-Gly^3-Val^4-Gly^5), i.e., poly(VPGVG), and its γ-irradiation crosslinked matrix and the poly(VPGVG)-derived hydrophobicity scale are noted. Also noted are the capacities of varying the bioactive role of bioelastic materials; that is, the bioelastic materials can be designed (1) to exhibit a range of elastic moduli, (2) to exhibit different rates of degradation, (3) for various modes of drug release, (4) to perform numerous free energy transductions, (5) to contain functional enzyme sites, and (6) to contain functional cell attachment sequences that promote growth to confluence.

Studies of cell attachment to bioelastic matrices containing either (1) REDV, an amino acid sequence within fibronectin shown to mediate melanoma cell adhesion and which as GREDVY attached to glycophase glass is reportedly specific for human endothelial cell spreading, or (2) RGDS, the primary cell adhesion site of fibronectin, are reported. Four different cell types were used; bovine ligamentum nuchae fibroblasts, bovine aortic endothelial cells, human umbilical vein endothelial cells and human malignant melanoma cells. None of four different REDV-containing sequences incorporated into X^{20}-poly(GVGVP) were found to promote cell adhesion. However, RGDS as GRGDSP incorporated into X^{20}-poly(GVGVP) promoted attachment and spreading of all cell types. Interestingly, this sequence within the crosslinked matrix functioned as a vitronectin-like sequence rather than a fibronectin-like sequence, being specific for the vitronectin cell surface receptor.

INTRODUCTION

Bioelastic materials have their origins in repeating sequences of the mammalian elastic protein, elastin. Because of the unique properties

Biotechnology and Bioactive Polymers, Edited by C. Gebelein
and C. Carraher, Plenum Press, New York, 1994

of certain elastin sequences, they can be designed to function and to perform roles that were not implicit in the parent protein. The result is a wide range of bioelastic materials that bear only peripheral resemblance to their elastin origins and whose active compositions and functions are not those of elastin, but whose retained compliance is a useful component.[1]

The most striking repeating elastomeric sequence of elastin is poly(Val^1-Pro^2-Gly^3-Val^4-Gly^5).[2-4] The polypentapeptide, poly(VPGVG), and its γ-irradiation crosslinked matrix have been shown to be extraordinarily biocompatible as demonstrated by the series of tests (1) mutagencity, (2) cytotoxicity, (3) systemic toxicity, (4) intracutaneous toxicity, (5) muscle implantation, (6) acute intraperitoneal toxicity, (7) systemic antigenicity, (8) dermal sensitization, (9) pyrogenicity, (10) blood clotting, and (11) hemolysis.[5]

The conformation of poly(VPGVG) is unique: it is a series of ß-turns which, on raising the temperature through a critical temperature range, fold into a hydrophobically assembled helical structure called a ß-spiral and several ß-spirals supercoil, again by optimizing hydrophobic interactions, to form twisted filaments.[6] This folding and assembly on raising the temperature, i.e., this increase in order on raising the temperature, is called an inverse temperature transition. The temperature, T_t, at which this inverse temperature transition is initiated depends on composition; increase hydrophobicity and the value of T_t decreases, and decrease hydrophobicity and T_t increases. This dependence of T_t on hydrophobicity has been used to develop a hydrophobicity scale.[7,8] It should be noted, however, that the compositional changes must be carried out with full knowledge of conformation. In particular, the only position in the pentamer that allows substitution of all the amino acids and additional derivatives is position four.[8] The hydrophobicity scale becomes a tool for designing new bioelastic materials with new functions.

By varying the composition within the allowable conformational constraints of the ß-spiral structure and using the developed hydrophobicity scale to maintain or readjust the temperature, T_t, of the folding and assembly transition, the bioelastic material (1) can be designed to exhibit a range of elastic moduli,[9] (2) can be designed to have a desired rate of degradation,[10] (3) can be designed for various modes of drug delivery,[11] (4) can be designed to perform free energy transduction involving the intensive variables of temperature, pressure, chemical potential, mechanical force and electrochemical potential,[8] (5) can be designed to contain functional enzyme sites, for example, for lysyl oxidase,[12] prolyl hydroxylase,[13] protein kinases and phosphatases,[14] and (6) can be designed to contain functional cell attachment sequences.[15]

The potential of combining the properties of biocompatibility,[5] variable compliance,[9] variable rate of degradation,[10] ability to introduce specific enzyme sites,[12-14] and most significantly being able to contain functional cell attachment sites where cells not only attach but grow to confluence[15] make bioelastic materials of potential use for tissue reconstruction. As one step in that direction, reported here is a study which identifies the integrin specificity (i.e., the cell surface receptor responsible for adhesion) that binds at the GRGDSP sequence when incorporated within poly(VPGVG) and examines a series of REDV-containing sequences for their capacity for cell adhesion in poly(VPGVG).

EXPERIMENTAL

1. CELLS

Bovine ligamentum nuchae fibroblasts (LNF) were provided by R. M. Senior and G. L. Griffin as newly explanted cultures. Stock LNFs at early passage were stored in liquid nitrogen. LNFs were cultured in Dulbecco's modified Eagle medium containing 10% fetal bovine serum, 2 mM L-glutamine, 0.1 mM nonessential amino acids, 100 u/mL penicillin and 100 μg/mL streptomycin. LNFs were used for experiment before passage 8.

Bovine aortic endothelial cells (BAEC), isolated as previously described,[16] were stored at early passage in liquid nitrogen. BAECs were cultured in Medium 199 containing 25 mM [N-2-hydroxyethylpiperazine-N'-2-ethansulfonic acid (HEPES), 10% fetal bovine serum, 5% Ryan's growth supplement, 100 u/mL penicillin and 100 μg/mL streptomycin. BAECs were used for experiment before passage 6.

Human umbilical vein endothelial cells (HUVEC) were obtained from Endotech Corporation, Indianapolis at passage 1. Early passage stock was stored in liquid nitrogen. HUVECs were cultured in Medium 199 containing 25 mM HEPES, 20% fetal bovine serum, 2 mM L-glutamine, 13.2 μl/mL Endo-Ret HI-GF (Endotech) endothelial cell growth supplement, 100 u/mL penicillin, 100 μg/mL streptomycin and 0.25 μg/mL amphotericin B. Plastic tissue culture flasks were precoated with 2% bovine gelatin at 7.5 μl/cm^2. HUVECs were also used for experiment at passage 8 or earlier.

A375 human malignant melanoma cells were obtained at passage 165 from the American Type Culture Collection, Maryland. Cells were cultured in Dubecco's modified Eagle medium containing 10% fetal bovine serum, 100 u/mL penicillin and 100 μg/mL strepomycin. A375 cells were used at passage 173 or earlier. All cells were cultured at 37°C in 5% CO_2 in air in a humidified incubator.

2. ADHESION ASSAY

The cell to matrix or precoated substratum adhesion assay has been described in detail.[15] Briefly, cells were harvested using 0.05% trypsin/0.53 mM EDTA, then treated with soybean trypsin inhibitor (0.2 mg/mL for 2 min). 50 μl of cells at 0.32 x 10^5 cells/mL were plated into glass cloning cylinders containing 50 μl of appropriate medium over the test matrix or other substratum. Cells were incubated and allowed to attach for 3 h or 20 h before being gently rinsed in phosphate buffered saline (PBS) pH 7.2 and then fixed using either 3.5% formaldehyde in PBS or 2.5% glutaraldehyde in PBS. Adherent cells were then counted and classified as well spread, poorly spread or rounded by phase contrast microscopy.

3. PEPTIDE INHIBITION OF ADHESION ASSAY

Inhibitory peptides were purchased from Telios Pharmaceuticals, San Diego. This assay was carried out as for a normal adhesion assay except that the 50 μl of cells were plated into 50 μl of medium containing the inhibitory or control peptide at 2 mM over the test substratum. Fibronectin (human, Calbiochem, San Diego) and Vitronectin (human, Telios Pharmaceuticals, San Diego) were coated at 10 μg/mL in PBS onto non-tissue culture plastic coverslips by placing 50 μl aliquots onto 10 mm square pieces of coverslip for one hour. The coating solution was then aspirated and replaced with medium prior to cell plating.

4. ELASTOMERIC MATRICES

These were prepared by the 20 Mrad γ-irradiation of coacervated polypeptide as previously described.[15,18,19]

5. PEPTIDE SYNTHESIS

The syntheses of poly(GVGVP) and poly[n(GVGVP)(GRGDSP)] have been previously described.[15,20] The peptides Boc-Val-Pro-OBzl, Boc-Gly-Val-Pro-OBzl, Boc-Val-Pro-Gly-Val-Gly-OH and Boc-Val-Pro-Gly-Phe-Gly-OH were prepared as previously described.[20,21] The two heptapeptides, Boc-Arg(Tos)-Glu(OCHx)-Val-Tyr(2,6-Cl$_2$Bzl)-Pro-Gly-OH and Boc-Arg(Tos)-Glu(OCHx)-Val-Phe-Pro-Gly-OH were synthesized by solid phase methods and the hexapeptide Boc-Gly-Arg(Tos)-Glu(OCHx)-Asp(OCHx)-Val-Pro-OH and the pentapeptide Boc-Arg(Tos)-Glu-(OCHx)-Asp(OCHx)-Val-Pro-OH were synthesized by the solution phase method. The protected peptides were characterized by carbon-13 nuclear magnetic resonance before polymerization to verify structure and purity. Thin layer chromatography (t.l.c.) was performed on silica gel plates obtained from Whatman, Inc., with the following solvent systems: R_f^1, CHCl$_3$:CH$_3$OH:CH$_3$COOH(95:5:3); R_f^2, CHCl$_3$:CH$_3$OH:CH$_3$COOH(90:10:3); R_f^3, CHCl$_3$:CH$_3$OH:CH$_3$COOH(85:15:5). The compounds on t.l.c. plates were detected by UV light, by spraying with ninhydrin, or by chlorine/tolidine spray. All Boc amino acids and HOBt were purchased from Advanced Chem. Tech. (Louisville, Kentucky). EDCI was obtained from Bachem, Inc. (Torrance, California).

5A. Solid Phase Synthesis of Boc-Arg(Tos)-Glu(OCHx)-Asp(OCHx)-Val-Phe-Pro-Gly-OH: (I) and Boc-Arg(Tos)-Glu(OCHx)-Asp(OCHx)- Val-Tyr (2,6-Cl$_2$-Bzl)-Pro-Gly-OH: (II).

The synthesis of the two heptapeptides, I and II, was carried out by the solid phase technique starting from the Boc-Gly-OCH$_2$-C$_6$H$_4$-Resin (2.0 g, 0.86 mmol each).[22] The Boc group was used for temporary N$^\alpha$-protection, and its removal was effected by 40% TFA/CH$_2$Cl$_2$. The neutralization step was carried out with 10% diisopropylethylamine in CH$_2$Cl$_2$. The coupling reaction was carried out with 3.0 equiv. each of Boc-amino acid, DCC and HOBt in CH$_2$Cl$_2$. When needed, the coupling step was repeated in 50% DMF/CH$_2$Cl$_2$ and uncoupled chains were finally acetylated. The side chain functions of Arg, Tyr, Glu and Asp were protected by Tos, 2,6-dichloro-benzyl and cyclohexyl groups, respectively. After completion of the sequence, the protected peptides were cleaved from the resin by transesterification with 2-dimethylaminoethanol at room temperature followed by treatment with water to yield the protected peptide acid.[23] The peptides were purified over a silica gel column using 15% methanol in chloroform to obtain 0.67 g (yield 63.2%) of I and 0.72 g (yield 59.5%) of II. The purity of the peptides were checked by C-13 NMR spectroscopy and t.l.c. For I: R_f^2, 0.21; R_f^3, 0.43 and for II: R_f^2, 0.25; R_f^3, 0.46.

5B. Solution Phase Synthesis of Boc-Arg(Tos)-Glu(OCHx)-Asp(OCHx)-Val- Pro-OH and Boc-Gly-Arg(Tos)-Glu(OCHx)-Val-Pro-OH Boc-Asp(OCHx)-Val-Pro-OBzl: (III)

Boc-Val-Pro-OBzl (8.09 g, 0.02 mole) was deprotected by stirring for 1.5 h in 4N HCl in dioxane. Excess HCl and dioxane were removed under reduced pressure, triturated with ether, filtered, washed with ether and dried (yield 100%).

A solution of Boc-Asp(OCHx)-OH(6.31 g, 0.02 mole) and HOBt(3.0 g, 0.022 mole) in DMF was cooled to -15°C with stirring and EDCI (4.2 g, 0.022 mole) was added. After 20 min a pre-cooled solution of the above hydrochloride salt and NMM (2.2 mL, 0.02 mole) was added and the reaction mixture stirred overnight at room temperature. The mixture was evaporated to a thick oil which was dissolved in $CHCl_3$. This solution was extracted with water, 10% citric acid, water 5% $NaHCO_3$, water, and then dried over Na_2SO_4. The solvent was removed under reduced pressure and the resulting oil was dissolved in ether and precipitated from petroleum ether. The solid was filtered, washed with petroleum ether and dried to obtain 10.2 g (yield 84.78%) of III: R_f^1, 0.52; R_f^2, 0.65.

5C. Boc-Glu(OCHx)-Asp(OCHx)-Val-Pro-OBzl: (IV)

Compound III (9.03 g, 0.015 mole) was deblocked with HCl/dioxane and coupled to Boc-Glu(OCH$_x$)-OH (4.89 g, 0.015 mole) using EDCI with HOBt in the same manner as that described for III to give 10.3 g (yield 85.3%) of IV: R_f^1, 0.45; R_f^2, 0.58.

5D. Boc-Arg(Tos)-Glu(OCHx)-Asp(OCHx)-Val-Pro-OBzl: (V)

Compound IV (10.1 g, 0.012 mole) was deblocked with HCl/dioxane and coupled to BOC-Arg(Tos)-OH(5.31 g, 0.012 mole) using EDCI in the presence of HOBt. The reaction was worked up the same as for Compound III to give impure product, which was purified by column chromatography on silica gel using 5% methanol in $CHCl_3$ to obtain 11.1 g (yield 80.1%) of pure product V: R_f^1, 0.39; R_f^2, 0.55.

5E. Boc-Gly-Arg(Tos)-Glu(OCHx)-Asp(OCHx)-Val-Pro-OBzl: (VI)

Compound V (6.0 g, 0.0054 mole) was deblocked using HCl/dioxane and coupled to Boc-Gly-OH (0.95 g, 0.0054 mole) using EDCI with HOBt as described in the preparation of III to obtain 5.2 g (yield 82.9%) of VI: R_f^1, 0.28; R_f^2, 0.48.

5F. Boc-Gly-Val-Gly-Val-Pro-ONp: (VII)

Boc-Gly-Val-Gly-Val-Pro-OBzl (15.2 g, 0.029 mole) was dissolved in glacial acetic acid (150 mL) and 1.5 g of 10% palladium on charcoal was added. This mixture was hydrogenated at 40 psi for 6 h. The resulting residue was triturated with ether, filtered, washed with ether and dried to obtain the acid.

This acid was dissolved in pyridine (120 mL) and reacted with bis-PNPC (1.5 equiv.). When the reaction was complete, as monitored by t.l.c., the solvent was removed under reduced pressure. The residue was taken into $CHCl_3$. The $CHCl_3$ was washed with water, 10% citric acid, water, 5% $NaHCO_3$, water and dried over Na_2SO_4. The solvent was removed under reduced pressure, triturated with ether, filtered, washed with ether and dried to obtain 12.21 g (yield 76.5%) of VII: R_f^1, 0.38; R_f^2, 0.63; R_f^3, 0.75.

5G. Boc-Arg(Tos)-Glu(OCHx)-Asp(OCHx)-Val-Pro-ONp: (VIII)

Compound V (5.0 g, 0.0045 mole) was converted into free acid and then reacted with bis-PNPC following the same procedure described for the preparation of VII to give 3.65 g (yield 70%) of VIII: R_f^1, 0.32; R_f^2, 0.38; R_f^3, 0.58.

5H. Boc-Gly-Arg(Tos)-Glu(OCHx)-Asp(OCHx)-Val-Pro-ONp: (IX)

Using the same procedure as described for VII, the compound VI (5.0 g, 0.0042 mole) was debenzylated and converted into ONp ester IX, 4.3 g (yield 84.4%); R_f^1, 0.24; R_f^2, 0.3; R_f^3, 0.55.

5I. Poly[16(Val-Pro-Gly-Val-Gly), 4(Val-Pro-Gly-Phe-Gly), (Arg-Glu-Asp-Val-Phe-Pro-Gly)]: (X)

Boc-Val-Pro-Gly-Val-Gly-OH (2.11 g, 0.004 mole), Boc-Val-Pro-Gly-Phe-Gly-OH (0.58 g, 0.001 mole) and compound I (0.31 g, 2.5 x 10^{-4} mole) were deblocked together with TFA, and a one-molar solution of the TFA salt in DMSO was polymerized for 12 days using EDCI (2 equiv.), as the polymerizing agent with HOBt (1 equiv.) and 1.6 equiv. of NMM as base. The polymer was dissolved in water, dialyzed using 3500 mol wt. cut-off dialysis tubing and lyophilized to obtain the protected polymer. This was deblocked by a high HF deprotection procedure.[24] The polymer was dissolved in water, dialyzed using 50 kD mol wt. cut-off dialysis tubing for one week and lyophilized to obtain 1.14 g (yield 50%) of X.

5J. Poly[16(Val-Pro-Gly-Val-Gly), 4(Val-Pro-Gly-Phe-Gly), (Arg-Glu-Asp-Val-Tyr-Pro-Gly)]: (XI)

Boc-Val-Pro-Gly-Val-Gly-OH (2.11 g, 0.004 mole), Boc-Val-Pro-Gly-Phe-Gly-OH (0.58 g, 0.001 mole) and compound II (0.35 g, 2.5 x 10^{-4} mole) were deblocked together with TFA and polymerized using the same procedure described for the compound X to obtain the protected polymer. This was deblocked by a low-high HF deprotection procedure, dialyzed and lyophilized to obtain 1.12 g (yield 48.5%) of XI.

5K. Poly[20(Gly-Val-Gly-Val-Pro), (Arg-Glu-Asp-Val-Pro)]: (XII)

Compound VII (4.33 g, 0.0067 mole) and compound VIII (0.38 g, 3.33 x 10^{-6} mole) were deblocked together using TFA, and a one-molar solution of the TFA salts in DMSO was polymerized for 14 days using 1.6 equiv. of NMM as base. The polymer was dissolved in water dialyzed using 3500 mol wt. cut-off dialysis tubing and lyophilized. The protected polymer was deblocked with a high HF deprotection procedure, base treated with 1N NaOH, dialyzed using 50 kD mol wt. cut-off dialysis tubing for one week and lyophilized to obtain 2.2 g (yield 75.1%) of XII.

5L. Poly[20(Gly-Val-Gly-Val-Pro), (Gly-Arg-Glu-Asp-Val-Pro)]: (XIII)

Compound VII (4.33 g, 0.0067 mole) and compound IX (0.4 g, 3.33 x 10^{-4} mole) were treated the same as for compound XII to obtain 2.47 g (yield 83.3%) of XIII.

The purity of the intermediates and of the final products was checked by thin layer chromatography, carbon-13 NMR spectroscopy and amino acid analysis. The amino acid analyses and carbon-13 NMR spectra are reported in Figures 1 and 2 for polymers X through XIII. The absence of extraneous peaks verifies the synthesis, and the ratio of relevant peaks verified the incorporation of ratios. Based on the amino acid analyses, the more correct statement of the formulae for the polymers with mixed pentamers would be X, Poly[0.78(Val-Pro-Gly-Val-Gly), 0.14 (Val-Pro-Gly-Phe-Gly), 0.08(Arg-Glu-Asp-Val-Phe-Pro-Gly)]; XI, Poly-[0.79 (Val-Pro-Gly-Val-Gly), 0.13(Val-Pro-Gly-Phe-Gly), 0.08(Arg-Glu-Asp-Val-Tyr-Pro-Gly)]; XII, Poly[0.92(Gly-Val-Gly-Val-Pro), 0.08(Arg-Asp-Val-Pro)]; XIII, Poly[0.89(Gly-Val-Gly-Val-Pro), 0.11(Gly-Arg-Glu-Asp-Val-Pro)]; and Poly[0.97(GVGVP), 0.03(GRGDSP)].

ABBREVIATIONS

Boc, tert-butyloxycarbonyl; OBzl, benzyl ester; EDCI, 1-ethyl-3-dimethylaminopropyl carbodiimide; HOBt, 1-hydroxybenzotriazole; bis-PNPC, bis(4-nitrophenyl)carbonate; TFA, trifluoroacetic acid; DMSO, dimethyl-sulfoxide; NMM, N-methylmorpholine; Tos, p-toluenesulfonyl; DCC, dicyclo-hexylcarbodiimide; OCHx, cyclohexyl; 2,6-Cl$_2$-Bzl, 2,6 dichlorobenzyl; DMF, dimethylformamide; HF, hydrogen fluoride; R (Arg), arginine; D (Asp), aspartic acid; G (Gly), glycine; E (Glu), glutamic acid; P (Pro), proline; F (Phe), phenylalanine; S (Ser), serine; Y (Tyr), tyrosine; V (Val), valine; X^{20}, 20 Mrad γ-irradiation crosslinked; PBS, physiological

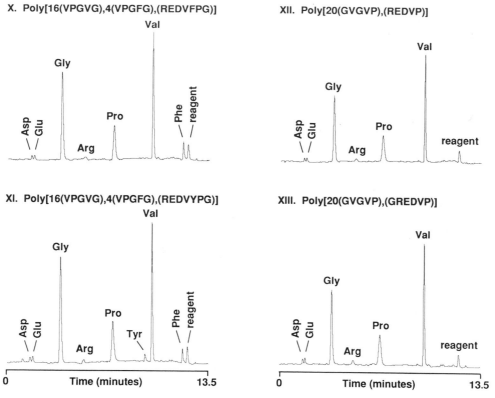

Figure 1. Amino acid analyses of the REDV-containing polymers used for the REDV-containing cell adhesion test matrices.

phosphate buffered saline (0.15 NaCl, 0.01 M phosphate, pH 7.2); HUVEC, human umbilical vein endothelial cells; LNF, ligamentum nuchae fibroblast; BAEC, bovine aortic endothelial cells; HEPES, N-2-hydroxyethyl/piperazine-N-2 ethanesulfonic acid; C^{13}NMR, C^{13} nuclear magnetic resonance; TC-hFN, tissue culture plastic coated with the human fibronectin; hFN, human fibronectin; hVN, human vitronectin; TC-O, uncoated tissue culture plastic; PS-O, uncoated non-tissue culture plastic; PShFN, non-tissue culture plastic coated with human fibronectin; TC-C, tissue culture plastic positive control; X^{20} stands for cross-linked nutrient using 20 μrads of γ-irradiation.

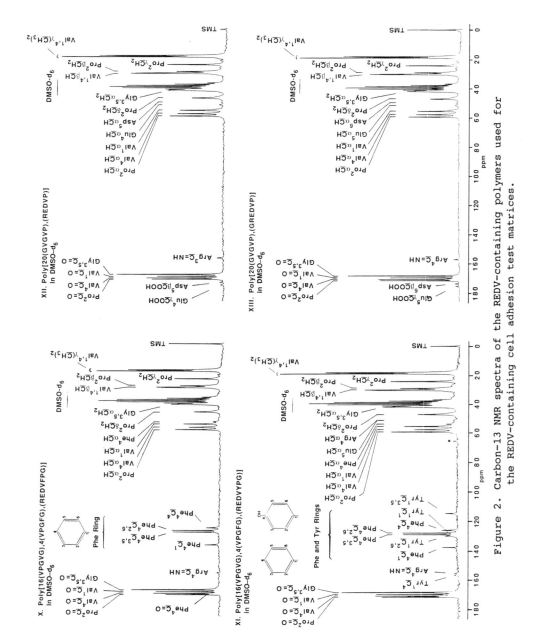

Figure 2. Carbon-13 NMR spectra of the REDV-containing polymers used for the REDV-containing cell adhesion test matrices.

RESULTS AND DISCUSSION

1. INCORPORATION OF RGDS PROMOTES CELL ADHESION

The cell adhesion properties of X^{20}-poly(GVGVP) can be varied by the covalent incorporation of RGDS,[15] the primary cell attachment sequence of fibronectin.[17] Unmodified X^{20}-poly(GVGVP) is a very poor support for cell attachment and spreading when tested in the absence of serum (Figure 3A). In the presence of serum sub-maximal cell adhesion and proliferation are obtained.[15] It has been shown that bovine endothelial cells and bovine ligamentum nuchae fibroblasts in the absence of serum adhere to X^{20}-poly[n(GVGVP)-(GRGDSP)] at maximal levels when n = 20 to 60.[15] Matrices having n = 100 supported cell adhesion but at a reduced level and matrices having n = 500 did not support cell adhesion. Thus; X^{20}-poly[n(GVGVP),(GRGDSP)] can be modulated between non-supportive to maximally supportive by varying n between 500 and 20. Also, when plated onto n = 20 to 60 matrices in the presence of 10% fetal bovine serum, cells attach and achieve growth rates similar to those of tissue culture controls (Figure 3B).

Human umbilical vein endothelial cells (HUVEC) were also examined by means of incubation for 4 h and 20 h on X^{20}-poly(GVGVP), X^{20}-poly[20(GVGVP).(GRGDSP)] and on uncoated and fibronectin coated plastic control surfaces. After 4 h or 20 h incubation (Figures 4A and 4B) the results were similar. The X^{20}-poly(GVGVP) provided very poor support for HUVEC adhesion whereas the X^{20}-poly[20(GVGVP),(GRGDSP)] was found to be a very good substrate for HUVEC adhesion, the level of cell attachment

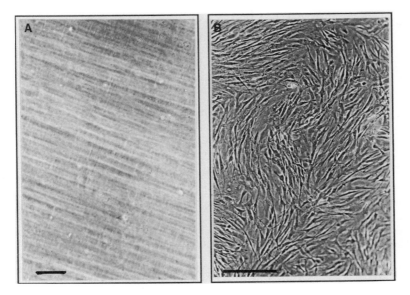

Figure 3. A: X^{20}-poly(GVGVP) 3 h after plating ligamentum nuchae fibroblasts at 1 x 10^4 cells/cm^2. B: X^{20}-poly[40(GVGVP),(GRGDSP)] 3 days after plating ligamentum nuchae fibroblasts at 1 x 10^4 cells/cm^2 in medium containing 10% fetal bovine serum. Scale bar represents 200 μm.

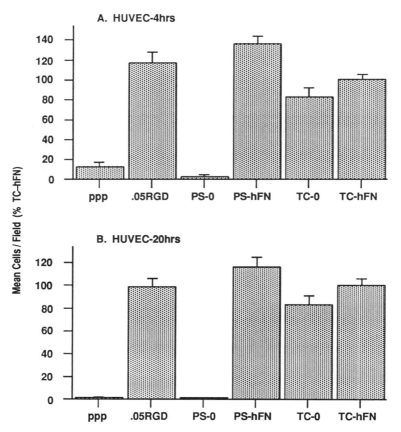

Figure 4. The numbers of HUVEC (mean ± s.e. expressed as a
percentage of those adhering to the TC-hFN control
adhering to test matrices after 4 hrs (A) and 20
hrs (B). Where PPP represents X^{20}-poly(GVGVP),
0.05% RGD represents X^{20}-poly[0.97(GVGVP),
0.03(GRGDSP), PS-0 represents uncoated plastic
cover slip, PS-hFN represents plastic cover slip
coated with human fibronectin, TC-0 represents
uncoated tissue culture plastic and TC-hFN
represents tissue culture plastic coated with
human fibronectin.

being intermediate between those of the two fibronectin coated control
surfaces at 4 h and equivalent to the fibronectin coated tissue culture
plastic at 20 h. Significantly, the X^{20}-poly[20(GVGVP),(GRGDSP)] also
supported cell spreading to levels similar to those of the fibronectin
coated controls. At 4 h, 95%, 84% and 95% of the attached cells were
classified as well spread on X^{20}-poly[20(GVGVP),(GRGDSP)], on fibronectin
coated plastic cover-slip and on fibronectin coated tissue culture
plastic, respectively. At 20 h, the levels were 85%,84% and 94%,
respectively.

2. INCORPORATION OF REDV DOES NOT PROMOTE CELL ADHESION

    REDV, (Arg-Glu-Asp-Val), has not been identified as an additional

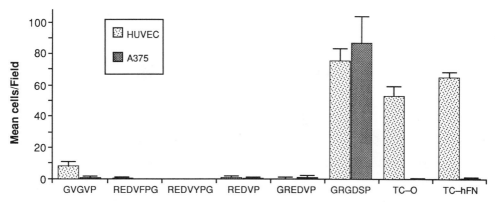

Figure 5. The numbers of HUVEC and A375 cells adhering to
REDV-containing elastomeric matrices after 4 and 7
hours, respectively, in serum-free medium. GVGVP
represents X^{20}-poly(GVGVP), REDVFPG, REDVYPG,
REDVP and GREDVP represent X^{20}-matrices containing
the different REDV peptides labeled as polymers X,
XI, XII and XIII as defined in the text, i.e.,
Poly[0.78(Val-Pro-Gly-Val-Gly), 0.14(Val-Pro-Gly-
Phe-Gly), 0.08(Arg-Glu-Asp-Val-Phe-Pro-Gly)]: (X);
Poly[0.79(Val-Pro-Gly-Val-Gly), 0.13(Val-Pro-Gly-
Phe-Gly), 0.08(Arg-Glu-Asp-Val-Tyr-Pro-Gly)]:
(XII); Poly[0.92(Gly-Val-Gly-Val-Pro), 0.08(Arg-
Glu-Asp-Val-Pro)]: (XII); and Poly[0.89(Gly-Val-
Gly-Val-Pro), (Gly-Arg-Glu-Asp-Val-Pro)]: (XIII),
respectively. GRGDSP represents X^{20}-poly
[0.97(GVGVP), 0.03(GRGDSP)]. TC-O represents
uncoated tissue culture plastic and TC-LNF
represents tissue culture plastic pre-coated with
10 μg/mL human fibronectin. The results are given
as mean cells per counted field and the vertical
bar represents one standard error.

cell adhesion site in fibronectin mediating melanoma cell adhesion.[25]
When attached to glycophase glass as GREDVY, it appears to serve as a
cell type specific site for human endothelial cell spreading.[26] The cell
receptor for REDV has been identified as the integrin α4β1.[27]

As described above, REDV was synthesized as REDVFPG, REDVYPG, REDVP
and GREDVP each of which was polymerized with GVGVP at a ratio of 20:1
then γ-irradiation crosslinked to form the REDV containing matrices. None
of these matrices were capable of supporting significant adhesion of
HUVEC or A375 melanoma cells (Figure 5). Both cell types showed high
levels of adhesion to the 20:1 poly[(GVGVP), (GRGDSP)]-containing control
matrix. One of the REDV-containing matrices, X^{20}-poly[0.78(Val-Pro-Gly-
Val-Gly), 0.14(Val-Pro-Gly-Phe-Gly), 0.08(Arg-Glu-Asp-Val-Phe-Pro-Gly)];
was also treated with LNF and BAECs but did not support the adhesion of
either (Figure 6). Thus, when incorporated into X^{20}-poly(GVGVP) matrices,
REDV appears incapable of supporting cell adhesion.

3. INVESTIGATION OF THE INTEGRIN BASIS OF CELL ADHESION TO RGDS
CONTAINING MATRICES USING SPECIFIC PEPTIDE INHIBITORS

In these experiments cells were plated onto X^{20}-

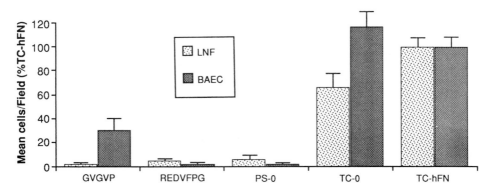

Figure 6. The number of LNF and BAEC cells adhering to X^{20}-
Poly[0.78(Val-Pro-Gly-Val-Gly), 0.14(Val-Pro-Gly-
Phe-Gly),0.08(Arg-Glu-Asp-Val-Phe-Pro-Gly)] matrix
after 3 hours in serum-free medium. PPP represents
X^{20}-poly(GVGVP); 20:1 REDVFPG represents X^{20}-
Poly[0.78(Val-Pro-Gly-Val-Gly), 0.14(Val-Pro-Gly-
Phe-Gly), 0.08(Arg-Glu-Asp-Val-Phe-Pro-Gly)]; PS-O
represents uncoated non-tissue culture plastic;
TC-O represents uncoated tissue culture plastic;
and TC-LNF represents tissue culture plastic
coated with human fibronectin at 10 μg/mL. The
results are given as mean cells per counted field
expressed as a percentage of the TC-LNF positive
control cell count. The vertical bar represents
one standard error.

(poly[20{GVGVP},{GRGDSP}] + poly{GVGVP}) matrix or onto fibronectin or
vitronectin coated plastic in the presence of 1 mM cell adhesion
inhibiting peptides. GRGDSP inhibits cell adhesion mediated by both
fibronectin and vitronectin receptors, GRGDdSP specifically inhibits only
fibronectin receptor mediated adhesion, GPenGRGDSPCA (cyclic RGDS)
specifically inhibits vitronectin receptor mediated adhesion and GRADSP
is an inactive control peptide.[28] Three widely different cell types were
used, bovine ligamentum nuchae fibroblasts (Figure 7), human umbilical
vein endothelial cells (Figure 8) and human A 375 malignant melanoma
cells (Figure 9), yet the matrix adhesion results were all similar
(Figure 10). All three cell types showed maximal adhesion to both the
control 40:1 (GVGVP):(GRGDSP) matrix and to the 40:1 matrix in the
presence of the 1 mM control peptide GRADSP. Both 1mM GRGDSP and 1mM
cyclic RGDS almost completely inhibited the attachment and spreading of
all three cell types to the 40:1 matrix. 1 mM GRGDdSP gave only a limited
inhibition of cell adhesion which was less than 50% for all cell types.
These results indicate that the cell adhesion to our RGDS-containing
matrix is mediated by the vitronectin receptor rather than, as might have
been expected, the fibronectin receptor.

When plated onto plastic substrata coated with human fibronectin in
the presence of the specific peptide inhibitors at 1 mM, no inhibition of
cell attachment was found for the fibronectin receptor specific GRGDdSP
nor for any of the other peptides (Figure 11). However, when the three
cell types were plated onto human vitronectin coated plastic substrata
both GRGDSP and the vitronectin receptor specific cyclic RGDS showed
strong inhibition of adhesion for all cell types (Figure 12). These few
cells which did adhere in the presence of GRGDSP and cyclic RGDS were
almost entirely round and not spread (data not shown; see Figures 7,8,9).

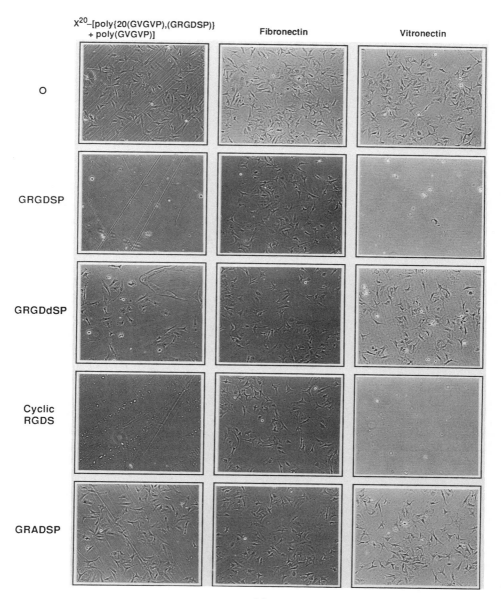

Figure 7. LNF adhesion to X^{20}-{poly[0.97(GVGVP), 0.03 (GRGDSP)] + poly(GVGVP)} and to 10 μg/mL human fibronectin coated and to 10 μg/mL human vitronectin coated plastic in the presence of 1 mM GRGDSP, 1 mM GRGDdSP, 1 mM cyclic RGDS, 1 mM GRADSP or in the absence of additional peptide (0).

As was the case for the 40:1 RGDS-containing matrix, the GRGDdSP did not strongly inhibit the adhesion of any of the cell types to vitronectin, the level of adhesion of LNFs stayed about same whereas the levels of adhesion of both HUVECs and A375s were higher for attachment to vitronectin than for the 40:1 RGDS-containing matrix. Thus, although the overall results clearly indicate that the major adhesive interaction of

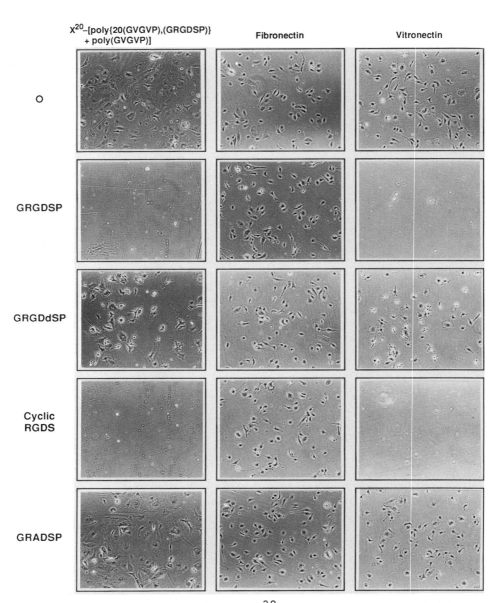

Figure 8. HUVEC adhesion to X^{20}-{poly[0.97(GVGVP), 0.03 (GRGDSP)] + poly(GVGVP)} and to 10 μg/mL human vitronectin coated plastic in the presence of 1 mM GRGDSP, 1 mM GRGDdSP, 1 mM cyclic RGDS, 1 mM GRADSP or in the absence of additional peptide (0).

cells with the 40:1 RGDS-containing matrix is based on their vitronectin receptors, it may be, for HUVEC and A375 cells, that there is a small contribution from their fibronectin receptors or other receptor.

The reason for the lack of inhibition of cell attachment to fibronectin by 1 mM GRGDSP and 1 mM GRGDdSP is not clear, but it may be that higher concentrations of peptides are necessary for the cell types

X^{20}–[poly{20(GVGVP),(GRGDSP)} + poly(GVGVP)]   Fibronectin   Vitronectin

O

GRGDSP

GRGDdSP

Cyclic
RGDS

GRADSP

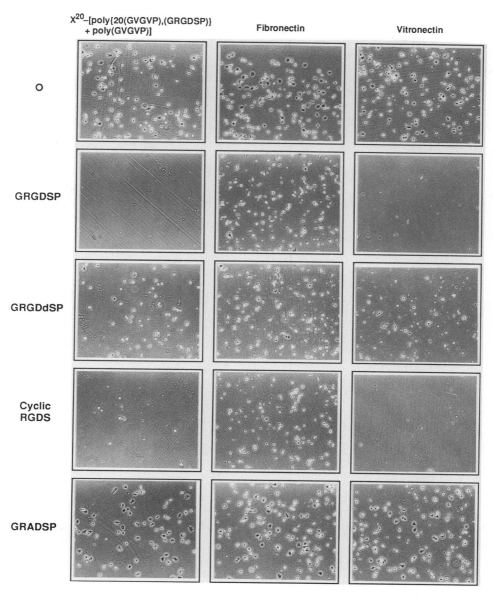

Figure 9. A375 melanoma cell adhesion to X^{20}-{poly-
[0.97(GVGVP), 0.03(GRGDSP)] + poly(GVGV)} and to
10 μg/mL human fibronectin coated and to 10μg/mL
human vitronectin coated plastic in the presence
of 1 mM GRGDSP, 1 mM GRGDdSP, 1 mM cyclic RGDS, 1
mM GRADSP or in the absence of additional peptide
(0).

used. Higher concentrations were not tried since at 1 mM GRGDdSP was
starting to inhibit cell attachment to vitronectin and thus higher
concentrations were unlikely to prove discriminatory for the cell types
used. The possible influence of coating solution concentration on the
response of the cells to peptide inhibition of adhesion was investigated
(Figure 13). At all levels of fibronectin coating solution concentration

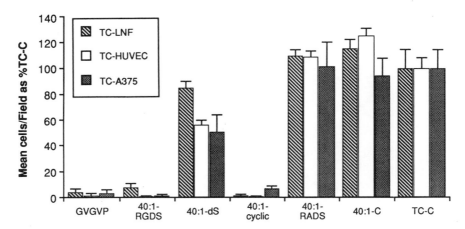

Figure 10. The adhesion of LNF, HUVEC and A375 cells to 40:1
(GVGVP:(GRGDSP)-containing matrix in the presence
of 1 mM GRGHDSP (40:1-RGDS), 1 mM GRGDdSP (40:1-
dS), 1 mM cyclic RGDS (40:1-cyclic) and 1 mM
GRADSP (40:1-RADS) and in the absence of
additional peptides (40:1-C). GVGVP represents
x^{20}-poly(GVGVP) and TC-C represents positive
tissue culture control which was tissue culture
plastic coated with human fibronectin for LNF and
HUVECs and was tissue culture plastic in the
presence of 5% fetal bovine serum for A375 cells.

tested (0.5 to 8.0 mg/mL) no effect of 1 mM GRGDdSP on LNF attachment was
found, even though at the lower concentrations the cells were
predominately poorly spread or rounded (data not shown). Conversely, at

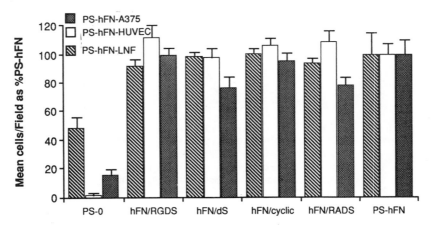

Figure 11. The adhesion of LNF, HUVEC and A375 cells to human
fibronectin coated non-tissue culture plastic in
the presence of 1 mM GRGDSP (hFN/RGDS), 1 mM
GRGDdSP (hFN/dS), 1 mM cyclic RGDS (hFN/cyclic),
1 mM GRADSP (hFN/RADS) and in the absence of
additional peptides (PS/hFN). PS-O represents
uncoated non-tissue culture plastic.

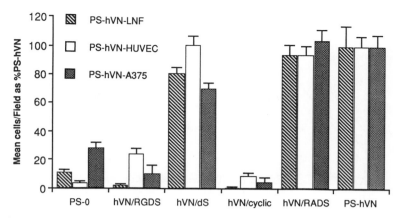

Figure 12. The adhesion of LNF, HUVEC and A375 cells to human vitronectin coated non-tissue culture plastic in the presence of 1 mM GRGDSP (hVN/RGDS), 1 mM GRGDdSP (hVN/dS), 1 mM cyclic RGDS (hVN/cyclic), 1 mM GRADSP (hVN/RADS) and the absence of additional peptides (PS-hVN). PS-O represents uncoated non-tissue culture plastic.

all levels of vitronectin coating solution concentration, 1 mM cyclic RGDS completely inhibited cell adhesion. Thus specific inhibition of LNF attachment to fibronectin could not be achieved by lowering the substratum concentration of fibronectin.

CONCLUSIONS

We have shown that the RGDS sequence, when incorporated as GRGDSP into elastomeric matrices based on GVGVP, behaves as a ligand for the vitronectin receptor rather than the fibronectin receptor. The vitronectin sequence recognized by the vitronectin receptor is RGDV.[29] The stereochemistry of the RGDS sequence has been demonstrated to affect its binding specifically in cell adhesion.[28] The specific peptide inhibitor, of vitronectin to vitronectin receptor based adhesion, is a cyclized sequence containing RGDS. Thus it is probable that the altered specificity of RGDS in the matrix is due to an altered stereochemistry. The failure of incorporated REDV to promote HUVEC and A375 cell adhesion could also be the result of an altered stereochemistry making the REDV unrecognizable by its receptor.

ACKNOWLEDGMENTS

This work was supported in part by grants HL-29578 and HL-41198 from the National Institutes of Health. The authors are pleased to acknowledge Richard Knight of the Auburn University Nuclear Science Center for carrying out the γ-irradiation crosslinking, Naijie Jing of the Laboratory of Molecular Biophysics for the NMR spectroscopy, and Robert M. Senior and Gail L. Griffin of Washington University Medical Center, St. Louis, for provision of fibroblasts.

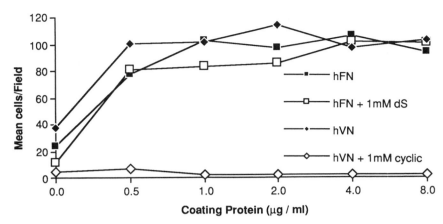

Figure 13. The effect of coating protein concentration on the
attachment of LNFs to human fibronectin and to
human vitronectin alone, and in the presence of 1
mM GRGDdSP (dS) and 1 mM cyclic RGDS (cyclic),
respectively.

REFERENCES

1. D. W. Urry, T. M. Parker, A. Nicol, A. Pattanaik, D. S. Minehan, D.
   C. Gowda, C. Morrow and D. T. McPherson, Am. Chem. Soc., Div. Polym.
   Mater.: Sci. and Eng., **66**, 399-402 (1992).
2. L. Sandberg, J. Leslie, C. Leach, V. Torres, A. Smith and D. Smith,
   Pathol. Biol., **33**, 266-274 (1985).
3. H. Yeh, N. Ornstein-Goldstein, Z. Indik, P. Sheppard, N. Anderson, J.
   Rosenbloom, G. Cicila, K. Yoon and J. Rosenbloom, Collagen and
   Related Research, **7**, 235 (1987).
4. L. B. Sandberg, N. T. Soskel and J. B. Leslie, N. Engl. J. Med., **304**,
   566-579 (1981).
5. D. W. Urry, T. M. Parker, M. C. Reid and D. C. Gowda, J. Bioactive
   Compatible Poly., **6**, 263-282 (1991).
6. D. W. Urry, in: *"Molecular Conformation and Biological Interactions,"*
   P. Balaram and S. Ramaseshan, Eds., Indian Acad. of Sci., Bangalore,
   India, 1991, pp. 555-583.
7. D. W. Urry, D. C. Gowda, T. M. Parker, C.-H. Luan, M. C. Reid, C. M.
   Harris, A. Pattanaik and R. D. Harris, Biopolymers, **32**, pp 1243-1250
   (1992).
8. D. W. Urry, Prog. Biophys. Molec. Biol., **57**, 23-57 (1992).
9. D. W. Urry, Mat. Res. Soc. Symp. Proc., **174**, 243-250 (1990).
10. D. W. Urry, D. C. Gowda, C. Harris, R. D. Harris and B. A. Cox, Poly.
    Preprints Am. Chem. Soc. Polym. Chem. Div., **32**, pp 84-85 (1992).
11. D. W. Urry, in: *"Cosmetic and Pharmaceutical Applications of
    Polymers,"* C. G. Gebelein, T. C. Cheng and V. C. Yang, Eds., Plenum
    Press, New York, 1991, 181-192.
12. H. M. Kagan, L. Tseng, P. C. Trackman, K. Okamoto, R. S. Rapaka and
    D. W. Urry, J. Biol. Chem., **255**, 3656-3659 (1980).
13. D. W. Urry, H. Sugano, K. U. Prasad, M. M. Long and R. S. Bhatnagar,
    Biochem. Biophys. Res. Commun., **90**, 194-198 (1979).
14. A. Pattanaik, D. C. Gowda and D. W. Urry, Biochem. Biophys. res.
    Comm., **178**, 539-545 (1991).
15. A. Nicol, D. C. Gowda and D. W. Urry, Biomed. Mater. Res., **26**, 393-
    413 (1992).

16. M. M. Long, V. J. King, K. U. Prasad, B. A. Freeman and D. W. urry, J. Cell. Phys., **140**, 512-518 (1989).

17. M. D. Pierschbacher and E. Ruoslahti, Nature, **309**, 30 (1984).

18. D. W. Urry, J. Protein Chem., **7**, 1 (1988).

19. D. W. Urry, B. Haynes, M. Zhang, R. D. Harris and K. U. Prasad, Proc. Natl. Acad. Sci. USA, **85**, 3407 (1988).

20. K. U. Prasad, M. A. Iqbal and D. W. Urry, Int. J. Pept. Protein Res., **25**, 408 (1985).

21. D. W. Urry and K. U. Prasad, in: "*Biocompatibility of Tissue Analogues*," D. F. Williams, Ed., CRC Press, Inc., Boca Raton, Florida, 89-116 (1985).

22. R. B. Merrifield, Biochemistry, **3**, 1385-1390 (1964).

23. M. A. Barton, R. U. Lemieux and J. Y. Savoie, J. Am. Chem. Soc., **95**, 4501-4506 (1973).

24. J. P. Tam, W. F. Heath and R. B. Merrifield, J. Am. Chem. Soc., **105**, 6442-6455 (1973).

25. M. J. Humphries, S. K. Akiyama, A. Komoriya, K. Olden and K. M. Yamada, J. Cell. Biol., **103**, 2637 (1986).

26. J. A. Hubbell, S. P. Massia, N. P. Desai and P. D. Drumheller, Biotechnology, **9**, 568 (1991).

27. A. P. Mould, A. Komoriya, K. M. Yamada and M. J. Humphries, J. Biol. Chem., **266**, 3579 (1991).

28. M. D. Pierschbacher and E. Ruoslahti, J. Biol. Chem., **262**, 17294 (1987).

29. S. Suzuki, A. Oldberg, E. G. Hayman, M. D. Pierschbacher and E. Ruoslahti, EMBO J., **4**, 2519 (1985).

EXPRESSION OF A SYNTHETIC MUSSEL ADHESIVE PROTEIN IN *ESCHERICHIA COLI*

Anthony J. Salerno and Ina Goldberg

Allied-Signal Inc.
101 Columbia Road
Morristown, NJ 07962-1021

Repetitious gene cassettes that encode the consensus decapeptide repeat of *Mytilus edulis* bioadhesive protein were designed, constructed, and expressed in *Escherichia coli*. The bioadhesive precursor (BP) with a MW of 25,000 was expressed from one 600-bp gene comprised of a 30-bp unit repeat that accounts for *E. coli* codon bias. In strains employing T7 RNA polymerase for induction, BP was produced at levels approaching 60% of total cell protein. BP forms intracellular inclusions and yet methionine was processed from the N-terminus of the purified protein as shown by amino acid composition and N-terminal sequencing to give an authentic consensus precursor protein. Although the repetitious gene containing 30-bp repeat units appeared stable in T7-based host/vector systems, it was less stable in a λP_L promoter-based host/vector system. Codon diversification was examined as a potential method to alleviate the problems by constructing a repetitious gene comprised of 120-bp repeats. This longer repeat unit failed to confer additional stability upon the repetitious gene.

INTRODUCTION

The marine mussel *Mytilus edulis* anchors itself to the environment by means of a byssus, which is an appendage composed of threads that emanate from the ventral base of its foot and end in an attachment plaque at the substrate surface.[1] The attachment plaque is a composite containing a number of proteins including collagen, catechol oxidase, and a bioadhesive polyphenolic protein.[2] The bioadhesive protein was found to be comprised, in large part, of about 75 repeats of the decapeptide sequence Ala-Lys-Pro-Ser-Tyr-Hyp-Hyp-Thr-Dopa-Lys.[3] Enzymatic processing of tyrosine to dihydroxyphenylalanine (dopa) and orthoquinone plays a critical role in development of adhesive and cohesive properties of the bioadhesive protein.[4] Impressive properties as a water-compatible adhesive have stimulated interest into its use in dental and medical applications.[4,5] It is also employed as a coating for cell attachment to tissue culture surfaces (e.g., Cell-Tack, Collaborative Research). In addition, this protein is also considered for industrial applications such as coatings and underwater adhesives. One requirement for such applications is an economical, large-scale supply of bioadhesive proteins. A recent report describing success toward this goal was the isolation of a partial cDNA clone encoding *M. edulis* bioadhesive protein and its expression in yeast to give a BP lacking the post-translational hydroxylations.[6]

In this work, we describe a somewhat different approach based on the construction and expression at high yields in *Escherichia coli* of synthetic genes encoding BP that is based on the consensus decapeptide repeat of the native protein.

EXPERIMENTAL

1. BACTERIAL STRAINS AND PLASMIDS

The bacterial strains and plasmids used in this study are listed in Table 1. *E. coli* AG1 and DC1138 were used for construction of pET and pAV7 derivatives, respectively. Strain IG110 was used for expression

Table 1. Bacterial strains and plasmids.

E. coli Strain	Genetic Markers/Characteristics	Source or Reference
DC1138	r^-m^+pro leu Δ(srlR-recA) 306:Tn10[λ]	Goldberg, et al 1989
AG1	hsdR17 recA1 endoAl gyrA96 thi supE44 relA	Stratagene
B121[DE3]	$r_B^-m_B^+$ ompT [λ(int::lacUV5 -T7gene 1)imm^{21}nin^5]	Studier, et al 1990
AS002	B121[DE3] [pLysS] Δ(srlR-recA) 306::Tn10	This study
IG110	rpoH165 lacZam trpam phoam supC^{ts} mal rpsL phe bio rel λ(srlR-recA)306::Tn10 [$\lambda\Delta$Bamrex::Km^{R}cI857Δ(cro-bioB)]	Goldberg, et al 1989
Plasmid		
pAV7	bla Δrop	This study
pET-3a	bla	Studier, et al 1990
pAG1-pAG8	pAV7 (StyI::*gag*150-600$_{30}$)	This study
pAG9	pET-3a (NdeI-BamHI::*gag*600$_{30}$)	This study
pAG13	pAV7 (StyI::*gag*120$_{120}$)	This study
pAG14, pAG15	pAV7 (StyI::*gag*720,600$_{120}$)	This study
pAG16	pET-3a (NdeI-BamHI::*gag*600$_{120}$)	This study

studies using the λP$_L$ promoter while the other strains were used in expression driven by the T7 promoter. The Δ(srlR-recA)306::Tn10 allele[7] was transferred into BL21[DE3](pLysS)[8] by P1 transduction[9] to give AS002.

Expression vector pJL6 served as the starting plasmid for pAV7.[10] The small EcoRV-PvuII DNA fragment of pJL6 was replaced with an SP6 promoter made from the following oligonucleotides: 5'-ATITAGGTGACACTATAGAATAGGGATCC-3' and 5'-GGATCCCTATTCTATAGGTGCACCTAAAT-3'. A BamHI site lies immediately adjacent to the 3' side of the promoter in the annealed oligonucleotides. This plasmid, denoted pAVO1, can produce anti-sense transcripts of DNA inserted into the cloning region from the SP6 promoter. The AvaI site of pAVO1 was digested, filled in using the Klenow fragment of DNA polymerase, and ligated to a T7 promoter made of the following oligonucleotides: 5'-TAATACGACTCACTATAGGGAGATCGCGA-3' and 5'-TCGCGATCTCCCTATAGTGACTCGTATTA-3'. A NruI site lies immediately adjacent to the 3' side of the promoter in the annealed oligonucleotides. This plasmid, denoted pAVO2, makes transcripts through the cloning region in the same direction as the λP$_L$ promoter. The ClaI cloning site of pAVO2 can be removed as an NdeI+HinfIII fragment, where the NdeI site lies within the initiation codon for the cII gene. Insertion of a synthetic StyI site into the NdeI+HindIII large fragment of pAVO2 yields pAV7.

2. GENERAL METHODS

DNA manipulation, transformation of *E. coli*, colony hybridization, DNA sequencing, and protein determination were all done using standard procedures.[11] Enzymes were purchased from commercial suppliers and used as directed. Glue-12, a decapeptide consensus sequence was chemically synthesized, cleaved, and purified as reported previously.[12] Glue-12 was polymerized with diphenylphosphoryl azide[13] to a mean MW of 35,000-40,000 and the trifluoroacetamide protecting group was removed from the lysinyl e-amino group in methanol: pyridine: water (7:2:1), lyophilized, and resuspended in water for immunization of rabbits. Initial immunization attempts failed, so the poly Glue-12 was conjugated to BSA and animals were reinjected with one rabbit developing a useful antiserum (Berkeley Antibody Co., Richmond, CA). For immunoblotting, proteins were transferred to nitrocellulose using a Trans-blot apparatus (Bio-Rad, Richmond, CA) according to manufacturers directions. The primary antibody was preadsorbed against an acetone powder prepared from AS002(pET-3a). This procedure as well as blocking of the nitrocellulose with blotto, binding the primary antibody (5 x 10^{-4} dilution), binding of goat anti-rabbit IgG conjugated to horseradish peroxidase, and detection using diaminobenzidine as a substrate are detailed elsewhere.[14] The detergent lysis method was used for large-scale plasmid preparation.[15] Stability of polydecapeptide glue analog gene (*gag*) cassettes was assessed by restricting the purified plasmid (10 μg) with StyI and electrophoresis on 5% polyacrylamide gels to size the population of *gag* cassettes. Coupled *in vitro* transcription-translation reactions were done using a commercial kit (Amersham) according to the manufacturers directions.

3. CONSTRUCTION OF *gag*

The method of construction of the 30-bp repeat *gag* cassette was disclosed previously.[16] Briefly, cassette oligonucleotides and linker oligonucleotides were heated and then allowed to anneal to each other respectively to form short duplex DNA (Figure 1A). The short duplex internal and external sequences were then combined in a 20 to 1 molar ratio, (8 μg total, reaction volume, 153 μL) and allowed to anneal by further cooling to form long duplex DNA, which was ligated, restricted with StyI, and fractionated by size-exclusion chromatography. Fractions containing oligomers >210 bp were ligated into the StyI site of pAV7. Following transformation, colonies carrying gag were identified by colony hybridization using the bottom oligonucleotide of Figure 1A as a probe.

The 600-bp gag comprised of 30-bp repeat units was moved as an NdeI+BamHI DNA fragment from pAG3 to similarly digested pET-3a. The 120-bp repeat unit gag cassette was constructed by mutually primed extension of two oligonucleotides of 90 and 91 bp in length by T7 DNA polymerase (Figure 1B). After digestion with StyI, the 120-bp gag cassette was purified on a low-melting agarose gel and ligated into StyI-digested pAV7. Colonies were screened by restriction analysis and then sequenced to identify the proper construct (pAG13). Tandem 120-bp gag cassettes in pAG14 and pAG15 were prepared by digesting pAG13 with StyI, recovering the vector and gag cassette, and ligating gag cassettes to form multimers. These were gel purified and then placed into the same vector, resulting in pAG14 (6 repeats) and pAG15 (5 repeats). The NdeI+BamHI gag-containing DNA fragment from pAG15 was moved into similarly digested pET-3a to give pAG16.

4. EXPRESSION AND DETECTION OF BP

Induction of strains containing T7-based expression systems[8] or a λP_L promoter system[17] was as described previously. Culture samples were prepared by centrifugation at 5,000 g for 5 min and pellets were suspended in 0.1 volume of solution A [50 mM TRIS HCl, 2% β-mercaptoethanol (β-SH), 0.5% cetyl trimethyl ammonium bromide (CTAB)]. Samples were freeze/thawed (once) and sonicated (3 x 10 s) prior to analysis. Cultures were grown at 37°C in LB medium (unless indicated otherwise) with ampicillin (100 μg/mL), tetracycline (12 μg/mL), kanamycin (50 μg/mL), and/or chloramphenicol (35 μg/mL) as required. For fractionation of AS002(pAG9) cells to detect the presence of inclusion bodies, a culture (8 x 10^8 cells/mL) was pelleted after 2 h of induction and resuspended in 0.1 volumes of solution B (40 mM TRIS, pH 8.0, 50 mM sodium chloride, 1 mM ethylenediaminetetraacetic acid, 1% β-SH). Samples were then subjected to three freeze/thaw cycles and insoluble material pelleted by a 15 min centrifugation in a microfuge. After suspending the pellet in an equal volume of buffer, the equivalent of 0.1 OD_{600} units of the original culture was used for gel analysis.

For the pulse-chase experiment; AS002(pAG9) was grown in M63 medium supplimented with 100 μg/mL ampicillin, 34 μg/mL chloramphenicol, 1 μg/mL biotin, 1 μg/mL thiamine, 0.2% glucose, and all amino acids except proline.[9] Isopropyl β-D-thiogalactopyranoside (IPTG) was added to the culture (5 x 10^8 cells/mL). After 1 h the culture was pulsed with [^{14}C]proline (2.5 μci/mL. Three min later cold proline chase was added (to 18 mM) and samples (0.75 mL) were taken immediately and at 10, 20, 50, 90, and 120 min. Cells were pelleted in a microfuge (20 s), washed with ice-cold M63 medium, resuspended in solution A, and sonicated 3x for 10 s each. After determining incorporated radioactivity, equal CPM were loaded in each lane for gel analysis.

Samples were analyzed for BP content on a gel adapted for basic proteins.[18] The protocol was modified to include CTAB in the buffer except that the CTAB concentration was increased to 0.05%.[19] Riboflavin was used for gel polymerization.[20] Samples were loaded after adding an equal volume of sample buffer (2 x: 5 M urea, 0.8 M acetic acid, 2% β-SH, 1% CTAB, 50 mM Tris HCl, 0.5 mg/mL methyl green) and heating at 50°C for 4 min. Gels were stained with fast green to detect proteins and BP was quantitated by scanning densitometry.[21]

A

Cassette

CCA	ACC	TAC	AAA	GCT	AAG	CCG	TCT	TAT	CCG			
			TTT	CGA	TTC	GGC	AGA	ATA	GGC	GGT	TGG	ATG
pro	thr	tyr	lys	ala	lys	pro	ser	tyr	pro	pro	thr	tyr

Linker   **Sty I**

CCA	ACC	TAC	AAA	GCC	AAG	GCT	TCT	TAT	CCG			
			TTT	CGG	TTC	CGA	AGA	ATA	GGC	GGT	TGG	ATG
pro	thr	tyr	lys	ala	lys	ala	ser	tyr	pro	pro	thr	tyr

Ligate, digest with Sty I

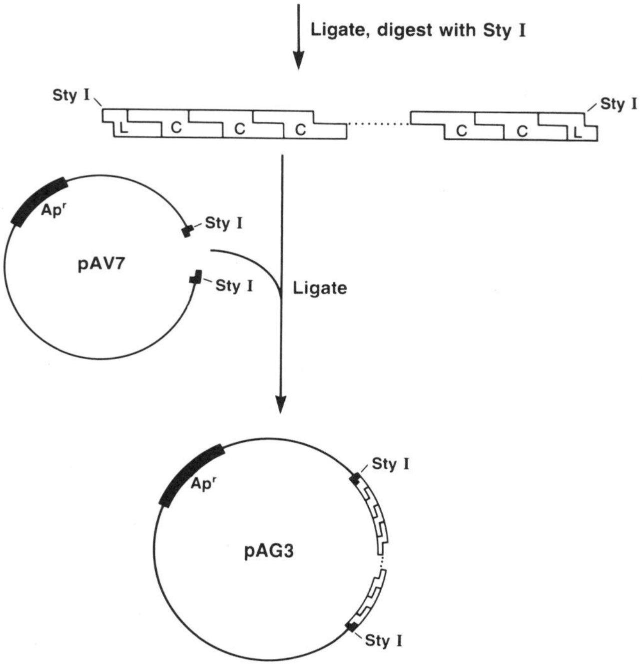

B

		10		20		30		40		50	

GCC	AAG	GCC	AGC	TAT	CCC	CCA	ACG	TAT	AAG	GCT	AAA	CCG	AGT	TAC	CCT	CCC	ACA	TAC
ala	lys	ala	ser	tyr	pro	pro	thr	tyr	lys	ala	lys	pro	ser	tyr	pro	pro	thr	tyr

60			70			80			90			100			110		

AAA	GCA	AAA	CCA	TCG	TAT	CCG	CCT	AAC	TAT	AAA	GCG	AAG	CCC	TCA	TAC	CCA	CCG	ACT
lys	ala	lys	pro	ser	tyr	pro	pro	thr	tyr	lys	ala	lys	pro	ser	tyr	pro	pro	thr

	120		

TAC	AAG	GCC	AAG
tyr	lys	ala	lys

Figure 1. (A) Construction of expression vector pAG3 containing a synthetic polydecapeptide gene cassette with a 30-bp repeat. (B) The DNA sequence of the 120-bp diversified decapeptide gene segment.

Cells were prepared for transmission electron microscopy by fixing pelleted cells with 2.5% glutaraldehyde in phosphate-buffered saline for 1 h. Cells were dehydrated in a graded alcohol series, transferred to 100% acetone, and embedded in Epon. Sections were stained in 4% uranyl acetate for 10-15 min prior to analysis.

5. PURIFICATION OF BP

One liter shake-flask cultures of AS002(pAG9) were harvested (5,000 g for 10 min), and resuspended in 0.01 volume of solution A followed by a freeze/thaw cycle (-70 to 37°C) and sonication to give a homogeneous, low-viscosity suspension. Alternatively, in some experiments the pellet was suspended in solution B and then subjected to three freeze/thaw cycles followed by three cycles of sonication (20 s each). Both procedures give good BP yields. CTAB was added to 0.5% (for lysates of solution B only), followed by urea to 2.5 M, and acetic acid to 0.8 M. The solution was incubated for 0.5 h. The pellet can be washed once with extraction solution to recover remaining BP. To the supernatant was added an equal volume of 0.8 M acetic acid. The BP-containing solution was then applied to a 2.5 x 7 cm cellex P column. The column was washed with 0.8 M acetic acid, 1.25 M urea until the baseline returned to zero. BP was then eluted with a linear gradient provided by 1 M potassium acetate, 4 M urea. Fractions containing BP were concentrated and dialyzed against 0.3 M ammonium acetate, pH 4.3. For structural studies, BP was further purified by size-exclusion chromatography using a 2.5 x 96 cm P-60 column and a buffer of 0.3 M ammonium acetate, pH 4.3. BP-containing fractions were pooled and lyophilized and then dissolved in water. BP was quantitated spectrophotometrically by tyrosine content at 276 nm and using a molar absorptivity of 1400 $M^{-1}cm^{-1}$.[22]

6. PHYSICAL ANALYSES OF BP

The N-terminal amino acid sequence of BP was determined by sequential Edman degradation using an automated protein sequencer (470; Applied Biosystems, Foster City, CA).[23] Amino acid composition analysis was performed by the post-column o-phthaldialdehyde method,[24] except for phenthiocarbamoyl derivatization of proline which was normalized with respect to alanine.[23]

RESULTS AND DISCUSSION

1. DESIGN AND CLONING OF *gag*

The decapeptide consensus repeat from the *M. edulis* bioadhesive protein was used as the basis for design of repetitious genes enabling the production of a polydecapeptide analog precursor protein.[3] Two gene cassettes with repeat unit lengths of 30- and 120 bp were designed in order to compare production levels and gene stability. The 30-bp repeat was comprised of codons optimized for *E. coli* expression;[25] multiple codons for Tyr, Lys, and Pro were used to minimize the introduction of nested repeat sites (Figure 1A). Replacement of Pro with Ala codons in repeats containing the StyI restriction site (Figure 1A) is consistent with the natural variability in the gene.[26] The 120-bp repeat represents the maximum length of unique sequence DNA that can be designed without the introduction of nested repeat sites at the region encoding Tyr-Lys (Figure 1B). It was necessary to incorporate a number of non-optimal codons to achieve this level of sequence diversity.

The vector used to clone *gag* cassettes is capable of expressing BP from either the T7 or λP_L promoter (Figure 1A). The StyI cloning site was designed to allow translateral control from the λcII Shine-Delgarno region to give BP carrying only an amino-terminal Met in addition to the polydecapeptide repeat.

A panel of plasmids were isolated that carried *gag* cassettes ranging in size from 120-720 bp (Table 1). Plasmids carrying 600-bp gene cassettes were chosen for detailed expression and gene stability studies.

2. EXPRESSION OF BP

Production of BP was examined using $-P_L$ promoter and T7 expression systems. An identical BP with a MW of approximately 25,000 is encoded by the 600-bp *gag* cassettes on the plasmids pAG3 and pAG9. Transcription is initiated from the λP_L promoter on pAG3 and from the T7 promoter on pAG9. Initial gel analysis of *in vitro* coupled transcription-translation reactions containing pAG3 showed a unique protein band compared to the vector control that was consistent with the predicted size (data not shown). However, in *E. coli*, little BP could be detected after induction using a λP_L promoter-based host vector system (Figure 2, lane 2) that was shown to accumulate a repetitive polypeptide analog to collagen.[17] In contrast, strain AS002(pAG9) synthesized BP at a very high rate (Figure 2, lane 1) and was selected for further analysis.

Figure 3 shows the time course of accumulation of BP. Strain AS002(pAG9) accumulates BP to levels of up to 60% of total cell protein. Levels of up to 5% of total cell protein have been reported for expression of a partial cDNA clone of *M. edulis* BP protein in yeast.[6] The drastic drop in viability is due to the almost complete recruitment of the cellular protein synthesis machinery by the T7 expression system.[8] Interestingly, BP accumulation in strain AS002(pAG16) was similar to that observed for AS002(pAG9). This result indicates that the less-than-optimal codon usage in the diversified *gag* (pAG16) did not lead to any decrease in the yield of BP. Since the promoter and Shine-and-Delgano sequence are identical on both plasmids, the level of gene expression must be predominantly controlled by these two regulatory regions.

Strains IG110(pAG3) and AS002(pAG9) were used in a pulse-chase experiment with [^{14}C]proline to assay BP turnover (Figure 4). As expected from the respectively observed yields, BP has a half life of approximately 10 min in IG110(pAG3) (Figure 4A) while it is stable in AS002(pAG9) (Figure 4B).

3. PURIFICATION OF BP

To develop an appropriate purification procedure for BP, induced cultures of AS002(pAG9) were harvested, lysed, separated into soluble and insoluble fractions, and analyzed for the distribution of BP. The

Figure 2. Induced production of BP in strains AS002(pAG9) (lane 1) and IG110(pAG3) (lane 2). Lane 3 contains purified BP standard.

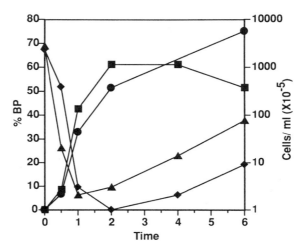

Figure 3. Intracellular accumulation of BP with time after induction and corresponding cell viability. Symbols: (■), AS002(pAG9); (●), AS002(pAG16); (▲), viable AS002(pAG9) cells x 10^{-5}; (◆), viable AS002(pAG16) cells x 10^{-5}.

majority of BP was present in the insoluble fraction, which indicates that BP forms intracellular inclusions (data not shown). However, about 33% of BP was present in the supernatant. To confirm the present of BP inclusions, induced AS002(pAG9) and vector control cells were examined by transmission electron microscopy (Figure 5). The cells producing BP were elongated and irregular with chromosomal material displaced to one side by amorphous, uniformly stained masses (Figure 5B). In contrast, cells containing the vector control were normal in size with centrally positioned chromosomal material and no uniformly stained amorphous masses (Figure 5A).

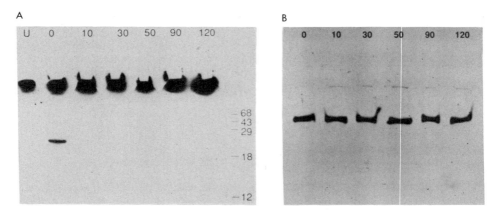

Figure 4. *In vivo* half life of BP in strains IG110(pAG3) (A) and AS002(pAG9) (B). Cells were pulsed with [^{14}C] proline and chased with unlabeled proline for the indicated periods of time.

A AS002(pET-3a)   **B** AS002(pAG 9)

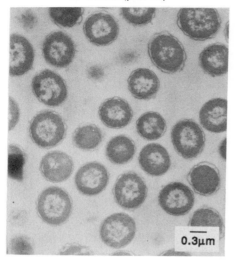
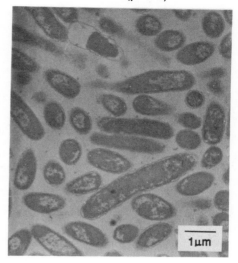

0.3μm   1μm

Figure 5. Electron microscopic analysis of polydecapeptide
overproduction in *E. coli* AS002. Cells containing
the vector alone (A) or the polydecapeptide gene
(B) were induced with IPTG and examined for the
presence of intracellular protein aggregates.

Since a significant proportion of the BP is soluble in the crude
lysate prepared from the culture, a purification strategy based on
selective extraction of BP was developed. BP is solubilized in a solution
containing 2.5 M urea, 0.8 M acetic acid and 0.5% CTAB. This solution was
designed to promote solubility of the highly basic BP by providing a low
pH and disrupting hydrophobic and ionic interactions. CTAB was required
for effective solubilization; in addition to its detergent action, it may
exchange at cationic binding sites with lysinyl-NH_4^+ groups. The presence
of the cationic detergent at low pH also precipitates nucleic acid and
some protein. Separation of the insoluble material results in supernatant
that is up to 90% pure with respect to total protein. BP can be further
purified by cation-exchange chromatography to give 93-95% purity with
respect to total protein (data not shown). Evenly spaced bands were
observed beneath purified BP which represent shorter forms of the protein
produced from partially deleted gene cassettes. These smaller forms can
be easily separated from the full length product by using a larger
gradient during cation-exchange chromatography since the various sizes of
BP are displaced from smallest to largest as the gradient proceeds. For
physical characterization, BP was further purified by size-exclusion
chromatography to yield full length protein of greater than 90% purity.

4. CHARACTERIZATION OF BP

Purified BP was shown to be authentic by amino acid composition and
sequencing analyses. The experimentally obtained amino acid composition
agrees well with the predicted composition based on the polypeptide gene
cassette (Table 2). However, no methionine was detected which suggested
the N-terminal Met was removed. Further evidence for Met processing came
from sequencing of BP. The N-terminal sequence was determined for 24
residues into the protein. The sequences agreed perfectly starting with

Table 2. Amino acid composition of bioadhesive precursor.

Amino Acid	Experimental[a]	Predicted
Methionine	ND[b]	0.5
Serine	7.6	9.8
Threonine	9.0	9.8
Alanine	11.4	11.3
Tyrosine	19.3	19.6
Lysine	20.7	20.1
Proline	32.1	28.9

(a) Average of duplicate analyses, expressed as mole %.
(b) ND, less than 0.01%.

the Ala following the Met residue (Figure 1A). The initial sequencing yield based on amino acid analysis was greater than 80%. Therefore, methionine aminopeptidase removes the methionine from BP to generate a protein devoid of any amino acids not normally found in the consensus decapeptide. Apparently, the aggregation of nascent BP into inclusions does not impede this process.

The structure of BP was further characterized by Western blotting. A polyclonal rabbit antiserum was raised against chemically synthesized decapeptide that contained 4-hydroxyproline at each position of the Pro-Pro dipeptide in the consensus sequence in order to mimic these post-translational modifications found in the natural protein. The serum detected purified BP and a unique identical band in induced cultures of strain AS002(pAG9) while not detecting any protein in the induced control culture (data not shown). These results demonstrate that the chemically synthesized analog and the recombinant BP share a common structure.

GENETIC STABILITY OF *gag* CASSETTES

It is well established that repetitious DNA (including satellite DNA tandem repeat DNA, VNTR DNA, hypervariable DNA) is subject to considerable fluidity or length polymorphism.[27] The tandem DNA of repetitious genes cloned in *E. coli* is also subject to analogous fluidity that is manifested as deletion of repeat units.[17,28-30] The *gag* cassettes described here are no exception. The BP was used as a model to investigate whether this problem could be alleviated by diversifying codon usage and thus increasing the unit repeat size. The rationale for this approach is the empirical observation that unique sequence DNA undergoes deletion events relatively infrequently compared to repetitious DNA.[31]

Gene cassettes comprised of short ($gag600_{30}$) and long ($gag600_{120}$) repeat units were placed in pAV7 and pET-3a vectors and deletions were observed after transformation and growth of various strains. A representative experiment is shown in Figure 6 where the size of $gag600$ was compared following growth of parallel cultures for 20 generations. Both repeat units showed significant deletions in the DC1138/pAV7 host vector pair (Figure 6, lanes 2 and 5), while no detectable deletions were observed for the AS002(pET-3a) host-vector pairs (lanes 3 and 4). Thus , the $gag600_{30}$ and $gag600_{120}$ repeat unit cassettes showed similar low levels of deletion. The T7 host/vector system appears to be a promising

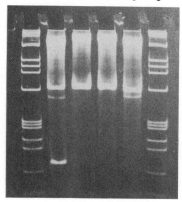

Figure 6. Deletion analysis of *gag*600 gene cassettes. Lanes: 1 and 6, molecular weight standards; 2, DC1138-(pAG15); 3, AS002(pAG16); 4, AS002(pAG9); 5, DC1138(pAG3).

candidate for scale-up experiments. Since $gag600_{30}$ did not show higher deletion levels, codon diversification is of little value in stabilizing *gag*.

ACKNOWLEDGEMENTS

The authors would like to thank M. Berenbaum, J. Williams, and D. Masilamani for their interest and support of this project, D. Piascik and M. Kuroiwa for technical assistance, Susan Anderson for peptide synthesis and Henry Lackland for conducting the amino acid analysis and protein sequencing. This work was funded in part by contract #N00014-89-C-0293 from the Office of Navel Research.

REFERENCES

1. J. H. Waite, Biol. Rev., **58**, 209 (1983).
2. J. H. Waite, Comp. Biochem. Physiol., **97B**, 19 (1990).
3. J. H. Waite, J. Biol. Chem., **258**, 2911 (1983).
4. J. H. Waite, Chemtech, **17**, 692 (1987).
5. J. B. Robin, P. Picciano, R. S. Kusleika, J. Salazar & C. Benedict, Arch. Ophthalmol., **106**, 973 (1988).
6. D. R. Filpula, S. M. Lee, R. P. Link, S. L. Strausberg & R. L. Strausberg, Biotechnol. Prog., **6**, 171 (1990).
7. D. K. Willis, B. E. Uhlin, K. S. Amini & A. Clark, Mol. Gen. Genet., **183**, 497 (1981).
8. F. W. Studier, A. H. Rosenberg, J. J. Dunn & J. W. Dubendorff, Meth. Emzymol., **185**, 60 (1990).
9. "*Experiments in Molecular Genetics*," J. H. Miller, Ed., Cold Spring Harbor Laboratory, Cold Spring Harbor, 1972.
10. J. A. Lautenberger, D. Court & T. S. Papas, Gene, **23**, 75 (1983).
11. "*Current Protocols in Molecular Biology*,"F. M. Ausubel, R. Brent, R. E. Kingston, D. D. Moore, J. J. Seidman, J. A. Smith & K. Struhl,

Eds., Greene Publishing Associates and Wiley-Interscience, New York, 1987.

12. M. D. Swerdloff, S. B. Anderson, R. D. Sedwick, M. K. Gabriel, R. J. Brambilla, D. M. Hindenlang & J. I. Williams, Int. J. Peptide Protein Res., **33**, 318 (1989).

13. T. Shioiri, K. Ninomiya & S, Yamada, J. Amer. Chem. Soc., **94**, 6203 (1972).

14. *"Antibodies: A Laboratory Manual,"* E. Harlow & D. Lane, Eds., Cold Spring Harbor Laboratory, Cold Harbor, 1988.

15. G. N. Godson & P. Vapnek, Biochim. Biophys. Acta, **229**, 516 (1973).

16. J. I. Williams, A. J. Salerno, I. Goldberg & W. T. McAllister, US Patent 5,089,406, 1992.

17. I. Goldberg, A. J. Salerno, T. Patterson & J. I. Williams, Gene **80**, 305 (1989).

18. S. Panyim & R. Chalkey, Arch. Biochem. Biophys., **130**, 337 (1969).

19. V. V. Shmatchenko & A. J. Varshavsky, Anal. Biochem., **85**, 42 (1978).

20. L. A. Marjanen & I. J. Ryrie, Biochim. Biophys. Acta, **371**, 442 (1974).

21. C. M. Willson, Anal. Biochem., **96**, 263 (1979).

22. *"Physical Biochemistry,"* D. Freifelder, Ed., W. H. Freeman and Co., San Francisco, 1976.

23. M. W. Hunkapiller & L. E. Hood, Meth. Enzymol., **91**, 486 (1983).

24. N. M. Meltzer, G. I. Tous, S. Gruber & S. Stein, Anal. Biochem., **160**, 356 (1987).

25. H. A. de Boer & R. A. Kastelein in: *"Maximizing Gene Expression,"* W. S. Reznikoff & L. Gold, Eds., Butterworth, Boston, 1986, Chapter 8, p. 225.

26. K. J. Maugh, D. M. Anderson, R. L. Strausberg, S. L. Strausberg, R. McCandliss, T. Wei & D. Filpula, US Patent 5,049,504, 1991.

27. P. G. Debenham, Trends Biotechnol., **10**, 96 (1992).

28. Y. Suzuki & Y. Ohshima, Cold Spring Harbor Symp. Quant. Biol., **42**, 947 (1978).

29. M. T. Doel, M. Eaton, E. A. Cook, H. Lewis, T. Patel & N. H. Carey, Nucleic Acids Res., **8**, 4575 (1980).

30. S. T. Case, Gene, **20**, 169 (1982).

31. R. M. Schaaper, B. N. Danforth & B, W, Glickman, J. Mol. Biol., **189**, 273 (1986).

EXPERIENCES WITH COLLAGEN-DERIVED BIOMATERIALS IN RESEARCH AND TREATMENT OF HUMANS

A. Stemberger, R. Ascherl, M.A. Scherer, F. Bader, W. Erhardt, B.C. Adelmann-Grill, K.H. Sorg, H.P. Thomi, M. Stoltz, J. Jeckle, T. Seaber[1], G. Blümel

Institut für Experimentelle Chirurgie der Technischen Universität München, Ismaninger Straße 22
8000 München 80
                    and
(1) Duke University Medical Center
Division of Orthopaedic Surgery
Durham, North Carolina 27710

Collagen derived biomaterials have been used in medicine for ages. Intact, native collagen fibrils can be prepared in high purity from bovine tendons by pepsin treatment. Formulation of foils and sponges by freeze-drying is an accepted procedure. In the body, collagen is resorbed by an active cellular mechanism. Porous collagen sponges serve as carriers for the fibrin adhesive system. In Germany, these materials are used to control bleeding in parenchymatic organs. A gentamicin-collagen compound is clinically used in bone and soft tissue infections.

INTRODUCTION

Collagen has been used as a biomaterial in medicine for ages. But now, modern methods of protein chemistry have made it possibe to produce native, pure collagen in form of sponges and foils. Our research group intended to clarify the process of preparation of collagen-derived materials and their use for hemostasis (also in combination with the fibrin adhesive system) and temporary wound covering, and their applicability as a drug carrier system in combination with antibiotics such as aminoglycosides. During a period of more than ten years, experimental investigations on bioavailability, resorption, tissue compatibility and mode of action have been performed in animal models and also - thoroughly controlled - in humans.

EXPERIMENTAL

During the approved method of isolation of native collagen from bovine tendons by pepsin treatment under acid conditions, non-collagenous proteins are removed.[3] Then sponges and foils are produced by lyophilisation or an air drying process. These methods can also be applied to a combination of collagen and aminoglycosides such as gentamicin and netilmicin. The gentamicin-collagen (genta-coll) compound of 10 x 10 x 1 cm consists of 280 mg collagen and 200 mg gentamicin sulfate. After sterilization with ethylene-oxide(EtO), these materials do not lose their antimicrobical potency. Sterilization by irradiation with a dosage of 15 and 25 kGy is in process. Unstable drugs such as fibrinogen were incorporated in pre-sterilized collagen materials under aseptic conditions (containing 2-3 mg fibrinogen per cm^3) and nitrogen and hydroxyproline determinations were used as quality controls. Also, the following tests were carried out: immunodiffusion for the detection of bovine protein, experiments on tensile strength, the testing of collagen-induced platelet aggregation, and scanning electron microscopy. To prove the hemostatic properties and resorption of a fibrinogen-collagen compound, hepatic lacerations were performed in pigs.

Bioavailability and tissue reaction of the genta-coll compound was investigated by its implantation into the femora of rabbits, including materials produced by different formulations and sterilization procedures (eto treatment, irradiation with 25 kGy). Then the gentamicin levels were analysed in the serum, urine, cortical bone and bone marrow by TDX (Abbott). Some specimens of bone were grinded under liquid nitrogen and eluted with an isotonic buffer system containing saccharose. Therapeutic efficacy was tested by inducing a foreign body associated osteomyelitis with staphylococcus aureus (10^7 ATTC 6538) in rabbits. Four weeks later, a clinically manifested osteomyelitis was diagnosed and treated with the genta-coll compound. The animals were observed during the following eight weeks: objective parameters in these studies were X-ray, technetium^{99} scintigraphy as well as microbiological and histological examinations. Another group of rabbits was investigated for the detection of collagen-related antibodies: Genta-coll was implanted into the soft tissue, and four weeks later a second implantation without any adjuvant was performed (all experiments were carried out according to the German laws for the protection of animals). Another four weeks later, the sera of these animals and the respective controls were investigated by ELISA method.

RESULTS AND DISCUSSION

Native collagen fibrils in pharmaceutically pure quality can be prepared from bovine tendons by pepsin treatment. As far as we are aware of the literature, reconstituted, purified collagen does not exhibit distinct antigenicity. Due to the development of specific production procedures, non-collagenous proteins are absent in final reconstituted collagen materials (Figure 1). The induction of collagen- and gelatin-related antibodies could not be demonstrated by in vivo experiments.

Pliable porous sponges can be prepared by lyophilisation. By scanning electron microscopic examinations it was possible to demonstrate reconstituted collagen fibrils (Figure 2). An aggregation of human platelets normally induced by collagen can be retained in ethylenoxide sterilized preparations. This method also applies to genta-coll combinations – on the contrary an irridation by 25 kGy completely abolishes the aggregating effect (Figure 3).

A quick and secure hemostasis of bleeding parenchymatous organs by hemostypic materials based on collagen can only be achieved in combination with the fibrin adhesive system containing thrombin and fibrinogen. The resorption is completed without significant foreign body reaction by

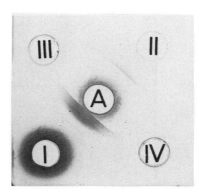

Figure 1. Immunodiffusion experiments. A: polyvalent antiserum against bovine proteins, I: control, extract
of ground bovine muscle protein, II: extract of
ground bovine tendon, III: pepsin treated collagen
slurry, IV: extract of freeze-dried collagen
sponge.

the third or fourth week. Scar formations on liver and spleen can not be
detected.

Collagen in combination with the fibrin adhesive system is accepted
in clinical routine for various clinical indications in abdominal and
trauma surgery and also for patients suffering from inborn and acquired
bleeding defects.

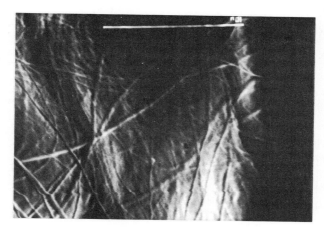

Figure 2. Scanning electron microscopy (6000 x) of the gentamicin-collagen sponge following EtO sterilization. The fibril structure of collagen is clearly
visible.

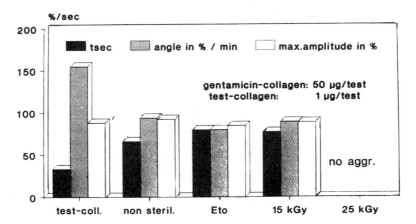

Figure 3. Platelet aggregation of gentamicin collagen, testing of EtO vs. irradiation. Control experiment
with test collagen. Non-sterilized material, EtO,
15 kGy (these experiments were performed with 50μg
pulverized genta-coll).

The hemostatic effect of a genta-coll compound can be proved after
implantation into the bone, irrespective of the sterilization method. As
shown in Figure 4, the aggregation of platelets on the other hand (performed according to Born's procedure),[1] is no longer possible after irradiation by 25 kGy. After in vivo implantation into the femora, no bleeding occured (see Figure 5). As demonstrated in Figure 6, large hematomas
could be detected after implantation of a gentamicin-containing drug
carrier system on the base of resorbable aliphatic polyester.

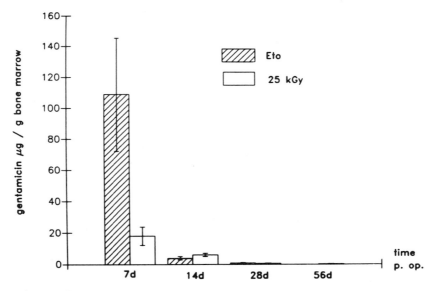

Figure 4. Levels of gentamicin in the bone marrow following
implantation of EtO and irradiated (25 kGy) genta-
coll.

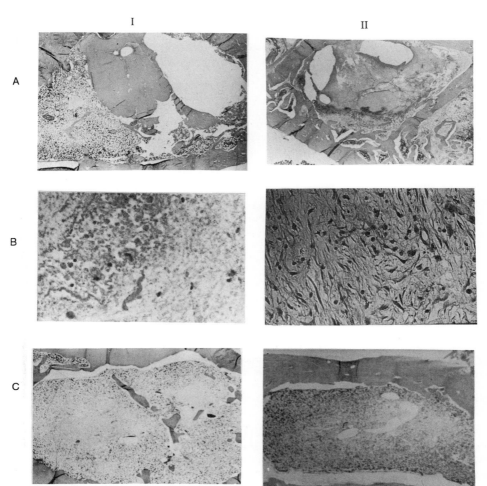

Figure 5. Resorption of genta-coll, comparison of EtO vs.
irradiated material. AI: EtO, 7 days p. op.: The
collagenous compound still clearly visible; granulocyte degradation process, typical with collagen
can be detected only in the marginal regions. AII:
25 kGy, 7 days p. op.: Within the whole region of
the implant accumulation of erythrocytes and
granulocytes with fibrous structures in between,
which cannot be clearly identified as collageuous
leftovers. BI: EtO, 14 days p. op.: Resembles AII;
fibrous structures surrounded by erythrocytes and
granulocytes; degradation process/resorption of
the EtO-sterilized collagen thus seems to be
delayed for 7 days compared to the irradiated
product. BII: 25 kGy, 14 days p. op.: The whole
region of implantation is filled with loose connective tissue; no granulocyte or fibrous structures visible; degradation process completed
(contrary to the eto-preparations). CI: and CII:
56 days p. op.: With both sterilization methods,
collagen is resorbable and the implantation region
is filled with bone marrow.

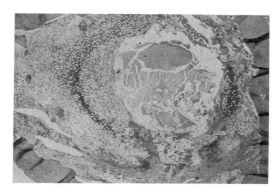

Figure 6. Following implantation of polyester-based gentamicin containing drug carrier into the femora of
rabbits: 7 days p. op., gross hemotoma.

Gentamicin concentrations in the cortical bone (13 μg gentamicin per
g bone) and in the bone marrow (110 μg gentamicin per g bone marrow) were
detected seven days after the implantation of genta-coll into the femora
of rabbits (Figure 4). At that time, an instillation of a corresponding
dosage of an injection solution of gentamicin revealed only traces of the
aminoglycoside in the bone. Toxic concentrations in serum and urine were
not observed. Two weeks p. op. the resorption of the collagen was completed in the soft tissue, and four weeks p. op. the resorption in the
bone. A poor local biocompatibility including adverse cellular reactions
did not arise (Figure 5).

In a prospective, controlled clinical study, genta-coll was applied
in revision surgery of septic loosening of total joint replacements.
These results corroborated and validated the data from the animal experi-
ents regarding gentamicin levels in urine, serum and drainage fluids. In
septic surgery, bone infections, and soft tissue infections like abdominosacral amputation, the genta-coll compound is an accepted resorbable
drug carrier.[2] Septic loosening of total hip replacement seems to be the
indication of choice for this therapeutic regimen. Recurrency in cases of

osteomyelitis and pseudarthrosis appears to be rather threatening but one
has to bear in mind that these patients had a mean of three revision
surgery and constitute a negative selection (see Table). Soft tissue
infections should only be considered a good indication if clinical experience shows a high rate of septic complications in spite of all routine
attempts to lower their incidence: Infection and pelvic abscesses following abdominosacral amputation are a good example.

In summary: Intact, native collagen fibrils can be prepared in high
purity from bovine tendons. As demonstrated by animal studies, an induction of collagen-related antibodies could be avoided in spite of the
repeated application of collagen sponges and foils. In the body, collagen
is resorbed by an active cellular mechanism. Porous collagen sponges are
carriers of the fibrin adhesive system. In Germany, this technique has
now been accepted as a blood staunching system in liver and spleen, and
collagen as a drug carrier for gentamicin is clinically used in septic
surgery. Research with silver impregnated collagen foils as temporary
wound covering of skin injuries and second degree burns are still in
progess.

Table 1. Clinical results of a longterm study with gentamicin-collagen. STI = soft tissue infection, OM = osteomyelitis, PSA = septic nonunion, sTHRL = septic loosening of total hip replacement

Dx	STI	OM	PSA	sTHRL
n Pts. tot.	33	18	13	26
>1 year p.op.	6	5	2	3
>2 years p.op.	-	3	1	0
>3 years p.op.	-	0	0	1
>4 years p.op.	-	1	1	0
>5 years p.op.	-	0	0	0
cumulated recurrency (%)	18,2	50	30,8	15,4

ACKNOWLEDGEMENTS

We are pleased to acknowledge partial support from the Sander Foundation, Neustadt/Donau FRG. The authors also thank Mrs. Annette Amper for preparing this manuscript.

REFERENCES

1. G. V. R. Born, J. Physiol. Lond., **162** 67-68 (1962). "Quantitative Investigations into Gengaggregation of Blood Platelets."
2. A. Stemberger, R. Ascherl, F. Lechner, G. Blümel G, Eds, "*Collagen as a Drug Carrier*," Franklin Scientific (1991).
3. R. L. Trelstadt in: "*Immunochemistry of the Extracellular Matrix*," Vol I, Methods, H. Furthmayr, Ed, CRC Press, Boca-Raton, Florida, 1982, p. 32-39. "Native Collagen Fractionation"

MODULATION OF PHOSPHOCHOLINE BILAYER STRUCTURES

Alok Singh, Michael A. Markowitz and Li-I Tsao

Center for Bio/Molecular Science and Engineering
Code 6900, Naval Research Laboratory
Washington, DC 20375

Vesicles derived from phosphocolines have been used in the encapsulation, retention and release of biomaterials. Diacetylenic phospholipids in aqueous dispersions are reported to transform into rigid, hollow cylindrical structures (0.5 - 0.8 μm diameter and length depending on the process) or tubules and helices. Tubules are used in the encapsulation and sustained release of bioactive materials. Synthesis of two diacetylenic phosphocholines, 1,2-bis-(9,16-dioxa-hexacosa-11,13-diynoic)-sn-3-phosphocholine and 1,2-bis(15-oxa-pentacosa-10,12-diynoic)-sn-3-phosphocholine, in which the diacetylene is linked to the acyl chain by an oxygen spacer is reported to modulate the morphology of bilayer structures and to understand the role of diacetylene in lipid-bilayer assembly. Lipid dispersions were characterized by calorimetric and film balance techniques and the structures were visualized by microscopic techniques. When both ends of the diacetylene were linked to the acyl chain by oxygen atoms, vesicles (diameters ranging from 0.3-3.4 μm) were observed. Linking only the terminal portion of the acyl chain to the diacetylene with an oxygen atom resulted in the formation of tubular microstructures with a diameter ranged from 0.4-4.7 μm. Transmission electron micrograph of the replica of freeze fractured dispersion revealed that the tubular structures consist of an aqueous core surrounded by a wall of lipid.

INTRODUCTION

Natural phosphatidylcholines and their synthetic analogues self-organize to form hollow, spherical, concentric, macromolecular assemblies or vesicles which have been extensively used as a model for cell membranes in encapsulation and sustained release of bioactive materials.[1,2] Incorporation of diacetylene in phosphatidylcholine acyl chains has resulted in the transformation of bilayer assemblies into helices and hollow, cylindrical microstructures composed of concentric bilayers.[3-5] While these structures claim broad technical applicability,[6-9] the mechanism of their formation is not well understood. Various theories have been suggested,[10-13] but none accounts for all the experimental observations, including reproducible diameter of cylinder, and formation of tubules from achiral or racemic lipids.[14,15] Most of the studies reported so far agree that the tubules and helices are the result of strong interaction between the diacetylenes that are present in the bilayer which influences the polar headgroup and hydrophobic interactions responsible for lipid self-assembly.[16-26] X-ray, FTIR and calorimetric analysis revealed a higher chain order in tubules[16,17,19-21] which is attributed to the presence of diacetylenic moiety. In our laboratory, along with other applications, we are exploring tubule technology in the development of environmentally safe antifouling coatings by sustained release of antimicrobial agents from tubules.[27-29] The current limitation on encapsulation application in tubules is the rigidity of cylinders which causes breaking of longer tubules into smaller ones and the fixed internal diameter which limits the size of the macromolecules which could be encapsulated.

This study is focussed on modulating the morphology of bilayer structures and developing flexible tubules which are less likely to break during processing and are able to entrap large molecules. The current approach probes the influence of changes in local chemical environment of diacetylenes on the resultant microstructures. Therefore, the diacetylenic moiety in phospholipid acyl chains has been chemically decoupled from rest of the acyl chain by an oxygen spacer (Figure 1). This led to the synthesis of lipids [2] and [3] and the formation of their macromolecular self-assemblies in the aqueous medium.

MATERIALS AND METHODS

Chloroform, methanol, ether, propargyl alcohol, 8-bromooctanoic acid, and decyl bromide were obtained from Aldrich Chemical Company. The chloroform and methanol were used as received. Ether and acetone were stored over $CaCl_2$ and Na_2CO_3, respectively. Triple distilled water was used in the lipid characterization studies. Published procedures were used in the coupling of acetylenes and lipid synthesis.[29-31] Synthetic intermediates and lipids were characterized by IR (Perkin-Elmer 1800 FTIR), and NMR (Brucker MSL-360 300 MHz) Spectroscopy. Purity of all the compounds was monitored by thin layer chromatography on silica gel (E. M. Merck).

1 SYNTHESIS

1A Synthesis of 9-Oxa-11-dodecynoic Acid [4]

An ether solution of 8-bromooctanoic acid (15.1 g, 65 mmol) was reacted with a five mole excess of propargyl alcohol dissolved in aqueous sodium hydroxide solution containing 15 mol excess of sodium hydroxide. After heating at 60°C for 24 hours, the reaction was quenched with hydrochloric acid. Ether extraction provided 16 g of an opaque liquid which was further purified by distillation under reduced pressure (160-163°C/0.075 mmHg). ^{1}H NMR spectrum accounted for expected chemical shifts. IR spectrum, as KBr pellet, showed the following absorption frequencies: 1709 (COOH), 2120 (-C≡C-acetylenic) and 3298 (-C≡CH) cm^{-1}.

1B Synthesis of 4-Oxatetradecyne [5]

Propargyl alcohol (28.0 g, 0.5 mol) was dissolved in a solution prepared from sodium hydroxide (60 g, 1.5 mol), water (90 mL), and etha-

$$CH_2\text{—}OC(O)\text{—}(CH_2)_7\text{—}X\text{—}C{\equiv}C\text{—}C{\equiv}C\text{—}Y\text{—}(CH_2)_n\text{—}CH_3$$
$$|$$
$$CHOC(O)\text{—}(CH_2)_7\text{—}X\text{—}C{\equiv}C\text{—}C{\equiv}C\text{—}Y\text{—}(CH_2)_n\text{—}CH_3$$
$$|$$
$$CH_2O(PO_3)\text{—}CH_2\text{—}CH_2\text{—}N(CH_3)_2$$

	n	X	Y
1	8	—CH$_2$—	—CH$_2$—
2	9	—OCH$_2$—	—CH$_2$O—
3	9	—CH$_2$—	—CH$_2$O—

Figure 1. Structures of diacetylenic phosphocholines.

nol (100 mL). Bromodecane (114 g, 0.51 mol) was then added and the mixture was refluxed for 5 hours. The reaction mixture was then acidified
and extracted with ether to recover 119 g of organic materials. Upon
distillation (82-84°C/7.5 mm Hg), 16 g (17%) [5] was obtained. ^{1}H NMR in
CDCl$_3$ revealed chemical shifts at δ 2.41 (d, J = 2.2 HZ, 1H, C≡CH), 3.51
(t, J = 6.6 Hz, 2H, -CH$_2$-O), and 4.14 (d, J = 2.2 Hz, 2H, O-CH$_2$-C≡C).
IR (KBr) showed the following absorption frequencies: 2120 (-C≡C-) and
3298 (-C≡CH) cm^{-1}.

1C Synthesis of 1-Iodo-4-oxa-tetradecyne [6]

Following the reported procedure, 13 g (66.2 mmol) [5] was reacted
with 73 mmol ethylmagnesium bromide in dry ether. The resulting alkynyl
magnesiumbromide upon reaction with iodine provided 15 g (71% yield) of
[6] as viscous liquid. Proton NMR in CDCl$_3$ showed the absence of chemical
shift at δ 2.41 which is due to -C≡CH. IR (KBr) revealed the following
absorptions: 2120 (-C≡C-) and absence of peak due to -C≡CH cm^{-1}.

1D Synthesis of 15-Oxa-pentacosa-10,12-diynoic Acid [7]

Undecynoic acid (7.15g, 39.2 mmol) was coupled with 1-iodo-4-oxa-
tetradecyne (13.9 g, 43 mmol) following the procedure described above to
yield 835 mg (5%) [9] after silica gel chromatography and recrystallization from hexane. M.P. 45°C. ^{1}H NMR chemical shifts were observed at δ
2.27 (t, 2H, -CH$_2$-COOH), 2.35 (t, 2H -CH$_2$-C≡C-); 3.50 (t, 2H, -CH$_2$-O),
and at 4.18 (s, 2H, O-CH$_2$-C≡C) indicated the formation of title compound. IR (KBr) revealed the absence of 2120 peak due to -C≡C- and the
presence of acid carbonyl at 1704 cm^{-1}. The acid was converted into its
anhydride (evidenced by the appearance of peaks at 1740 and 1810 cm^{-1})
employing dicyclohexyl carbodiimide (DCC) in methylene chloride.

1E Synthesis of 9,16-dioxa-hexacosa-11,13-diynoic Acid [8]

Following the reported procedure for coupling diacetylene, 3.2 g, 10
mmol [6] was coupled with [4] (1.98 g, 10 mmol), to yield 1.2 g [8] in
31% yield. ^{1}H NMR (CDCl$_3$) revealed chemical shifts at δ 2.27 (t, 2H,
-CH$_2$-COOH), 3.50 (t, 4H, -CH$_2$-O), and 4.18 (s, 4H, O-CH$_2$-C≡C). The acid
was converted into its anhydride by treating with DCC.

1F Synthesis of Phospholipid [2]

Phospholipid [2] was prepared by reacting 1.15 g (1.5 mmol) of the
anhydride of [9] with 226 mg, 0.5 mmol glycerophosphorylcholine-cadmium
chloride complex (GPC CdCl$_2$) in the presence of 180 mg, 1.5 mmol 4-
dimeth-ylaminopyridine (DMAP). After the usual workup, 300 mg (60% yield)

of pure [2] was obtained. ^{1}H NMR (CDCl$_3$) revealed the following chemical shifts: δ 0.88 (t, 6H, C$\underline{H}_3$); 1.2-1.38 (Br singlet, 40H, -(C$\underline{H}_2$)-); 1.55-1.62 (m, 12H, -(C$\underline{H}_2$)-); 2.30 (m, 4H, -C$\underline{H}_2$-COOH); 3.31 (s, 9H,-N$\underline{Me}_3$); 3.49 (t, J = 6.5 Hz, 8H, -C$\underline{H}_2$-O); 3.71-3.90 (m, 6H, -C$\underline{H}_2$-O-CO, -C$\underline{H}_2$-N); 4.18 (s, 8H, O-C$\underline{H}_2$-C≡C); 4.2-4.4 (m, 2H, -P-O-C$\underline{H}_2$-); 5.19 (s, 1H, -C$\underline{H}$O-).

1G Synthesis of Phospholipid [3]

Reaction of 696 mg (0.97 mmol) of the anhydride of [8] with 137 mg (0.31 mmol) GPC. CdCl$_2$ in the presence of 119 mg (0.97 mmol) DMAP provided 153 mg (51% yield) of [3]. ^{1}H NMR (CDCl$_3$) δ 0.88 (t, 6H, C$\underline{H}_3$); 1.2-1.75 (Br singlet, 56H, -(C$\underline{H}_2$)-); 2.23-2.44 (m, 8H, -C$\underline{H}_2$-COOH, -C$\underline{H}_2$-C≡C-); 3.36 (s, 9H,-N$\underline{Me}_3$); 3.49 (t, J=6.5 Hz, 4H, -C$\underline{H}_2$O); 3.71-4.12 (m, 6H, -C$\underline{H}_2$-O-CO, -C$\underline{H}_2$-N); 4.18 (s, 4H, O-C$\underline{H}_2$-C≡C); 4.2-4.42 (m, 2H, -P-O-C$\underline{H}_2$-); 5.23 (s, 1H, -C$\underline{H}$O-).

2 LIPID MICROSTRUCTURE PREPARATION AND VISUALIZATION

Microstructures were prepared by hydrating the lipid in water (2mg/mL) above the T_m of the lipid followed by vortex mixing. The dispersion is then slowly cooled to room temperature (1°C/min). Dispersion was initially examined by optical microscopy using Nikon model diathote inverted optical microscope with a temperature controlled stage. Transmission electron microscopy (TEM) was accomplished with a Zeiss EM-10 microscope. Samples were applied onto carbon-coated copper grids, air dried and observed immediately. All samples were stained with 2% aqueous UO$_2$. For freeze-fracture, samples were transferred to a Balzers copper specimen plates, equilibrated at room temperature and quickly frozen by plunging into melting nitrogen. The samples were then transferred to a Balzers 360 freeze-fracture device, fractured and then replicated with a Pt-C film at -100°C and 10^{-6} torr. The replicas were floated off onto triply distilled water and then picked up on coated grids and examined.

3 THERMAL BEHAVIOR OF LIPID DISPERSION

Differential scanning calorimetric (DSC) studies were performed with a Perkin-Elmer DSC-7 differential scanning calorimeter. Samples were prepared by weighing a known amount of the lipid (3-4 mg) into the DSC pan and then injecting 60 μL H$_2$O. The DSC pan was immediately sealed, equilibrated at 70°C for 3 hrs to ensure complete hydration, cooled to room temperature and transferred to the calorimeter. Samples were scanned following a cooling and heating cycle at a rate of 1°/min until reproducible scans were obtained. The melting (T_m) and cooling transition (T_c) values were determined from the peak temperature of the transitions. Enthalpies were determined from the area under the peak as determined by the calorimeter. Using this procedure, a T_m of 41.7°C and an enthalpy of 7.8 kcal/mole were obtained for DPPC.

4 LANGMUIR FILM BEHAVIOR

Force-area isotherms were recorded on a thermostated Wilhelmy plate film balance (NIMA) with an automatic data collection unit. The monolayers were spread from CHCl$_3$ solutions of the lipids (1 mM), using a Hamilton syringe, onto a subphase of water. 2.9-5.2 x 10^{-8} moles of lipid were used in each experiment. The monolayers were compressed to their collapse pressures at a rate of 50 cm^2/min. The data for each experiment was compiled from at least two isotherms which were reproducible to within 1 Å^2/molecule.

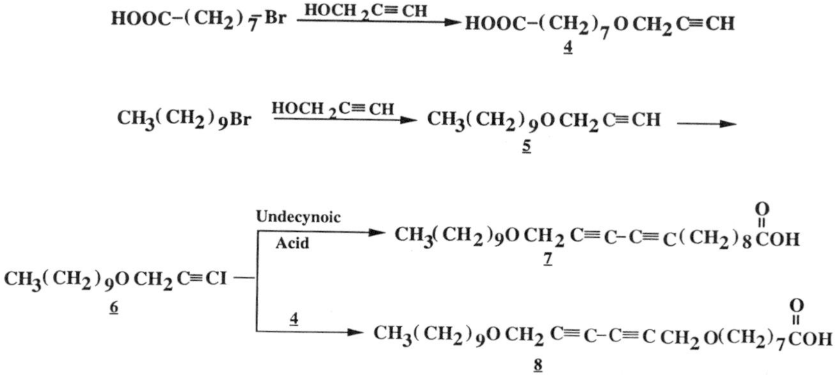

$$HOOC-(CH_2)_7Br \xrightarrow{HOCH_2C\equiv CH} HOOC-(CH_2)_7OCH_2C\equiv CH$$
$$\underline{4}$$

$$CH_3(CH_2)_9Br \xrightarrow{HOCH_2C\equiv CH} CH_3(CH_2)_9OCH_2C\equiv CH \longrightarrow$$
$$\underline{5}$$

$$CH_3(CH_2)_9OCH_2C\equiv CI \underbrace{\begin{array}{c} \xrightarrow[\text{Acid}]{\text{Undecynoic}} CH_3(CH_2)_9OCH_2C\equiv C-C\equiv C(CH_2)_8\overset{O}{\overset{\|}{C}}OH \\ \underline{7} \\ \\ \xrightarrow{4} CH_3(CH_2)_9OCH_2C\equiv C-C\equiv CCH_2O(CH_2)_7\overset{O}{\overset{\|}{C}}OH \\ \underline{8} \end{array}}$$
$$\underline{6}$$

Scheme 1. Synthesis of diacetylenic acids.

RESULTS

The modified phospholipids were synthesized following the synthetic route depicted in Scheme 1. Diacetylenic acids were prepared by hetero-coupling of [6] with [4] as well as undecynoic acid following essentially the published procedure.[27] Ether linked acetylenic molecules, [4] and [5] were synthesized by reacting corresponding bromo analog with propargyl alcohol in the presence of excess of aqueous sodium hydroxide. When sodium hydride was used instead of aqueous sodium hydroxide, the formation of [4] or [5] was not observed. Phospholipids were prepared by reacting corresponding acid anhydride with GPC-CdCl$_2$ in the presence of DMAP by following the methods reported earlier.[25]

The thermal behavior of lipid [2] and [3] is studied by differential scanning calorimetry. The results are compiled in Table 1. Two melting transitions at 15.1°C and 18.3°C were observed for [2]. Upon cooling, a single transition was observed at 9.4°C. Repeated scanning of the sample did not alter the thermogram profile. The enthalpy of the melting transitions was 9.52 kcal/mole while that of the cooling transition was 10.42 kcal/mole. Single melting transition and cooling transitions were observed for [3] at 23.5°C (enthalpy 16.8 kcal/mole) and 15.1° C (enthalpy 15.0 kcal/mole), respectively.

Table 1. Thermal behavior of lipids [2] and [3].

Lipid	Yield %	T_m °C ΔH Kcal/m	T_c °C ΔH Kcal/m	Morphology	Molecular Area Å^2/mol
1	–	43.1 (23)	38.0 (−20.0)	Tubules	52
2	60	15.1,18.3 (9.52)	9.4 (−10.42)	Vesicles	87
3	51	23.5 (14.62)	15.1 (−15.0)	Tubular	89

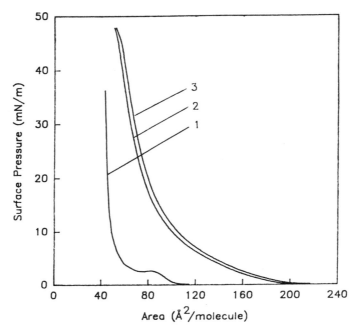

Figure 2. Force area isotherms (23°C) of [1], [2] and [3].
Monolayers were spread from CHCl$_3$ solutions of the
lipids and compressed up to their collapse
pressures at a rate of 50 cm^2/min.

The force-area curves of lipids [1], [2], and [3] on a subphase of
water are presented in Figure 2. Water temperature was maintained at
23°C. The monolayer behavior of [1] is comparable to that previously
reported as evidenced by the molecular area = 52 Å^2/molecule and the
collapse pressure = 36.5 mN/m.[28] In contrast to the monolayer behavior
of [1], liquid expanded monolayers of [2] and [3] were obtained upon
compression. The molecular area of [2] and [3] were 87 Å^2/molecule and
89 Å^2/molecule, respectively, and the collapse pressure of [2] and [3]
was 46 mN/m.

The self-assembled structures from lipids [2] and [3] were compared
with that of [1] (Figure 3), an extensively studied former tubule.[3] The
optical micrographs reveal that [2] produced vesicular structures (Figure
4). Lipid [3] produced predominantly tubular structures along with some
large vesicles (Figure 5). Repeated thermal cycling did not produce any
change in microstructure morphology for either lipid. The diameter of
the vesicles formed from [2] ranged from 0.3 μm to 3.4 μm while the
diameter of the tubular microstructures formed from [3] ranged from 0.4
μm to 4.7 μm and the length was greater than 110 μm.

Transmission electron micrographs of the lipid dispersion formed
from [3] have a smooth appearance with an average diameter of 1.2 - 2.0
μm. TEM of replicas of a freeze fractured sample of 3 reveal that the
tubular microstructures consist of a thin lipid coat around an aqueous
core (Figures 6 and 7).

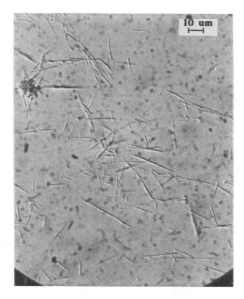

Figure 3. Optical micrograph of aqueous dispersion of [1].

DISCUSSION

Both, vesicles (closed, spherical, hollow structures made of concentric bilayers) from natural phosphatidylcholines (PC) and tubules (rigid, hollow, open ended, cylindrical structures consisting of concentric

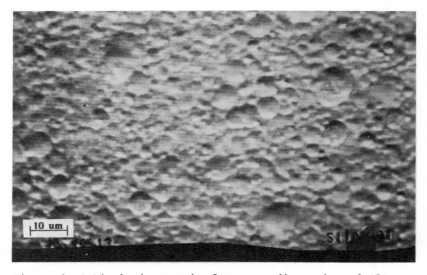

Figure 4. Optical micrograph of aqueous dispersion of [2].

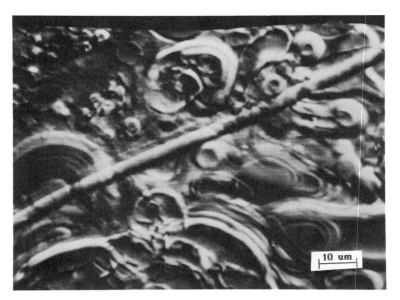

Figure 5. Optical micrograph of aqueous dispersion of [3].

bilayers), have many common structural features which are technologically useful.[1-3] Formation of the tubules is mainly caused by the following three interactions: head-group, hydrophobic, and inter diacetylene inter-actions. The first two interactions are common to all lipid systems. Because of its ability to promote tubule formation, the diacetylene has an important function as a morphology modulator in lipid self-assembled bilayers. The goal of this study was to modify the diacetylenic environ-ment to first disrupt the tubule formation and then systematically fine tune chemical changes to achieve a control over bilayer morphologies.

The complete interruption of tubule formation can be achieved by making structural changes in the polar headgroup region (responsible for long range organizational order) and/or by changing the diacetylene microenvironment in the acyl chains which in turn can influence the hydrophobic interactions. The formation of tubules with variable diame-ter has been reported by chemically altering the headgroup structure from choline to hydroxyalkanol.[23-26] That system is, however, complex in that it has many variables, e.g., pH and ionic strength of dispersion medium, and the presence of counterions, which influenced the lipid self-assembly and consequently the morphology of microstructures. Keeping the choline moiety in headgroup intact and changing the local chemical environment of diacetylene has the advantage of attributing the results only to a single parameter i.e., chemical changes in the acyl region.

The introduction of ether linkages on both sides of the diacetylene disrupted the rigidity of the acyl chain and rendered it unable to prop-erly align for close packing. This was demonstrated by the thermal (Table 1) and monolayer behavior of the lipid, [2] (Figure 4 and Table 1). The lower transition temperatures and enthalpy of the lipid [2] as compared to that for [1] demonstrate the increased fluidity of the acyl chains. The liquid-expanded to liquid-condensed transition, which occurs as a result of headgroup realignment and close chain packing, observed for the [1] monolayer was in contrast to the monolayer of [2] which was liquid expanded up to its collapse pressure.

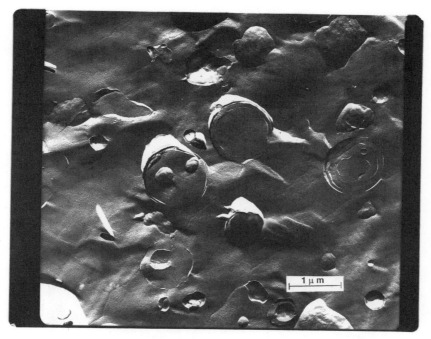

Figure 6. Transmission electron micrograph of replica of
freeze fractured sample of [3]. Bar represents
1 μm.

When only the terminal portion of the acyl chain was linked to the
diacetylene by an oxygen, cylindrical as well as vesicular microstructures were observed to coexist by optical and transmission electron
microscopy. The diameter of the tubular microstructures was up to 3
times greater than that reported for tubules formed from [1] (internal
diameter, 0.5 μm). The larger diameter of the cylinders formed from [3]
(internal diameter, 1.0-1.4 μm) might be attributed to the strong lateral
diacetylenic interaction and presence of fluid acyl chains. The decreased order in acyl chain packing is evident in the relatively low
melting transition and enthalpy of the chains as compared to corresponding non-ether linked PC [1]. The more fluid packing of the chains is
also demonstrated by the liquid expanded monolayer as opposed to the
transition to a liquid condensed state that occurs as [1] is compressed.
The ability of [3] to form tubular microstructures can be explained by
the order of the acyl chains of [3] as seen in the calorimetric results
(higher enthalpy). The increased disorder was demonstrated by the lower
melting transitions and enthalpy for [2] as well as the presence of two
melting transitions for [2]. This suggests that the diacetylene interaction of [3] is only partially disordered and that the chain retains
enough rigidity to allow formation of tubular microstructures. The
differences in the T_m values and the enthalpies of the two lipids can be
explained by the decreased hydrophobicity of the acyl chain with the
diacetylene connected by two oxygen linkages as opposed to one. The
presence of two melting transitions for [2] as opposed to one melting
transition for [3] indicates that the additional oxygen linkage affects
the molecular packing of the chains. An additional explanation may be
that the two melting transitions for [2] result from a decoupling of the
acyl chain segments by the ether linkage on the glycerol side of the
diacetylene.

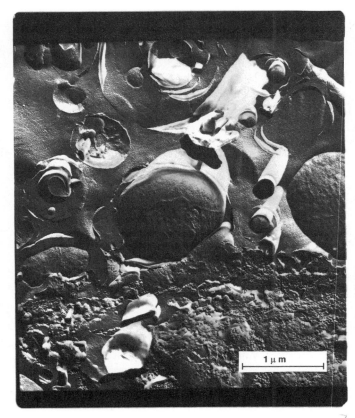

Figure 7. Transmission electron micrograph of replica of
freeze fractured sample of [3]. Bar represents
1 μm.

CONCLUDING REMARKS

In this study the role of diacetylene in the tubule formation is
confirmed. By changing microenvironment of diacetylenes by incorporation
of ether linkages ß to the diacetylene in phospholipid acyl chains lipid
bilayers can be modulated. These results have direct implications on the
development of new applications including controlled release of biomaterials. The formation of less rigid tubular structures from [3] has many
practical aspects, e.g. tubules may encapsulate more materials and will
not break during processing steps. To understand the results further, a
careful examination of the formation characteristics of the tubular
microstructures and the encapsulation behavior is needed

ACKNOWLEDGEMENT

Financial support from Office of Naval Research for an NRL program
on Molecular Design of Microstructures is gratefully acknowledged.

REFERENCES

1. D. Chapman, Biological Membranes, Academic Press, London (1968).
2. R. R. C. New, "*Liposomes, a Practical Approach*," Oxford University Press, New York (1990).
3. P. Yager and P. E. Schoen, P.E., Mol. Cryst. Liq. Cryst. **106**, 371 (1984).
4. J. H. Georger, A. Singh, R. R. Price, J. M. Schnur, P. Yager and P. E. Schoen, J. Amer. Chem. Soc., **109**, 6169 (1987).
5. P. Yager, P. Schoen, C. Davies, R. Price and A. Singh, Biophys. J., **48**, 899 (1985).
6. B. P. Gaber, J. M. Schnur and D. Chapman, Eds., "*Biotechnological Applications of Lipid Microstructures*," Plenum Press, New York, pp. 305-320 (1988).
7. T. G. Burke, A. Singh and P. Yager, Ann. N. Y. Acad. Sci. **507**, 330 (1987).
8. J. M. Schnur, P. Yager, R. Price, J. M. Calvert, P. E. Schoen and J. H. Georger, U.S. Patent 4,911,981 (1990).
9. F. Behroozi, M. Orman, W. Stockton, J. Calvert, F. Rochford and P. E. Schoen, J. Appl. Phys., **68**, 3688 (1990).
10. P. G. de Gennes, C.R. Seances Acad. Sci., **304**, 259 (1987).
11. W. Helfrich and J. Prost, Phys. Rev., **A38**, 3065 (1988).
12. J. N. Israelachvili, "*Intermolecular and Surface Forces*," Academic Press, New York (1985).
13. J. C. Chappel and P. Yager, Chem. Phys. Lipids, **58**, 253 (1991).
14. A. Singh, T. G. Burke, J. M. Calvert, J. H. Georger, B. Herendeen, R. R. Price, P. E. Schoen and P. Yager, Chem. Phys. Lipids, **47**, 135 (1988).
15. A. Singh, P. E. Schoen and J. M. Schnur, J. Chem. Soc., Chem. Commun., 1222 (1988).
16. S. L. Blechner, W. Morris, P. E. Schoen, P. Yager, A. Singh and D. G. Rhodes, Chem. Phys. Lipids, **58**, 41 (1991).
17. T. G. Burke, A. S. Rudolph, R. R. Price, J. P. Sheridan, A. W. Dalziel, A. Singh and P. E. Schoen, Chem. Phys. Lipids, **48**, 215 (1988).
18. A. Singh, B. P. Singh, B. P. Gaber, R. Price, T. G. Burke, B. Herendeen, P. E. Schoen, J. M. Schnur and P. Yager, in: "*Surfactants in Solution*," K. L. Mittal, Ed., Plenum Press, New York, **8**, 467 (1989).
19. A. S. Rudolph, B. P. Singh, A. Singh and T. G. Burke, Biochim. Biophys. Acta **943**, 454 (1988).
20. P. Yager, R. R. Price, J. M. Schnur, P. E. Schoen, A. Singh and D. Rhodes, Chem. Phys. Lipids **46**, 171 (1988).
21. P. Schoen, P. Yager, J. P. Sheridan, R. Price, J. M. Schnur, A. Singh, D. G. Rhodes and S. L. Blechner, Mol. Cryst. Liq. Cryst. **153**, 357 (1987).
22. A. Singh and M. A. Markowitz, in: "*Membrane Structure and Functions—The State of the Art*," B. P. Gaber and K. R. K. Easwaran, Eds., Adenine Press, New York, 1992, p 37.
23. A. Singh and S. Marchywka, Polym. Mat. Sci. Eng., **61**, 675 (1989).
24. M. Markowitz and A. Singh, Langmuir, **7**, 16 (1991).
25. M. A. Markowitz, J. M. Schnur and A. Singh, in: "*Synthetic Microstructures in Biological Research*," J. M. Schnur and M. Peckarar Eds., Plenum Publishing, New York, 1992, in press.
26. M. A. Markowitz, J. M. Schnur and A. Singh, Chem. Phys. Lipids, **62**, 193 (1992).
27. R. Price and M. Patchan, J. Microencap., **8**, 301 (1991).
28. R. Price and M. Patchan, J. Microencap., (1992) in press.
29. A. S. Rudolph, A. Singh, R. Price, B. Goins and B. Gaber, in:

"Fundamentals and Applications in Targeting of Drugs," G. Gregoridis, et al., Eds, Plenum Press, New York, p 103 (1990).

30. A. Singh, J. Lipid Res., **31**, 1522 (1990).

31. D. Johnston, S. Sanghera, M. Pons and D. Chapman, Biochim. Biophys. Acta, **602**, 57 (1980).

32. A. Singh and J. M. Schnur, Synth. Commun., **16**, 847 (1986).

33. D. J. Johnston, L. R. Mclean, M. A. Wittman, A. D. Clark and D. Chapman, Biochemistry, **22**, 3194 (1983).

SPECIAL FUNCTIONAL GROUP TRIGLYCERIDE OIL BASED INTERPENETRATING POLYMER NETWORKS

L. H. Sperling and L. W. Barrett

Chemical Engineering Department
Materials Science and Engineering Department
Materials Research Center
Center for Polymer Science and Engineering
Whitaker Laboratory 5
Lehigh University
Bethlehem, PA 18015-3194

Castor oil, vernonia oil, and a few other triglyceride oils contain special functional groups, such as hydroxyl or epoxide, which can be polymerized to make soft elastomers with glass transitions as low as $-50°C$. These can be combined with either amorphous polymers, such as polystyrene, to make IPNs, or with poly(ethylene terephthalate)PET, a crystallizable polymer, to make semi-IPNs. The latter compositions were investigated, finding that the triglyceride oils can improve the crystallization rate of the PET, and castor oil and vernonia oil networks can also serve to rubber-toughen the PET.

INTRODUCTION

1. TRIGLYCERIDE OIL CHEMISTRY

While most triglyceride oils contain only double bond functionality, a few special oils contain hydroxy, epoxy, or other chemical groups. Of course, all of the triglyceride oils are triesters between glycerine and the various acid residues. The special functional group triglyceride oils can be polymerized with either dibasic acids or diisocyanates, depending on the group present, to form elastomers with glass transition temperatures in the range of -40 to $-50°C$.

As a special plus for developing nations, oils such as castor oil or vernonia oil are renewable resources, containing hydroxyl or epoxide groups, respectively. Since these oils are crops rather than petrochemical based, all nations should be interested, since sooner or later the large quantities of petroleum oil now consumed may run low, or be voluntarily curtailed due to environmental restrictions. While castor oil is a commercial product, widely used in paints, adhesives, and urethane foams, vernonia oil is still technically a wild oil, but currently being examined for possible development. Both castor oil and

Biotechnology and Bioactive Polymers, Edited by C. Gebelein
and C. Carraher, Plenum Press, New York, 1994

vernonia oil are tropical or semi-tropical crops. However, lesquerella oil, another hydroxyl-bearing oil, is obtained from a wild flower native to Arizona and New Mexico. The chemical structure of these three oils is shown in Table 1. It is noted that lesquerella oil has two additional CH_2 groups per acid residue over that of castor oil. Castor oil and vernonia have the same number of carbon atoms.

Table 1. Chemical structures of functional group triglyceride oils.

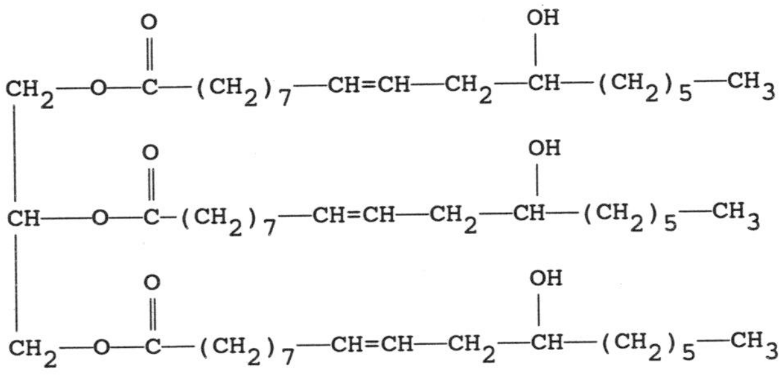

$$CH_2-O-\overset{\overset{\displaystyle O}{\|}}{C}-(CH_2)_7-CH=CH-CH_2-\overset{\overset{\displaystyle OH}{|}}{CH}-(CH_2)_5-CH_3$$

$$CH-O-\overset{\overset{\displaystyle O}{\|}}{C}-(CH_2)_7-CH=CH-CH_2-\overset{\overset{\displaystyle OH}{|}}{CH}-(CH_2)_5-CH_3$$

$$CH_2-O-\overset{\overset{\displaystyle O}{\|}}{C}-(CH_2)_7-CH=CH-CH_2-\overset{\overset{\displaystyle OH}{|}}{CH}-(CH_2)_5-CH_3$$

CASTOR OIL

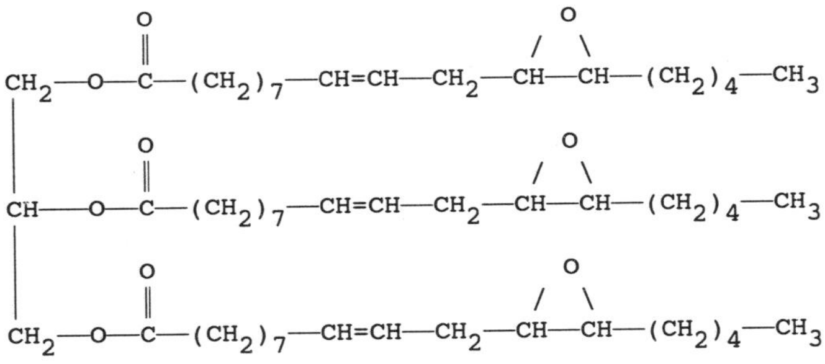

$$CH_2-O-\overset{\overset{\displaystyle O}{\|}}{C}-(CH_2)_7-CH=CH-CH_2-\overset{\overset{\displaystyle O}{\diagup\diagdown}}{CH-CH}-(CH_2)_4-CH_3$$

$$CH-O-\overset{\overset{\displaystyle O}{\|}}{C}-(CH_2)_7-CH=CH-CH_2-\overset{\overset{\displaystyle O}{\diagup\diagdown}}{CH-CH}-(CH_2)_4-CH_3$$

$$CH_2-O-\overset{\overset{\displaystyle O}{\|}}{C}-(CH_2)_7-CH=CH-CH_2-\overset{\overset{\displaystyle O}{\diagup\diagdown}}{CH-CH}-(CH_2)_4-CH_3$$

VERNONIA OIL

$$CH_2-\overset{\overset{\displaystyle O}{\|}}{C}-(CH_2)_9-CH=CH-CH_2-\overset{\overset{\displaystyle OH}{|}}{CH}-(CH_2)_5-CH_3$$

$$CH-\overset{\overset{\displaystyle O}{\|}}{C}-(CH_2)_9-CH=CH-CH_2-\overset{\overset{\displaystyle OH}{|}}{CH}-(CH_2)_5-CH_3$$

$$CH_2-\overset{\overset{\displaystyle O}{\|}}{C}-(CH_2)_9-CH=CH-CH_2-\overset{\overset{\displaystyle OH}{|}}{CH}-(CH_2)_5-CH_3$$

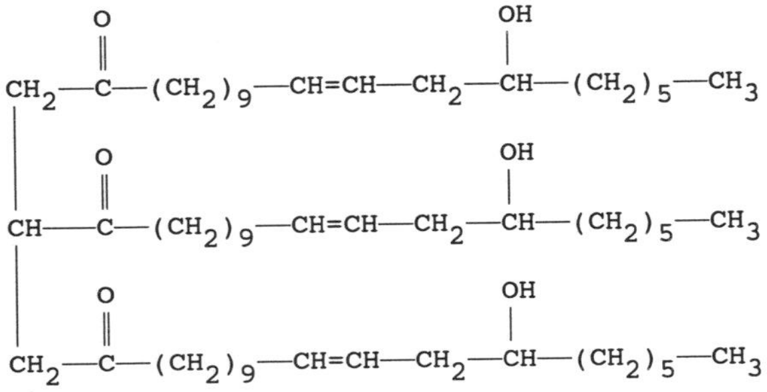

LESQUERELLA OIL

Using these oils, it is possible to make a 100% natural product elastomer. For example, pyrolytic decomposition of castor oil constitutes the commercial source of sebacic acid. Sebacic acid or its acyl chloride derivative can then be reacted with castor oil to form a crosslinked elastomer. Similarly, vernonia oil can be used to form suberic acid, and hence the corresponding polyester elastomers.[1,2]

While most of these elastomers are rather soft and weak as homopolymers, in combination with various plastics they serve to increase impact resistance, and in general form tough plastics or reinforced elastomers. A good way to combine these new elastomers with plastics is through the formation of interpenetrating polymer networks.

2. INTERPENETRATING POLYMER NETWORKS

An interpenetrating polymer network, IPN, is a combination of two crosslinked polymers. Usually, one or both are synthesized or crosslinked in the immediate presence of the other. There are several kinds of IPNs. The sequential IPNs are made by first forming one polymer network, then swelling in the second monomer, crosslinker, and initiator, and polymerizing *in situ*. A simultaneous interpenetrating network, SIN, is formed by simultaneously polymerizing both monomers by non-interfering routes, such as by chain and stepwise kinetics. Other methods of preparing IPNs include emulsion polymerization, suspension polymerization, etc. Semi-IPNs are cases where one polymer is crosslinked, and the other is linear. Gradient IPNs have macroscopic variation in either the overall composition or the crosslink level from location to location. Of course, the polymers may be elastomeric, glassy, or crystalline. The field of IPNs has been reviewed several times recently.[3-6]

Research on triglyceride oils in Sperling's laboratory can be divided into several stages, historically. Through an international program supported by NSF in the U.S. and Colciencias in Colombia, South America, first sequential,[7] then SIN[8-11] materials based on castor oil were synthesized and characterized. Usually, polystyrene was the second polymer. Materials tough enough to form shoe heels outlasting their mate leather soles were made at the time.

Taking the simultaneous interpenetrating networks of castor oil and polystyrene as an example, Figure 1 illustrates the synthetic detail.[9] Castor oil prepolymers made with sebacic acid were heated and added to diisocyanate, styrene, divinyl benzene, and benzoyl peroxide. Stirring was required to obtain phase inversion, see Figure 2.[8] Just before the gel point of the oil polymer, the material was poured into the mold, heated further, and eventually postcured. The diisocyanate is added to balance the castor oil prepolymers stoichiometrically as it is difficult to remove water beyond the gel point when only diacid is used. Figure 3 illustrates the stress-strain behavior of such materials.[11] Here, several SINs of castor oil and polystyrene are compared with a sequential composition. The SINs are all technically elastomers, while the sequential composition is technically a plastic, having a much higher modulus, and a yield point. Composition 5 was used for the shoe heels mentioned above.

Later, research was initiated on the wild oils vernonia[12] and lesquerella.[9,13] As usual, the size of the phase-separated domains determines the clarity of the resulting material. In Figure 4, epoxidized lesquerella palmeri oil is made into both SINs and IPNs with crosslinked polystyrene.[9] The finer morphology obtained via the

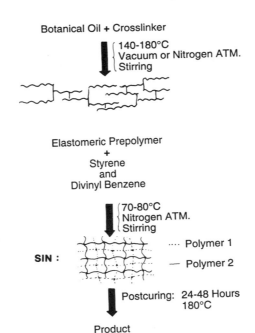

Botanical Oil + Crosslinker

140-180°C
Vacuum or Nitrogen ATM.
Stirring

Elastomeric Prepolymer
+
Styrene
and
Divinyl Benzene

70-80°C
Nitrogen ATM.
Stirring

SIN :

···· Polymer 1

— Polymer 2

Postcuring: 24-48 Hours
180°C

Product

Figure 1. Synthesis scheme for a simultaneous interpenetrating network based on a functional group triglyceride oil and styrene.[9] The oil plus crosslinker
(diacid or diisocyanate) are reacted to form a
prepolymer. Then, styrene and divinyl benzene are
added, and reaction allowed to proceed to just
before the gel point. Then, the product is poured
into the mold. The last stage is postcuring at
elevated temperatures.

sequential route makes more transparent materials. Probably, the SIN
phase separated before gelation, allowing phase domains to be much larger
than the distances between crosslinks.

Other teams picked up the research during this period. Important
contributions were made by Tan and Xie,[14] Patel and Suthar,[15] Liang, et
al.,[16] Rajalingham and Radhakrishnan,[17] Song and Donghua,[18] and Ayorinde,
et al.[1,2] More recent contributions will appear in reference.[3]

3. VERNONIA OIL-POLY(ETHYLENE TEREPHTHALATE) SEMI-IPNS

Most recently, EniChem America has supported a research program at
Lehigh for the study of semi-IPNs based on both castor oil and vernonia
oil semi-IPNs with poly(ethylene terephthalate), PET.[19-28] The intent of
the program was to produce both faster crystallization of the PET and to
toughen the product. The PET utilized by EniChem was a low molecular
weight material used in injection molded glass fiber composites, and was
brittle as a neat resin. Research on IPNs containing a crystalline
component has received only limited attention up until now. Four basic
parameters have been identified which control various properties,
depending on their extent and/or the time order of their appearance:[19-23]

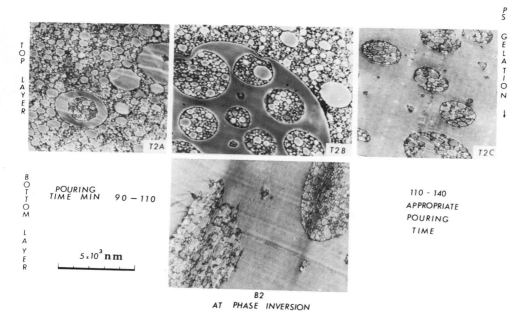

TOP LAYER

T2A   T2B   T2C

PS GELATION ↓

BOTTOM LAYER

POURING TIME MIN   90 — 110

110 - 140 APPROPRIATE POURING TIME

5×10^3 **n m**

B2

AT PHASE INVERSION

Figure 2. The phase inversion of a castor oil-polystyrene 10/90 simultaneous interpenetrating network can be followed by electron microscopy. In this case, first the oil is continuous, then both phases are cocontinuous, then the polystyrene is continuous.[8] When the polystyrene is continuous, the morphology of a phase-within-a-phase-within-a-phase resembles that of high impact polystyrene.

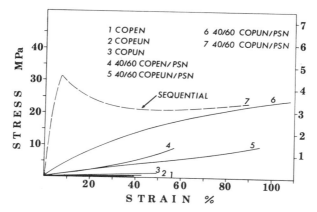

1 COPEN
2 COPEUN
3 COPUN
4 40/60 COPEN/PSN
5 40/60 COPEUN/PSN
6 40/60 COPUN/PSN
7 40/60 COPUN/PSN

SEQUENTIAL

STRESS MPa

STRAIN %

Figure 3. A comparison of several castor oil-polystyrene SINs with a corresponding sequential IPN. The latter has a higher modulus, and a yield point.[11]

EPOXIDIZED

LESQUERELLA PALMERI OIL

POLY[(ELP,SA(0.70))- SIN -(S,DVB(5%))] 50/50

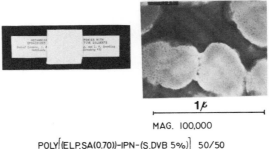

1 μ

MAG. 100,000

POLY[(ELP,SA(0.70))-IPN-(S,DVB 5%)] 50/50

Figure 4. The size of the domains controls the clarity of multiphase materials. Here, the larger domains of the SIN make the material opaque, while those of the sequential IPN are nearly clear.[9]

1. Crystallization of the PET.
2. Gelation of the vernonia or castor oils on polymerization.
3. Phase separation of the two components.
4. Bond interchange between the oil and the PET.

Figure 5 illustrates a typical heating, mixing, and cooling cycle for a vernonia oil-PET semi-IPN.[26] Poly(ethylene terephthalate) of about 12,700 g/mol was dried at 120°C in a vacuum oven for at least 48 hours prior to use. Refined vernonia oil was obtained from Rhone-Poulenc, Inc., which consisted of approximately 75% vernolic acid-containing triglycerides. PET and vernonia oil were mixed in a flask with magnetic stirring at 300°C. With continued stirring, the initially immiscible mixture becomes miscible, as evidenced by its clarity. Later, this was ascertained to be caused by bond interchange reactions, which take place at these temperatures. After about 10 minutes at 300°C, the mixture was cooled to 280°C, and the required amount of sebacic acid was added, which reduced the temperature to about 250°C. Under these conditions, the sebacic acid reacted with the vernonia epoxide groups to form a polyester network, and reached the gel point in about 15 minutes, however, after only five minutes of heating, the mixture was poured into a preheated mold and allowed to cool, during which time the PET crystallized. Finally, the material was placed in a vacuum oven for 18 hours at 160°C to complete the vernonia polyester network formation. While these conditions produce only a slightly bond-interchanged composition, different ranges of temperatures and times may be employed.

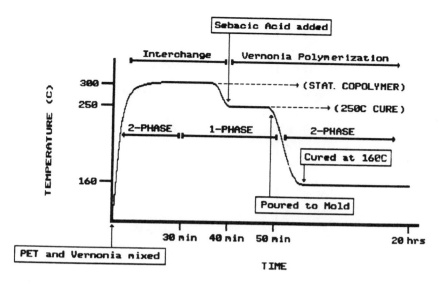

Figure 5. Schematic diagram of the semi-IPN synthesis process
for a vernonia oil-PET composition, axes not to
scale.[26] The solid line shows a typical time-
temperature course taken to produce the
compositions of interest. The times at the
various temperatures can be varied to make a range
of materials.

The crystallization of the PET and the gelation of the oils on
polymerization separately are relatively well-understood phenomena.
Figures 6 and 7 illustrate the consequences in the morphology of first
crystallization or first gelation.[26] In systems which crystallize before
gelation, well-developed spherulites appear (Figure 6). If gelation
happens first, crystallization produces only poorly defined spherulites
at best (Figure 7).

Bond interchange was a most unexpected phenomena. The three ester
groups connecting the acid residues to the glycerine can participate, as
well as the hydroxyl on the castor oil, or the epoxy on the vernonia.
The hydroxyl group reacts fastest of all with the ester groups in the
PET. Thus, when the oil is first mixed with the PET, it does not
dissolve. However, after a period of time, bond interchange occurs
sufficiently so that a single phase is formed. On cooling, the mix may
phase separate first (upper critical solution temperature), or
crystallize first, depending on the extent of interchange and cooling
rate. Naturally, bond interchange tends to disrupt the crystallization,
because the PET is copolymerized with the oil to a greater or lesser
extent.

The extent of bond interchange can be followed by FTIR, or by the
change in the glass transition temperature. The change in T_g in the
oil-rich phase of a 50/50 vernonia/PET mix with cooking time is shown in
Figure 8.[27]

One of the objectives of the research program was to produce a more
rapid crystallizing PET. Commercially, 1% by weight of the chemical
nucleating agent sodium benzoate is often used, for it is very effective
at improving the onset and peak crystallization temperatures of PET.

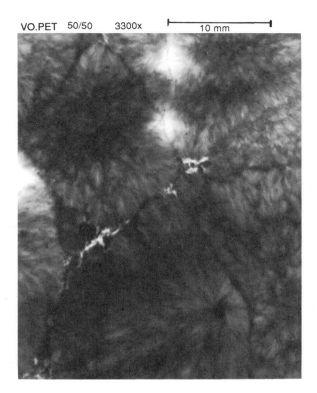

VO.PET 50/50   3300x |———— 10 mm ————|

Figure 6. Transmission electron micrograph showing well-
defined spherulites, which occurs when
crystallization of the PET precedes the gelation
of the vernonia oil polymer.[26]

While castor oil by itself is very effective, see Figure 9,[22]
combinations of the two were also tried. The castor oil and sodium
benzoate may be mixed together with PET in a simple manner, or the sodium
benzoate may be bonded directly to the castor oil to make a new molecule,

using castor oil monosodium terephthalate (COMSTA). In Figure 9, DSC
crystallization scans of PET compositions containing the bonded materials
are compared with similar compositions wherein the two are just mixed
together. Depending on whether the mixture is being cooled from the melt
or heated from the quenched glass, one or the other is the best under the
conditions tried.

Table 2 shows the temperatures and heats of crystallization for
several compositions, both cooled from the melt, and heated from the
quenched glass. The heat of crystallization provides a measure of the
total crystallinity induced.

Stress-strain curves, Figure 10,[22] show that the toughest materials
occur for midrange semi-IPN compositions. Thus, while the 100% oil-based
elastomer is soft and weak, and the PET at the low molecular weight
investigated is brittle, the semi-IPN is relatively tough and highly
energy absorbing.

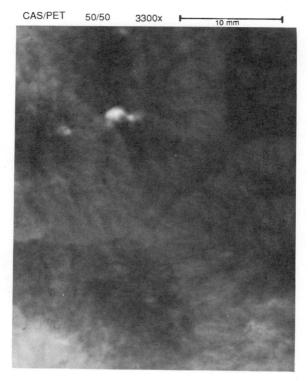

CAS/PET   50/50   3300x   ├────── 10 mm ──────┤

Figure 7. Transmission electron micrograph of a castor oil-
PET composition.[26] In this case, gelation of the
castor oil polymer and crystallization are more or
less simultaneous, resulting in poorly developed
spherulites. If gelation is first, spherulites
may fail to form, the product remaining amorphous.

Table 2. PET crystallization data for castor oil and/or
sodium benzoate compositions.[22]

Composition	ΔH^+_m	T^+_m	$\Delta H^\pm_m$	$T^\pm_m$
PET	45.9	202.2	36.5	133.6
10% Castor oil	58.1	215.0	40.6	109.7
1% Sodium Benzoate	52.4	222.5	33.3	117.2
10% Castor +				
1% Sodium Benz.	49.4	221.8	*	*
10% COMSTA	55.9	219.5	38.7	102.1

* This sample could not be quenched effectively enough for
  comparable data.
+ Cooled from the melt.
± Heated from the quenched glass.

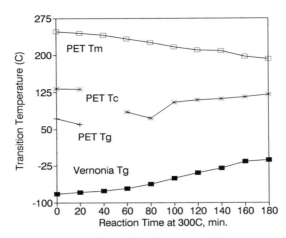

Figure 8. As the reaction time between the vernonia oil and
the PET is allowing to proceed, the PET melting
temperature decreases, while the vernonia oil
glass transition temperature increases. Both
changes suggest mixing, caused by copolymerization
of the two components.[27]

DISCUSSION

Of each of the phenomena discussed above, the most interesting from
a chemist's point of view is the bond interchange reaction. The oil and
PET initially are immiscible, but become more so as reaction time
together elapses. If reaction time would be allowed to progress
indefinitely, a statistical copolymer would result.

However, at limited cooking times a more miscible material emerges.
This product has two phases on cooling. The PET-rich phase retains PET
sequences long enough to crystallize rapidly. The oil network-rich phase
has its glass transition temperature raised somewhat, but is still useful
for toughening the plastic.

This research program originally had two objectives: (1) By adding
castor oil or vernonia oil or their respective prepolymers, could the
relatively slow crystallization of neat PET be quickened? (2) By
polymerizing and crosslinking the oil afterwards, could the product be
rubber-toughened? Thus, two improvements in the product could be
achieved with the modest addition of one component.

From a biotechnology point of view, each of these oils can easily be
grown as a crop. An important point, however, is that locations should
be chosen wherever possible that do not interfere with land usage for
food crops. In so far as vernonia oil grows in relatively arid
locations, this may be a significant advantage.

From an economic point of view, the cost still remains too high for
the oils to be used except for significantly value increased products.
Roughly speaking, the cost of castor oil is approximately three times
that of butadiene on a weight basis. While the cost of producing
vernonia oil or lesquerella oil commercially is still unknown, they may
be in the same range as castor oil (given the same economics of scale).
Thus, if (and when!) the price of petroleum products increases roughly a

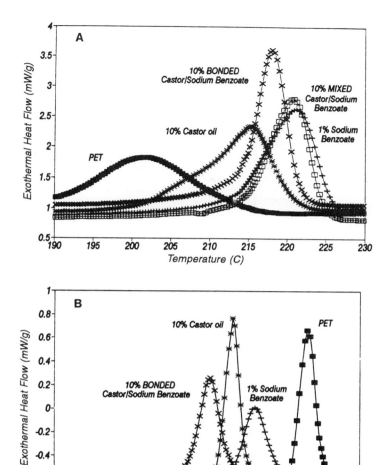

Figure 9. Differential scanning calorimetry curves showing crystallization from the melt (A), and from the glass (B). The latter also shows the glass transition temperatures of the PET near 70°C.[22]

factor of three, these products will easily compete on the market. During the oil crises of the 1970s, these conditions were almost met for brief periods of time.

CONCLUSIONS

Castor oil, vernonia oil, and similar special-functional group triglycerides occupy a unique position in nature. While ordinary double-bond functional oils can be polymerized only through free-radical oxygen attack, these oils can be polymerized to form a series of polyester or

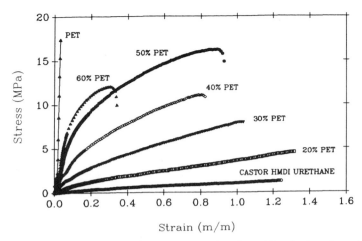

Figure 10. Stress-strain curves showing the behavior of a range of castor oil-urethane PET semi-IPNs.[22] The 50/50 composition is toughest.

polyurethane networks, and subsequently IPNs and semi-IPNs. The mechanical behavior of the products is encouraging, and now a world-wide interest in triglyceride-based oil IPNs has developed.

REFERENCES

1. O. A. Afolabi, M. E. Aluko, G. C. Wang, W. A. Anderson, and F. O. Ayoride, J. Ameri. Oil Chem. Soc., **66**, 7 (1989).
2. F. O. Ayoride, G. Osman, R. L. Shepard, and F. T. Powers, J. Amer. Oil Chem. Soc., **65**, 11 (1988).
3. D. Klempner, L. H. Sperling, and L. A. Utracki, Eds., "*Interpenetrating Polymer Networks*," ACS Books, Washington, D.C., publication expected 1993.
4. L. H. Sperling, "*Interpenetrating Polymer Networks and Related Materials*," Plenum, New York, 1981.
5. D. Klempner and K. C. Frisch, Eds., "*Advances in Interpenetrating Polymer Networks*," Technomic, Lancaster, PA, Vol. I, 1989, Vol. II, 1990; Vols. III and IV expected 1993.
6. J. J. Fay, C. J. Murphy, D. A. Thomas, and L. H. Sperling, Polym. Eng. Sci., **31**, 1731 (1991).
7. G. M. Yenwo, L. H. Sperling, J. Pulido, J. A. Manson, and A. Conde, Polym. Eng. Sci., **17**, 251 (1977).
8. N. Devia, J. A. Manson, L. H. Sperling, and A. Conde, Macromolecules, **12**, 360 (1979).
9. L. H. Sperling, J. A. Manson, and M. A. Linne , J. Polym. Mater., **1**, 54 (1984).
10. L. H. Sperling, J. A. Manson, S. Qureshi, and A. M. Fernandez, I&EC Prod. Res. & Dev., **20**, 163 (1981).
11. N. Devia, J. A. Manson, L. H. Sperling, and A. Conde, Polym. Eng. Sci., **19**, 878 (1979).
12. A. M. Fernandez, J. A. Manson, and L. H. Sperling, in: "*Renewable Resource Materials: New Polymer Sources*," C. E. Carraher, Jr. and L. H. Sperling, Eds., Plenum, New York, 1986.
13. M. A. Linne , L. H. Sperling, A. M. Fernandez, S. Qureshi, and J. A. Manson, in: "*Rubber-Modified Thermoset Resins*," C. K. Riew and J. K. Gillham, Eds., ACS Adv. Chem. 208, 1984.

14. P. Tan and H. Zie, Hecheng Xiangjiao Gongye, **7**, 180 (1984).
15. M. Patel and B. Suthar, J. Appl. Polym. Sci., **33**, 67 (1987).
16. J. L. Liang, H. T. Liu, W. H. Ku, and G. M. Wang, Abstracts, IUPAC International Symposium on Polymers for Advanced Technologies, Jerusalem, Israel, August, 1987.
17. P. Rajalingam and G. Radhakrishnan, Abstract for poster session, "International Symposium on Multiphase Macromolecular Systems," ACS Div. Polym. Chem., 14th Biennial Meeting, San Diego, CA, November 19-23, 1988.
18. Ma Song and Zhang Donghua, Plastics Industry (P.R.C.), **2**, 42 (1987).
19. L. W. Barrett and L. H. Sperling, Polym. Mat. Sci. Eng. Prepr., **65**, 345 (1991).
20. L. W. Barrett and L. H. Sperling, Polym. Mat. Sci. Eng. Prepr., **66**, 395 (1992).
21. L. W. Barrett and L. H. Sperling, Polym. Prepr., **33 (1)**, 948 (1992).
22. L. W. Barrett, L. H. Sperling, J. Gilmer, and S. Mylonakis, accepted, "*Interpenetrating Polymer Networks*," ACS Advances in Chemistry Series, 1993.
23. L. W. Barrett and L. H. Sperling, accepted, Polym. Eng. Sci., 1992.
24. L. W. Barrett and L. H. Sperling, Polym. Mater. Sci. Eng. Prepr., **65**, 180 (1991).
25. L. H. Sperling, C. E. Carraher, Jr., S. P. Qureshi, J. A. Manson, and L. W. Barrett, in: "*Biotechnology and Polymers*," C. G. Gebelein, Ed., Plenum Press, NY, 1991.
26. L. W. Barrett, O. L. Shaffer, and L. H. Sperling, accepted, J. Appl. Polym. Sci.
27. L. W. Barrett and L. H. Sperling, accepted by J. Polym. Sci., Part A, Chem. Ed.
28. L. W. Barrett, L. H. Sperling, J. Gilmer, and S. Mylonakis, accepted, J. Appl. Polym. Sci.

BIODEGRADABLE POLYMERS IN THE ENVIRONMENT: ARE THEY REALLY BIODEGRADABLE?

Graham Swift

Rohm and Haas Company
Spring House, PA 19477

Biodegradable polymers in the environment represent a cause for concern, depending how one defines and establishes the degree of biodegradation. In this paper, I will define biodegradable polymers, explain why they are used, how the degree of biodegradability is estimated and what the environmental implications are for incomplete biodegradation. Tests available are not yet totally satisfactory and further development needs are indicated. Some possible approaches are raised.

INTRODUCTION

The interest in biodegradable water-soluble polymers and plastics is due to the environmental concerns and issues raised by the disposal of conventional plastics and polymers. The waste-management of plastics has several options in addition to biodegradables, including recycle, incineration, and continued landfilling. The entrenched position of current plastics and their cost advantages make it unlikely that biodegradables will be considered, in the short-term, for applications beyond niche markets where the other waste-management options are not economically attractive, such as agricultural film, marine and fresh water fishing line, and diaper sheeting. In all these areas, plastic recovery is difficult and biodegradation becomes an attractive alternative to maintain a clean environment. In contrast to plastics, water-soluble polymers offer few alternatives to biodegradation, since once they enter the aqueous environment, recovery is not readily available or commercially attractive. The only real alternative is complete recalcitrance if that can be guaranteed to be non-harmful to the environment. Before biodegradable polymers can be claimed as the route to environmental salvation, there is a simple question that must be asked, and answered in the affirmative:

Are biodegradable polymers really biodegradable and environmentally safe?

This is very necessary because the term "biodegradable" has been widely used and interpreted to the point where confusion reigns. Therefore, in this paper, I will address the definition of biodegradable polymers, how to measure the degree of biodegradability and whether this

Biotechnology and Bioactive Polymers, Edited by C. Gebelein
and C. Carraher, Plenum Press, New York, 1994

is acceptable for environmental waste-management. In closing, I will indulge in a few comments on what I see as some of the future needs to complete our understanding of biodegradable polymers and their effects on the environment.

POLYMERS IN THE ENVIRONMENT

As mentioned in the introduction, polymeric materials in the environment are of two distinct types: plastics (generally water-insoluble) and water-soluble polymers. Plastics are used widely in the packaging industry, and ifimproperly disposed of, enter the environment as litter and present a real danger to wild-life as well as being offensive to the natural beauty of the landscape. Even disposal in landfills is a problem since the volume occupied by plastics is large; they contribute heavily to the depletion of available space. The synthetic plastics shown in Table 1 represent 10-15 billion pounds production per year.[1] Polyethylene used in agricultural film, polypropylene used in diaper stock, and nylon used in fishing gear are likely targets for biodegradable substitutes.

Water-soluble polymers enter the environment surreptitiously, unseen in aqueous solution, and represent real or imaginary concerns which are difficult to substantiate, but must be addressed. They, as plastics, represent a large volume of material, probably over 10 billion pounds per year, entering the environment. Unlike plastics they are composed of synthetics and a large proportion of natural polymers, either as isolated or chemically modified. Examples are shown in Table 2. The natural, unmodified, polymers represent no problems in the environment; however, modification brings uncertainty and these polymers may or may not be biodegradable.

Figure 1 indicates the several options available to the waste-management of water-soluble polymers and plastics. Degradation pathways, with the exception of photodegradation, are the common options. Photodegradation is unlikely to contribute to the degradation of water-soluble polymers. The common denominator for all options is fate and this the major problem in dealing with polymers in the environment.

If polymers enter the environment, they can, as shown in Figure 2 by their degradation pathways, leave residues, or completely biodegrade (mineralize). The issue than becomes how to establish that residues biodegrade further or are recalcitrant and innocuous. This may be more difficult than developing and establishing totally biodegradable polymers. At issue is the fate of the original polymer in the environment, and how far and to what it has (bio)degraded.

Table 1. Synthetic plastics.

Polyethylene
Polypropylene
Polystyrene
Poly(ethylene terephthalate)
Poly(vinyl chloride)
Nylon 66

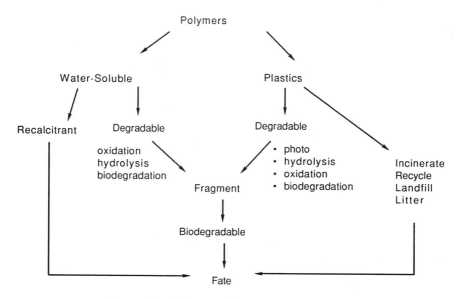

Figure 1. Polymers in the environment.

The importance of defining biodegradability in an environmental sense is very apparent. All degradation pathways lead to physical/chemical changes in the original polymer, but not necessarily removal from the environment. Biodegradation is necessary for this step. Biodegradation can be defined as shown below or schematically as shown in Figure 3.

Table 2. Water soluble polymers.

NATURAL POLYMERS
Cellulose
Starches
Pullulan
MODIFIED NATURAL
Carboxylated celluloses
Hydroxyethyl celluloses
Starch grafts
Chitosan
SYNTHETICS
Poly(vinyl alcohol)
Poly(alkylene glycols)
Polyacrylates

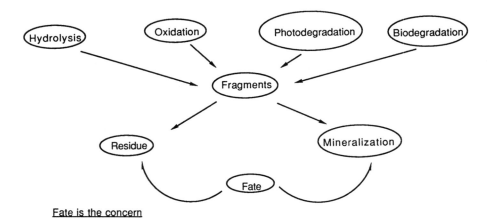

Fate is the concern

Figure 2. Degradation pathways - significance of biodegradation.

Molecular degradation, which is promoted by enzymes, may occur under
aerobic and anaerobic conditions leading to complete or partial removal
from that environment. Therefore, as we have seen, a polymer need only
partially biodegrade in a given environment to be identified correctly as
biodegradable. If the fragments are demonstrably innocuous or are known
to occur in nature, this would be acceptable for polymer waste-
management. Hence, we see the subtle difference between stating a polymer
is biodegradable and is acceptable in a waste-management sense, one must
establish that degradation fragments are innocuous. In other words,
chemical fate is the issue in regard to the acceptability of
biodegradable polymers, just as in any chemical exposure in the
environment and testing protocols should reflect this.

Biodegradation = Fate in the Environment

CONSIDERATION FOR BIODEGRADATION TESTING PROTOCOLS

Consideration in establishing test protocols are shown in Table 3,
and are equally applicable to water-soluble polymers and plastics,
despite their physical differences.

The measurement of biodegradability relates back to the fate issue
and chemical changes. To measure biodegradability truly, all the polymer
charged to the environment must be accounted for as gaseous byproducts,
cellular matter increase (biomass) and residues or unchanged polymer.
After biodegradation has ceased, any residue must be established as non-
toxic or harmless in the environment. Figure 4 expresses the requirements
for measuring biodegradation in an mathematical sense. This very clearly
delineates the problem of establishing the biodegradability of polymers.
It is readily extendible (to include other elements) from the
carbon/hydrogen polymer shown . Physical changes and rate of change may
also be important with respect to a plastic life-time in a given
application, e.g. fishing line, where performance may be lost at a
specific point of biodegradation. Tests shown in Table 4 are already

164

aerobic

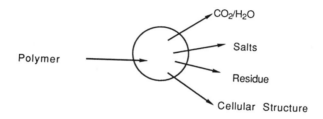

anaerobic

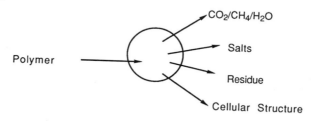

Figure 3. Biodegradation defined.

established as ASTM standards and can be used for plastics exposed to a given environment and retrieved for testing over a predetermined time period. This will permit an estimate of rate of biodegradation under specified conditions.

Environments are a very key part of biodegradation testing; they may be natural or simulated and pertain to use or disposal. The whole crux of

Table 3. Biodegradation test protocol considerations.

MEASUREMENT
Physical Changes
Chemical Changes
Rate of Changes

ENVIRONMENT
Natural
Simulated
Accelerated
Use
Disposal

POLYMER
Concentration
Form

Biodegradation of a polymer in a given aerobic[*] environment may be expressed as:

$$Ce = CO_2 + Cc + Cr$$

Where:

Ce = Polymeric carbon charged to the environment
CO_2 = Carbon dioxide evolved
Cc = Carbon incorporated into cellular matter
Cr = Polymeric carbon residual in the environment

When:

Cr = 0, Total biodegradation
Cr > 0, Incomplete biodegradation
Cr = Ce, No biodegradation

If:

Cr > 0, Fate becomes an issue

[*] Anaerobic, $Ce = CO_2 + CH_4 + Cc + Cr$

Figure 4. Measurement of biodegradability.

testing is the choice of environment. Using natural exposure may require long-term testing which is not practical for new polymer development, yet accelerated tests may not be predictive. The inoculum chosen may not contain the microbes that are active on the polymer under test. The only solution is to test widely or use a very rich inoculum for screening such as sewage sludge. Even so, such are the vagaries of biodegradation testing that a negative result never confers the status of non-biodegradability on a polymer. Repeated testing in different environments is the order, until one finally decides further testing is a waste of time, and accepts that a polymer is not likely to biodegrade.

Table 4. Physical test requirements for plastics.

MECHANICAL PROPERTY LOSS
Tensile strength (ASTM, D882-83)
Impact strength (ASTM, D1709-85)
Tear strength (ASTM, D1922-67)
VISUAL CHANGES
Bacterial colony growth (ASTM, G22-76)
Fungal growth (ASTM, G21-70)
Disintegration
PHYSICAL CHANGES
Weight loss

Plastic form or pretreatment, degree of crystallinity, etc., may be important and may influence rate of biodegradation. The concentration of water-soluble polymers can also greatly influence their biodegradability: too high may be toxic; too low may be difficult to measure.

TEST PROTOCOLS AVAILABLE

Currently available test protocols are shown in Table 5. Water-soluble polymers require only quantitative tests where a mass balance is the ultimate goal. Plastics require an additional test for physical property change during degradation (as referred to earlier), as this may influence use and disposal considerations.

Water-soluble polymers fit well into the aerobic and anaerobic test protocols established for surfactants and water-soluble chemicals by SDA[2] and OECD.[3] Preliminary screening is done by biological oxygen demand (aerobic) determination (BOD) with progression through carbon dioxide evolution (plus methane for anaerobic test) and residual carbon analysis to radiolabelled polymers. The inoculum is sewage sludge or other of choice. Test costs increase rapidly in the progression, but information obtained has significantly more value for establishing biodegradability. The carbon dioxide (methane)/residual carbon test gives good accountability as demanded by the equations discussed earlier. Radiolabelling permits testing at realistic exposure concentration in the environment.

Recently, ASTM developed protocols based on the above SDA and OECD aerobic and anaerobic test for plastics.[4] These are expected to be reproducible and quantitative, but round-robin tests have not yet been done to establish this. Work is also underway to develop simulated environments such as compost, sea and fresh water, and landfill (active) in which to test the physical degradation of biodegradable plastics in the laboratory.

The final link to connect the degree of biodegradation established in the aforementioned quantitative biodegradation tests to fate in the environment has yet to be made. ASTM is beginning to develop bridging protocols for plastics where residues, after biodegradation is complete, will be evaluated for toxicity towards suitable flora and fauna. Recently N. Scholz reported similar tests for water-soluble polymers,[5] indicating the importance and recognition being given to this biodegradation-fate relationship in that area.

Table 5. Tests available.

WATER-SOLUBLE POLYMERS
Quantitative
PLASTICS
Quantitative Qualitative (physical)

CONCLUSIONS

This active and interesting research area has not yet provided a satisfactory answer to the question: "Biodegradable polymers in the environment: are they really biodegradable?" However, I believe that the question is understood to mean fate in the environment and the right approaches have been established to answer it. Experimentation is difficult, but the question must be answered. Therefore, it will take time before biodegradable polymers will be widely accepted and used. The conclusions are:

Biodegradable plastics and water-soluble polymers have a part in environmental waste-management.

Recovery difficulties with conventional polymers will govern the urgency and area of use.

Fate will need to be established for those polymers not totally biodegradable.

Test protocols are available to establish whether a plastic or water-soluble polymer is biodegradable. But extent of biodegradation and fate need more attention.

CLOSING COMMENTS

The currently available tests for polymers are adequate for establishing some degree of biodegradation. However, we have to move beyond this level to determine the fate of any residual fragments in the environment. This can be done in several ways, for example the bridging toxicity test methods are a beginning. Ultimately, however, identification of the degrading microbes and enzymes is preferred so that the metabolic pathway can be established for a polymer's degradation. A clear picture of environmental fate will then be established to everyone's satisfaction.

REFERENCES

1. A. M. Thayer, Chem. Eng. News, Jan. 30, 1989.
2. R. D. Swisher, *Surfactant Biodegradation*," Marcel Dekker, Inc., New York, Basel, 1987.
3. OECD, *Guidelines for Testing Materials, Section Degradation and Accumulation*," 1981, #301A-E, 302A, B, C, 303A, 304A.
4. ASTM Standards D5909, D5910.
5. N. Scholz, Tenside Surfact. Det., **18**, 277-281 (1991).

FOUR YEAR ABSORPTION STUDY OF THREE MORPHOLOGICALLY DIFFERENT FORMS OF POLY-L-LACTIDE IN RABBITS

Balkrishna S. Jadhav & Deger C. Tunc*

Johnson & Johnson Orthopaedics
325 Paramount Drive
Raynham, Massachusetts 02052-0350

The rate of absorption of Poly-L-Lactide (PLLA) in the body, among other factors, is influenced by its morphology. In this study the absorption of implants made from three distinctly different morphological forms of PLLA, was investigated in rabbits. The three morphological forms of PLLA studied were mostly amorphous, mostly crystalline, and orientruded (extruded and drawn). Mostly amorphous and mostly crystalline PLLA implants were placed surgically in the lumbar muscles and orientruded PLLA implants were placed in the femoral intra-medullary canal of the rabbits. Molecular weight degradation and loss of mass of these implants were monitored as a function of time of implantation. The molecular weight of all three forms of PLLA implants was substantially degraded in the initial six months of implantation. There was no significant difference in the degradation rate of molecular weight of these three forms of PLLA among one another. However the rate of absorption of each form of PLLA was significantly different from the other two forms. Orientruded PLLA had the most rapid absorption and mostly crystalline PLLA was the slowest to absorb. At the four year time period both orientruded and mostly amorphous PLLAs were completely absorbed and only 77% of the mostly crystalline PLLA was absorbed.

INTRODUCTION

In recent years there has been considerable increase in the interest in Poly-L-Lactide (PLLA) for orthopaedic applications. Among its properties that are attractive for this application are: (1) absorbability, (2) non-toxicity, all its degradation products are natural metabolites and (3) high mechanical strength and sufficient retention of mechanical strength when implanted in the body.

---

* Present address: Union Carbide Corporation, P.O. Box 670, Bound Brook, NJ 08805

Biotechnology and Bioactive Polymers, Edited by C. Gebelein
and C. Carraher, Plenum Press, New York, 1994

One of the attractive characteristics of PLLA is that it is slow to crystallize and thus it lends itself for morphological manipulation offering wide range of properties which can potentially be utilized for a variety of applications. However this morphological manipulation also influences its absorption significantly. The objective of this study was to investigate the absorption of three distinctly different morphological forms of PLLA implants as a function of time of implantation in rabbits.

THREE FORMS OF PLLA

The three morphological forms of PLLA that were studied are as following:

(1) Mostly Amorphous Form: This is primarily obtained by injection molding or extrusion of PLLA in which the polymer is basically melted, shaped and then quenched. This form of PLLA is a tough and ductile material.

(2) Mostly Crystalline Form: This is obtained by annealing amorphous PLLA at a high temperature above its glass transition temperature in an inert atmosphere. This form of PLLA has higher yield strength and higher modulus as compared to amorphous PLLA.

(3) Orientruded PLLA:[1] This is obtained by injection molding or extrusion of PLLA into a cylindrical rod and subsequently drawing this rod uniaxially to impart molecular orientation in the polymer. Orientruded PLLA has very high tensile strength and tensile modulus. During orientrusion PLLA undergoes a profound morphological transformation. It changes essentially from a single phase structure to a two phase, highly oriented crystalline/semicrystalline composite structure.

PROTOCOL

In this study we used fifty New Zealand rabbits. The average weight of these rabbits was 3.7 kg. Each rabbit received one implant of each of the three morphological forms of PLLA. Each form of PLLA implant was placed surgically at one particular site of the three sites namely left anterior lumbar musculature, left posterior lumbar musculature and left femoral intra-medullary canal. About four to eight rabbits were sacrificed and the implants retrieved at 26, 52, 95, 104, 156 and 208 weeks time periods. The retrieved implants were rinsed with 70% ethanol and distilled water and then dried under vacuum at room temperature to constant weight.

Absorption of each implant was measured by determining the loss in its original mass due to implantation at each time interval. Degradation of molecular weight of each form of PLLA implant was measured by the change in its inherent viscosity in chloroform at 25°C. Histology was performed at the four year time period on tissues from the three implantation sites.

PREPARATION OF THREE FORMS OF PLLA

Ultra high molecular weight PLLA was synthesized in our laboratories by bulk polymerization of L(-) Lactide. The inherent viscosity of this PLLA was about 7.6 dl/g. This PLLA was ground and purified and subsequently pelletized. Pelletized PLLA was injection molded to obtain mostly amorphous form of PLLA. Part of this mostly amorphous PLLA was annealed at 122°C for 24 hours to obtain mostly crystalline form. PLLA pellets were also extruded into a circular rod shape and drawn uniaxially to a draw ratio of about 4.4 to obtain orientruded form of PLLA.

These three morphological forms of PLLA were characterized by Wide Angle X-ray Diffraction (WAX) for percent relative crystallinity, by inherent viscosity measurement for molecular weight and by Thermo-Gravimetric Analysis for residual monomer content. Cylindrical implants were machined from these three forms of PLLA. The following Table 1 lists the properties of three forms of PLLA implants and their respective implantation sites.

RESULTS

1. DEGRADATION OF MOLECULAR WEIGHT OF THREE FORMS OF PLLA

Figure 1 shows the inherent viscosity of each morphological form of PLLA as a function of time of implantation in rabbits. All three forms of

Table 1. Properties of three forms of PLLA implants and their implantation sites.

	Injection Molded	Inj. Molded & Annealed	Orientruded
Relative crystallinity	Mostly Amorphous*	Mostly crystalline**	53%
Inherent Viscosity dl/g	3.0	2.5	2.5
% Residual Monomer	<1	<1	<1
Size of the implant	3.5mm Dia x 10mm length	3.5mm Dia x 10mm length	3.5mm Dia x 25mm length
Avg. weight of the implant	0.110g	0.112g	0.285
Implantation site	Left anterior lumbar musculature	Left posterior lumbar musculature	Left femoral Intra-medullary canal

* WAX detected no crystallinity
** WAX detected no amorphous region

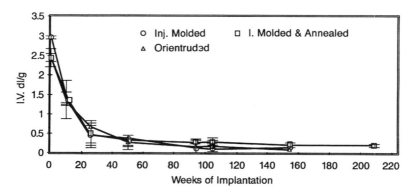

Figure 1. Change in the inherent viscosity of each of the three forms of PLLA implants as a function of time of implant in rabbits.

PLLA were degraded substantially in the initial 26 weeks of implantation. There was no significant difference in the rate of degradation of molecular weight of any one form of PLLA as compared to the other two forms.

2. ABSORPTION OF THREE FORMS OF PLLA

Figure 2 shows the absorption of each morphological form of PLLA. Absorption again was measured by the loss in the mass of the implants as a function of time of implantation in rabbits. Orientruded PLLA was the fastest to absorb and injection molded and annealed (mostly crystalline) PLLA was the slowest to absorb. At the four year time period, both orientruded and injection molded (mostly amorphous) PLLA samples were completely absorbed whereas only about 77% of injection molded and annealed (mostly crystalline) PLLA was absorbed.

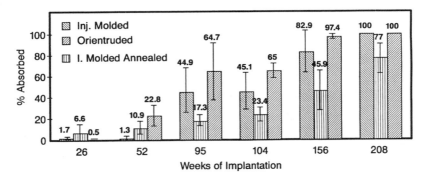

Figure 2. Absorption by the loss of mass of each of the three forms of PLLA implants as a function of time of implant in rabbits.

HISTOLOGY

Histology was performed on the tissue from three implantation sites at four year time period. Tissue from intramuscular implantation sites were embedded in paraffin wax and then sectioned. These sections were stained with hematoxylin and eosine for qualitative histopathologic evaluation. Tissue from femoral intramedullary implantation site were embedded in PMMA. Undecalcified sections were prepared and surface stained and evaluated qualitatively by light microscopy.

Figure 3 shows the intramuscular implantation site of the mostly amorphous PLLA. It shows that no implant material was present. It shows a presence of macrophages which was believed to be related to the absorption of PLLA. Surrounding muscle and fat are normal in appearance.

Figure 4 shows the intramuscular implantation site of the mostly crystalline PLLA. It shows the presence of implant material surrounded by the macrophages. The presence of macrophages was again believed to be related to the absorption of PLLA.

Figure 5 shows the femoral intramedullary canal site where orientruded PLLA was implanted. It shows no presence of any implant material. Macrophages were detected in the marrow cavity indicating the absorption of PLLA. Surrounding bone was normal in appearance.

DISCUSSION

The absorption data of three morphological forms of PLLA was statistically analyzed and compared with one another at each time period. This analysis showed that between the time period of one year to three years, absorption of each form of PLLA was significantly different from the other two forms. It also showed that at the four year time period, absorption of injection molded and annealed (mostly crystalline) PLLA was significantly lower than the other two forms.

Factors other than morphology which could have contributed to the differences in the absorption profile of these three forms of PLLA, but actually did not in the opinion of the authors, include the following:

(a) Presence of non-polymeric impurities such as residual monomer and catalyst: These three morphological forms of PLLA were obtained by the processing of an identical batch of PLLA. The residual monomer content was found to be less than 1% in all three forms of PLLA. Thus the possibility of the presence of different levels of impurities could not have been a factor which would have contributed to the significant difference in the absorption of three forms of PLLA.

(b) Molecular weight of three forms of PLLA: Molecular weights of these three forms of PLLA, as measured by the inherent viscosity measurement, were very close to one another. So this factor could also be eliminated.

(c) Surface area to weight ratio of implants: Although implants made from orientruded PLLA were larger than those from the other two forms of PLLA, the surface area to weight ratio of all three forms were very close to one another.

173

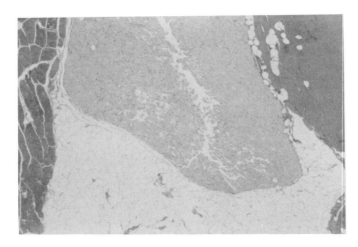

Figure 3. Histological slide of the left anterior lumbar
muscle site at four year time period where mostly
amorphous PLLA was implanted in the rabbit.

(d) Implantation site dependency: An assumption was made that as far as
the intramuscular implantation site and the femoral intramedullary
canal implantation site were concerned, the absorption of PLLA was
site independent. This assumption was based on the absorption data
obtained in our previous rabbit study.[2] In this study ultra high
molecular weight PLLA was implanted subcutaneously and also in the
femoral intramedullary canal of the rabbits. The absorption data
showed no significant difference in the absorption of PLLA at those
sites. Similar finding was reported by Miller, et al.[3]

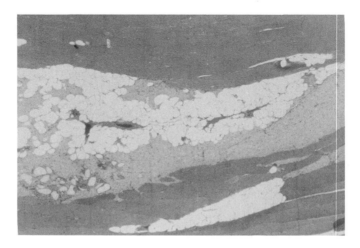

Figure 4. Histological slide of the left posterior lumbar
muscle site at four year time period where mostly
crystalline PLLA was implanted in the rabbit.

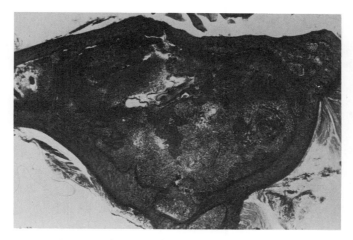

Figure 5. Histological slide of the left femoral intra-
medullary canal site at four year time period
where orientruded PLLA was implanted in the rabbit.

Hence the significant difference in the absorption of three
morphological forms of PLLA was indeed due to the differences in their
morphology. A plausible explanation for these differences in the
absorption profiles could be described in terms of the accessibility of
the body fluids to the bulk of the implants.

In the absorption of PLLA under physiological conditions hydrolysis
by water was the most dominant degradative reaction. The absorption of
PLLA was influenced by the rate of hydrolysis as well as by the rate of
diffusion of water into the bulk of PLLA. At the physiological pH of ~7.3
or less, the rate of diffusion of water plays a more dominant role than
the rate of hydrolysis, which probably would explain why PLLA absorbs by
bulk erosion rather than by surface erosion.

In the case of orientruded PLLA, two phases were formed which in
combination with the possibility of formation of microvoids during
orientrusion appeared to accelerate the diffusion of water into the
implant. This acceleration of water diffusion would probably explain the
fast absorption of orientruded PLLA. In the case of injection molded and
annealed PLLA, the high level of relative crystallinity, was probably
responsible for its slow absorption. It can be argued that high
crystallinity caused significant resistance to both water diffusion and
hydrolysis due to the necessity to overcome the lattice energy of the
crystalline regions.

CONCLUSIONS

The results of this absorption study lead to the conclusion that at
the four year time period both orientruded and mostly amorphous forms of
PLLA were completely absorbed. Orientruded PLLA was absorbed most
rapidly. Annealing of PLLA was shown to slow down significantly the
absorption of the polymer.

REFERENCES

1. D. C. Tunc and B. S. Jadhav, in: *"Progress in Biomedical Polymers,"* C. G. Gebelein and R. L. Dunn, Eds., Plenum Press, New York, 1990, p. 239.
2. D. C. Tunc, M. W. Rohovsky, B. S. Jadhav, W. B. Lehman, A. Strongwater and F. Kummer, in: *"Advances in Biomedical Polymers,"* C. G. Gebelein, Ed., Plenum Press, New York, 1987, p. 87.
3. R. A. Miller, J. M. Brady and D. E. Cutright, J. Biomedical Research, **11** (5), 711-9 (1977).

NON-AQUEOUS POLYMERIC SYSTEMS (NAPS) FOR DIAGNOSTIC APPLICATIONS

A. F. Azhar, A. D. Burke, J. E. DuBois, and A. M. Usmani

Boehringer Mannheim Corporation
9115 Hague Road
Indianapolis, IN 46250

Water-borne enzymatic coating based dry chemistries have been used for more than 20 years, especially by diabetics, for self-monitoring of blood glucose. Until recently it has been believed that enzymes work only in water and not organic solvents. Synthesis of a hydroxylated acrylic polymer and the novel concept of dispersing enzymes have enabled us to design non-aqueous coatings. These coating-films gave excellent dose response and dynamic range because of their ability to be ranged by antioxidants. The color signals were independent of the reaction time and the chemistry was thermally stable. Molecular forces and thermodynamical considerations have been used to explain the performance of such dry chemistries.

INTRODUCTION

Self-monitoring of blood glucose levels for diabetics is essential, and this has become possible for the past 20 years due to the advent of dry chemistries. In a typical glucose measuring reagent, glucose oxidase (GOD) and peroxidase (POD) enzymes, along with a suitable indicator, e.g., tetramethyl benzidine (TMB) are dissolved or dispersed in a latex or a water soluble polymer. This water-borne enzymatic coating is applied to a lightly TiO_2 pigmented plastic film and dried to a thin film. The coated plastic, cut to about 0.5 cm x 0.5 cm size, and mounted on a plastic strip is the dry reagent. The user applies a drop of blood and allows it to react with the dry reagent for about 60s or less. The blood is either wiped off manually or removed by capillary force and the developed color is then read visually or by an instrument. The chemistry, technology, design and characterization of dry chemistries have been discussed earlier by us.[1-2] Biochemical reactions in glucose determination are shown below.

Until now, dry chemistries have exclusively been water-borne because of the belief that enzymes function effectively only in water. We have researched non-aqueous polymers and coatings for dry chemistries to which red blood cells will not adhere in addition to giving quick end-point reaction.[3] These new coatings have shown superior thermostability as well. In this paper we will describe polymer synthesis, coating design, reaction stability and a proposed stability mechanism.

EXPERIMENTAL

1. SYNTHESIS OF HIGHLY HYDROXYLATED ACRYLICS

The non-aqueous diagnostic polymer must impart particular desirable properties to the resulting coating-film. Most importantly, hydrophilicity and hydrogel character to the film must be imparted, thus allowing intimate contact with aqueous whole blood sample. Furthermore, the polymer should provide some enzyme stabilizing effects.

An acrylic resin comprising of 65 wt% 2-hydroxyethyl methacrylate (HEMA), 33 wt% butyl methacrylate (BMA) and 2 wt% dimethylaminoethyl methacrylate (DMAEMA) was made by solution polymerization at 40% solid in xylene/1-methoxy-2-propanol (1/1). The polymerization was done at 90°C for 6 h using 1% azobisisobutyronitrile initiator. Another acrylic polymer, HEMA/BMA/t-butylaminoethyl methacrylate (65/34/1) was also made similarly. In fact a very large number of polymers were made - the above compositions are typical among the best.

2. ENZYME DISPERSION

The glass transition temperature (Tg) of both glucose oxidase (GOD) and peroxidase (POD) is 50°C.[4] In organic solvents, these enzymes become extremely rigid and can be dispersed with ease. Dispersions were made by grinding GOD and POD in xylene/methoxy propanol with or without a surfactant using an Attritor mill (2-4 h) or a ball mill (24 h). Grinding was continued until dispersions of <1 μm were obtained. Upon completion of dispersion, grinding media was strained and the dispersion stored at 4°C. The composition of a typical dispersion is 1.876 g GOD, 4.298 g POD, 11.79 g sodium dodecyl sulfate, 41.06 g xylene and 41.06 g 1-methoxy-2-propanol.

Table 1. Composition of a typical non-aqueous coating.

	wt%	Actual Batch, g
HEMA/BMA/DMAEMA(65/33/2)[a]	33.29	1.68
TMB	2.38	0.12
GOD	1.17	0.0589
POD	2.68	0.1355
SDBS[b]	3.28	0.1656
xylene	26.53	1.339
1-M-2-p[c]	26.53	1.339
Mica[d]	1.96	0.099

(a) 40% solid
(b) sodium dodecyl benzene sulfonate
(c) 1-methoxy-2-propanol
(d) cosmetic grade, C-4000, ultra fine

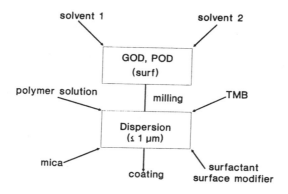

solvent 1   solvent 2

GOD, POD
(surf)

polymer solution   milling   TMB

Dispersion
(≤ 1 µm)

mica   coating   surfactant
surface modifier

Figure 1. NAPS process technology.

3. COATING COMPOSITIONS AND PROCESS TECHNOLOGY

A generalized method for preparation of the non-aqueous enzymatic coating is shown in Figure 1. The composition of a typical non-aqueous coating useful for low range blood glucose detection is shown in Table 1.

Many surfactants and surface modifiers were investigated and some useful combinations are sodium dodecyl sulfate alone or with Flow Tone 4 (sulfonated castor oil), and sodium dodecyl benzene sulfonate. In order to improve resolution in measurement of glucose levels in the high range (220-800 mg/dL), many hindered phenol antioxidants were found useful. These antioxidants that function as ranging compounds can be post-added to the non-aqueous systems. Typical ranging compounds are indicated in Table 2.

4. PREPARATION OF COATING-FILMS

The coatings were applied on a lightly TiO_2 pigmented polycarbonate plastic film at a wet film thickness of 100 µm and dried in an air-forced oven at 50°C for 15 min. A laboratory coater was used for applying coating onto polycarbonate at 2 m/min rate.

Table 2. Ranging compounds useful in NAPS.

Polymer	Ranging Compound	Compound TMB Mole Ratio
HEMA/BMA/DMAEMA (65/33/2)	APAC[a]	1:20
HEMA/BMA/TBAEMA (65/34/1)	APAC	1:15
HEMA/BMA/DMAEMA (65/33/2)	BHT[b]	1:2.5
	BHT/PG[c] (85/15)	1:3.0

(a) 3-Amino-9-(aminopropyl)-carbozole dihydrochloride
(b) butylated hydroxy toluene (2,6-di-tert-butyl-p-cresol)
(c) propyl gallate

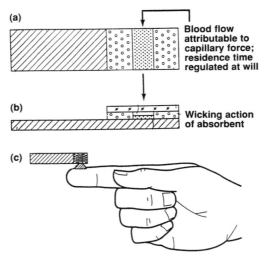

Figure 2. Touch and Drain Model. (a) Dry Coated Surface, (b) Cross Section of Dry Coated Surface, Adhesive, and Cover Piece, and (c) Contact with Blood Drop Results in Blood Filling the Cavity. After Desired Reaction Time, Blood is Drained Off by Touching End of Cavity with an Absorbent.

5. REACTIONS OF COATING-FILMS

For dose response and other studies with whole blood, the coated-films were incorporated into a touch and drain test device as shown in Figure 2. In this device, the blood flows across the surface of the coating-film. The residence time can thus be regulated at will. A McBeth 1500 visible spectrometer was used for measuring color signals.

RESULTS

1. COMPARISON OF AQUEOUS AND NON-AQUEOUS DIAGNOSTIC COATING SYSTEMS

The characteristics and performance of aqueous and non-aqueous diagnostic coatings differ. The composition and structural differences are indicated in Table 3.

2. DOSE RESPONSE AND CHARACTERISTICS OF NAPS

The dose response with low glucose blood is shown in Figure 3. Most water-borne diagnostic coating-films do not lend themselves to ranging by antioxidants. Our new coatings can easily be ranged however (Figure 4). The color signals were independent of the blood's residence time in the device from 10 to the 60s. The generated color remained stable over extended periods. Uric acid (30 mg/dL) produced interference by lowering the color signal. This interference can be prevented by buffering the coatings to 5.5 pH. Long-term stability of our coating-films under elevated temperature/moderate humidity was outstanding. Actual whole blood glucose concentrations in the 5 to 900 mg/dL and the reflectance

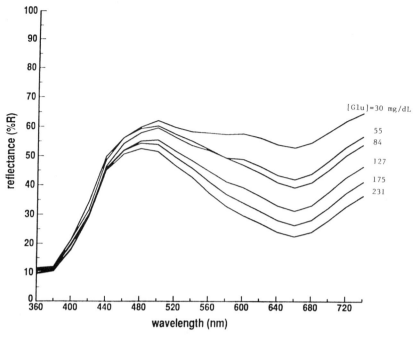

Figure 3. Low range whole blood glucose response.

readings correlated very well. We have thus been able to design a non-aqueous diagnostic coating-film that is superior to water-borne diagnostic coatings.

Table 3. Comparison of conventional (aqueous) and NAPS.

Feature	Aqueous[1]	Non-Aqueous[5]
Continuous phase	water	organic solvents
Discontinuous phase		
Enzymes (GOD,POD)	disolved	dispersed
Indicator (TMB)	dispersed; adsobed onto latex particles due to micellar forces	dissolved; therefore coats enzyme dispersions very uniformly
Micelles (surfactant)		
Antitoxidant	dispersed; adsorbed onto latex particle due to micellar forces	dissolved
Polymer binder	latex polymer of very high molecular weight	polymer solution of high molecular weight

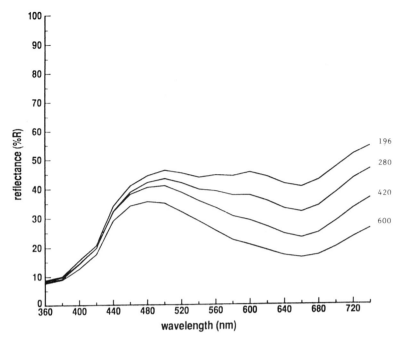

Figure 4. Dose response for high range blood glucose.

DISCUSSION

The continuous phase in aqueous diagnostic coatings is water. The enzymes, GOD and POD, are soluble in water whereas the dispersed indicator TMB is adsorbed on polymer due to micellar forces. In non-aqueous coatings, the continuous phase is organic, the enzymes are dispersed and solvent soluble TMB coats the enzyme dispersion. The structure of the micelles is also reversed in non-aqueous systems.

The biocatalytic activity of enzymes, dispersed in select organic solvents, remained intact and was actually enhanced due to the polymer. Our GOD/POD dispersions are extremely rigid. The conformation of the enzyme molecule in solutions is determined by a complicated network of both hydrogen bonds and electrostatic as well as hydrophobic interactions. To sustain native conformation, the enzyme molecule must have a definite hydration sphere. Introduction of organic solvents in the dispersions and coatings does not eliminate or distort the hydration shell as evidenced by catalytic activity remaining intact. In fact, the hydration shell acts as a "lubricant or flexibilizer" in non-aqueous medium thus allowing the catalytic reactions to occur.

The GOD molecule is about 50Å in size whereas a typical diagnostic emulsion polymer is 200-500 times that in size. The emulsion particle cannot therefore compress the protein. In non-aqueous systems, molecular and micellar forces can produce added stability. The coils of the acrylic polymer surround the protein, pack well, compress the protein and thus improve thermal stability (Figure 5).

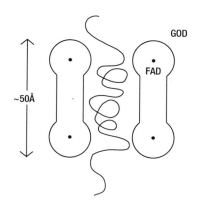

Figure 5. Compression of enzyme by polymer produces stability.

CONCLUSION

Non-aqueous diagnostic coatings, using properly designed polymers, showed superior performance when compared to aqueous coatings in thermal stability, quick end-point and regulating biochemical reactions. NAPS chemistry should be useful for aqueous analytes, e.g., glucose, lactate but even more so for lipid analytes, e.g., cholesterol, triglyceride.

REFERENCES

1. E. Diebold, M. Rapkin and A. Usmani, CHEMTECH, **21**, 462 (1991).
2. A. Burke, J. DuBois, A. Azhar and A. Usmani, CHEMTECH, **21**, 547 (1991).
3. A. Azhar, A. Burke, J. DuBois, M. Skarstedt and A. Usmani, U.S. Patent pending.
4. J. E. Kennamer and A. M. Usmani, J. App. Polym. Sci., **42**, 3073 (1991).
5. A. Azhar, A. Burke, J. DuBois and A. Usmani, Polym. Mater. Sci. Eng., **66**, 432 (1992).

PROTEIN PURIFICATION BY SELECTIVE ADSORPTION ON IMMOBILIZED PYRIDINIUM

LIGANDS

That T. Ngo[*]

BioProbe International Inc.
Tustin, California

Avid AL is an affinity gel with synthetic low molecular weight affinity ligands capable of selectively purifying IgG from the serum of all animals tested. By packing the gel in a radial flow column, a high flow rate can be applied and 12 g of highly purified IgG can be obtained within 5 hours directly from the serum. Immunoglobulins bound to Avid AL can be eluted with 0.05 M sodium acetate at pH 3 or at pH 7.4 with a neutral buffer containing "electron-rich" compounds. The affinity gel appears to be suitable under a varieties of storage and depyrogenation conditions.

INTRODUCTION

Selective adsorption of proteins to immobilized affinity ligands and subsequent elution of the adsorbed proteins are the basis of protein purification by affinity chromatography. The adsorption process involves a precise interaction of a specific region of the protein ligate with an equally specific area of an affinity ligand immobilized on the solid phase. A well known example of such a specific interaction is the binding of the Fc fragment of an IgG to Protein A.[1] This specific interaction is exploited in the affinity purification of IgG using an immobilized protein A gel.

Porath, et al.,[2] recently discovered a novel protein adsorption to immobilized non-ionic sulfone-thioether ligands. The interaction is termed "thiophilic." The proposed structure of the immobilized ligand was presented as agarose-$CH_2CH_2SO_2CH_2CH_2SCH_2CH_2OH$. This gel exhibits extraordinary selectivity in adsorbing IgG in the presence of "water-structuring" salts. The process is, however, different from the conventional hydrophobic interactions because thiophilic gel does not bind serum albumin, a hydrophobic protein, which, under similar conditions, is adsorbed to hydrophobic gels. A one-step purification of monoclonal antibodies using "thiophilic" adsorbent has been

---

* Address for correspondence: 20 Sandstone, Irvine, CA 92714

demonstrated.[3] The adsorption mechanism is thought to be the permanent
sulfone dipole and the free electron pair of the thioether acting in
concert as electron acceptor and donor sites, respectively, and that the
2-carbon proximity of the thioether to the sulfone group may be a
structural requirement for expression of the thiophilic character.[4] A
ring structure could be envisaged between the thioether and sulfone
moieties of the thiophilic ligands and the electron acceptor and donor
sites on the protein by an electron donor-acceptor or charge transfer
process.[4]

Ngo, et al, have recently discovered another interesting mode of
protein adsorption to a surface having immobilized pyridinium (IP)
ligands.[5-8] The proposed structures of the immobilized ligands are
depicted in Figure 1. We refer to the adsorption as "Aza-arenophilic"
adsorption (Aza for nitrogen and areno for aromatic). These gels show
extraordinary protein adsorption selectivity and binding capacity.[6] Such
a gel was able to purify immunoglobulins to a very high degree of purity
in one step directly from a phosphate buffered saline diluted serum. The
adsorption of proteins, in particular the immunoglobulins from the
unfractionated serum takes place in phosphate buffered saline (0.15 M
NaCl) only without the presence of "water-structuring" salts. The bound
proteins can be eluted with acid buffers or with an "electron-donor"
containing buffer at neutral pH.[8] The selectivity of protein adsorption
to IP gels can be varied by varying the composition and the salt
concentration of the loading buffer. The selectivity of these gels can be
further augmented by using selective elution buffers. It is also possible
to alter the selectivity of the gel by changing the halogen and
nucleophilic-substituents of the immobilized ligands (see Figure 1).

Studies on the preparation, proposed structure of the immobilized
ligands, protein adsorption properties, applications in immunoglobulin
purification, binding parameters of immunoglobulin and fragments
therefrom and a proposed mechanism of protein adsorption are herein
summarized.

PREPARATION OF IMMOBILIZED PYRIDIUM GELS (IP-GELS)

The immobilized pyridinium gels were prepared by first reacting
Sepharose Cl-4B, under anhydrous conditions, with penta-halopyridine and
4-dimethylaminopyridine (DMAP) and then with a nucleophile solution, such
as mercaptoethanol, ethanolamine, ethylene glycol or glycine.[5-8]

PROPOSED STRUCTURE OF THE IMMOBILIZED LIGANDS

Nucleophilic substitution reactions of mono-halogenated pyridines
occur most readily at the 2- or 4-halogenated carbon by an addition-
elimination mechanism.[9,10] The 2-step addition-elimination reactions with
nucleophiles occurs almost exclusively at carbon-4.[11-13] Chambers, et
al.,[14] showed that nucleophilic attacks by pyridine on perhalogenated
pyridine also took place at position-4 of the perhalogenated pyridine to
give a pyridinium adduct. By this analogy, the reaction of DMAP and a
perhalogenated pyridine forms a pyridinium adduct which subsequently
reacts with the hydroxyl groups of Sepharose and then with nucleophiles.
A total of 16 gels have been prepared and the protein adsorption
properties of these gels have been investigated in varying degrees of
detail.

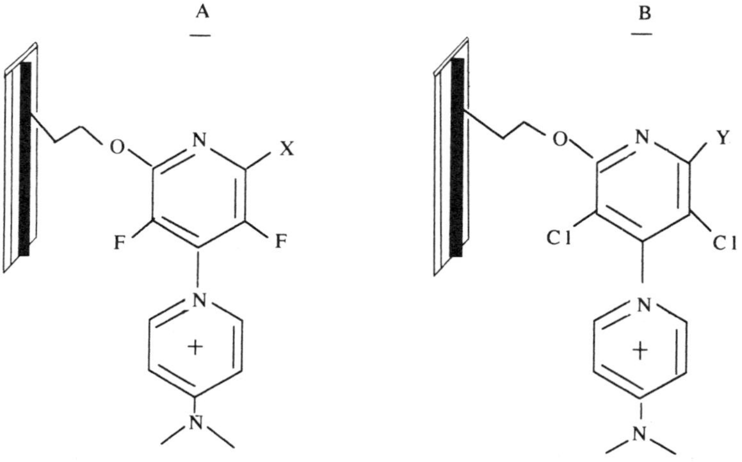

Nucleophilic substituents:

$$X = - O - CH_2 - CH_2 - OH \quad \text{or} \quad - HN - CH_2 - C \overset{O}{\underset{OH}{\diagdown}}$$

$$Y = - S - CH_2 - CH_2 - OH$$

Figure 1. Proposed structure for immobilized pyridinium gels.

ADSORPTION OF PROTEINS TO IMMOBILIZED PYRIDINIUM GELS (IP-GELS)

We have shown that in 20 mM phosphate buffer, pH 7.4, almost all serum proteins were adsorbed to IP gels. The addition of 0.5 M potassium phosphate caused a desorption of serum albumin. By decreasing the pH of the eluant to 5, 4, and 2.8, the desorbed results were mainly immunoglobulins.[6] The results of the analysis of eluted proteins, by enzyme-linked immunosorbent assay (ELISA), for the distribution of serum albumin, IgG, IgA and IgM revealed that: (1) in 20 mM phosphate, pH 7.4, virtually all serum proteins were bound; (2) the fraction of protein eluted by buffer containing 0.5 M potassium sulfate (2nd pooled fraction) was consisted of mostly albumin with very little amount of the immunoglobulins; (3) the 3rd pooled fraction desorbed by 0.1 M glycine, pH 5, was composed of mostly IgG with only a small amount of albumin; and (4) the 4th and 5th pooled fractions desorbed, at pH 4 and 2.8, consisted of all the immunoglobulins classes (IgG, IgA and IgM) and some albumins. The amount of proteins contained in these fractions was less than that of the pooled 2nd and 3rd fractions. The results of SDS-gradient (10-15%) polyacrylamide gel electrophoresis showed proteins other than albumin and immunoglobulins were adsorbed and desorbed in different fractions.[6] For example, appreciable amount of proteins with the same molecular weight as transferrin were detected in the 2nd pooled fraction and to a small degree in the 3rd pooled fraction.

PURIFICATION OF IMMUNOGLOBULINS WITH IP-GEL

A total of 16 different gels were synthesized and screened for IgG purification. The results from screening these gels showed that gel B (see Figure 1 for the structure), with dichloro-and hydroxyethylthio-substitutes, appeared to be ideal for IgG purification. Using this gel, the serum needs only to be diluted 5-fold with PBS. There is no need to add a high concentration of salt to the dilution buffer or to the serum. The bound IgG can be eluted by using 0.05 M sodium acetate, with the pH adjusted to 3 with concentrated HCl, not with glacial acetic acid. The purity of the isolated goat IgG was checked by using SDS-gradient polyacrylamide gel electrophoresis under both reducing and non-reducing conditions. The purity of IgG isolated from serum of mouse, rabbit and human was also examined with SDS-gradient electrophoresis. The results indicated a high degree of purity for the isolated IgG with only a minor amount of other proteins. It must be emphasized that the use of a proper elution buffer is critical for the successful purification of the antibody. For example, when the pH of the 0.05 M sodium acetate buffer was adjusted to 3 with acetic acid (not with concentrated HCl), the yield of IgG was very low. It was also noticed that 0.05 M acetate is more effective than 0.1 M in eluting the bound IgG. For more strongly adsorbed IgGs, the inclusion of 20% ethylene glycol or glycerol in the elution buffer is necessary. Glycine buffer (0.1 M), pH 3, can also elute the adsorbed antibodies. However, the antibodies desorbed by using glycine buffer is less pure than that obtained by using acetate buffer.

Several "electron-rich" compounds such as triethylamine, mercaptoglycerol, dithiothreitol, acetonitrile, etc, were able to desorb bound IgG at near neutral pH.[8] The ability to elute bound antibodies at neutral pH is an advantage of the IP-gel in purifying acid-labile antibodies. It is also shown that neutral pH elution buffer consistently yielded antibodies with higher specific binding activity than those prepared by using acid elution buffer.[8] We have used IP-gel to purify IgG from serum of 12 different animals. All IgGs tested so far adsorb to the gels. The degree of purity of the isolated IgG varied from one animal species to the next. The goat IgG isolated by this method has the highest degree of purity.

ADSORPTION OF IgG AND ITS PROTEOLYTIC FRAGMENTS TO IP-GELS

The adsorption of intact IgG molecule or its proteolytic fragments, i.e., the Fc and Fab fragments to IP-gel B, appeared to follow a simple Langmuir adsorption isotherm. The dissociation constants for IgG, Fab and Fc were determined to be 17.4 μM, 23.9 μM and 8.8 μM, respectively. The adsorption capacities were, respectively, 14, 5.3 and 1.8 mg/mL gel.[8] The data showed that, in addition to binding IgG, the gel does also bind both Fab and Fc fragments albeit with different affinities and capacities. It thus appeared that the simultaneous presence of both Fab and Fc provided a cooperative adsorption to the gel.

PROPOSED MECHANISM OF PROTEIN ADSORPTION TO IP-GEL

The IP-gel described here showed unusual protein adsorption selectivity and capacity. At low ionic strength most serum proteins were adsorbed to IP-gels. However, serum albumin, a "hydrophobic" protein, was desorbed by the presence of a "water structuring" salt, such as potassium sulfate, at 0.5 M.[6] This is contrary to the conventional hydrophobic interaction where the adsorption of albumin is promoted by "water structuring" salt. The adsorption of IgG to IP-gels takes place in either low or high ionic strength buffers with or without a "water structuring" salt. Therefore, the mode of adsorption of proteins to an IP-gel can neither be totally attributed to "pure" hydrophobic interactions nor to ionic interactions.

The adsorbed IgG can be selectively desorbed by a low pH buffer or by a neutral pH buffer containing "electron-rich" compounds, such as amines, nitriles or thio-compounds. Some immobilized pyridinium gels do not contain the element sulfur and yet they all are capable of adsorbing IgG with or without the presence of a "water structuring" salt. In this respect the IP-gel is different from the thiophilic gel which requires the presence of sulfur, either as a sulfone or a thioether, in the ligand and the definite presence of a "water structuring" salt to promote the adsorption of IgG.[2,4,15] The pyridinium ring, such as that found in the ligands of IP-gels, can serve as an electron acceptor in charge-transfer interactions.[16-21] The relevant electron-rich ligands in a protein molecule are most likely the aromatic amino acid side chains. In this regard, tryptophan was estimated to be the strongest electron donor. This was followed by tyrosine and phenylalanine.[22] Cilento and Guisti showed the transfer of an electron from indole nucleus to the pyridine coenzyme.[16]

The oxidized pyridine coenzymes and their model, 1-benzyl-3-carbozamide-pyridium chloride, form charge transfer complexes in a 1:1 ratio with tryptophan and other indole derivatives in aqueous buffer systems.[17,21] The quaternary nitrogen of the pyridinium ring can also interact with aromatic ring of amino acid side chain via ion-dipole interaction. For example, recent studies have shown that highly hydrophobic molecules built up from ethenoanthracene units display a strong and fairly general affinity for quaternary ammonium compounds.[23] This complexing effect can be ascribed to an ion-dipole attraction between the positively charged quaternary a ammonium ion and the electron-rich systems of the cyclophane. Furthermore, neutral molecules with electron-deficient π-systems are preferentially bound by the macrocyclic cyclophane. This binding suggests the operation of favorable donor-acceptor π-stacking interactions.

A synthetic acetylcholine receptor comprising primarily aromatic rings has been prepared.[24] This synthetic receptor provides an overall hydrophobic binding site capable of recognizing the positive charge of the quaternary ammonium group of acetylcholine through a stabilizing interaction with electron-rich π-systems of the aromatic rings (cation-π interaction). The biological receptors of cationic ligates, such as acetylcholine and phosphocholine, appear to comprise also primarily aromatic amino acid residues. A choline binding site of an antibody to phosphocholine comprises aromatic amino residues at close proximity to the quaternary ammonium ion.[25,26] Two anionic residues, aspartate and glutamate, also participate in the binding of the cationic ligate but are located further away than are the aromatic residues. The hydrophobic environment created by the aromatic amino acid residues could result in a low dielectric constant space, which in turn could result in a higher effective local charge than might be predicted from the nearby acidic amino acid side chains.

A recent study on the three-dimensional atomic structure of an acetylcholinesterase has shed light on the structure and chemical make-up of the choline binding site.[27] The active site of this enzyme lies at the bottom of a gorge that reaches halfway into the protein. The quaternary ammonium ion appears not bound to a negatively charged "anionic" site, but rather to some of the 14 aromatic residues lining the gorge.[27] Of these 14 aromatic amino acid residues there are 5 tryptophan, 5 tyrosine and 4 phenyalanine residues. Thus the structural studies on phosphocholine binding antibody,[25,28] and acetylcholinesterase,[27] are in agreement with the theory put forward by Dougherty and Stauffer,[24] that the π-electrons of aromatic rings attract a quaternary ammonium ion with greater affinity than isosteric uncharged ligands. Such attraction is thought to proceed via a stabilizing cation-interaction. An analysis of 33 high-resolution protein crystal structures revealed a significant tendency for positively charged amino groups of lysine, arginine, asparagine, glutamine and histidine to lie within 6Å to the electron clouds of aromatic amino acid side chains, where they make van der Waals' contact with the $d(-)\pi$-electrons of the aromatic ring.[29] This interaction of a charged amino group with the face of an aromatic ring contrasts the attraction of electron-rich oxygen or sulfur atom to the electron-poor edge of such rings.[30]

The underlying attractive forces that drive the adsorption of proteins to immobilized pyridinium ligands should include at least four factors: (1) the immobilized pyridinium ligands with the extensive delocalized π-electron system allows the operation of a favorable π-stacking type of interactions; (2) the pyridinium moiety of the ligand serves as an electron acceptor in the formation a charge transfer complex with an electron donor such as an aromatic amino acid residue; (3) hydrophobic interactions by removal of bound water molecules, an entropy driven process; and (4) the quaternary ammonium ion of the pyridinium group can interact with aromatic rings of a protein via cation-π interaction. The presence of N-dihalogenated-pyridyl substituent on the pyridinium ring greatly enhances the electron accepting character of the immobilized ligands.

In the case of IP-gel B, which showed excellent utility for immunoglobulin purification, an additional factor should be considered. The gel with a 2'-hydroxyethylthio-substituent on the pyridine ring may exhibit considerable "thiophilic" character as described by Porath, et al.,[2,22,31] The coupling of a mercaptopyridine to a divinylsulfone-activated gel gave an adsorbent that is more thiophilic than that obtained by coupling mercaptoethanol to a divinylsulfone-activated gel or a gel obtained by coupling mercaptopyridine to an epoxide-activated gel.[22] In an analysis of 36 high-resolution protein crystal structures, Reid, et al.,[30] found that sulfur-aromatic interactions are commonly observed in the hydrophobic core of proteins and that about half of all electronegative sulfur atoms from cyst(e)ine and methionine residues are in contact with an aromatic ring (phenylalanine, tyrosine and tryptophan). The sulfur atoms express an affinity toward the edge of the aromatic ring, thus, the electronegative sulfur atom is attracted by the positively charged aromatic hydrogens and avoids the region above the ring in the vicinity of the π-electrons.

It is clear that aromatic amino acid residues do interact intramolecularly with proximal methionine and cyst(e)ine residues. The interactions between the electronegative sulfur of a thioether and aromatic rings may partially provide the driving force in the thiophilic adsorption of proteins to a thiophilic gel. Porath and Belew proposed a hypotheses on the mechanism of thiophilic adsorption, which assumed a simultaneous operation of both electron donor and acceptor sites in the ligand molecule and interacts cooperatively and respectively with electron acceptor and donor sites on the protein molecule.[31] The electron donor site could be the sulfur atom of the thioether group, and the electron acceptor site could be either the methylene group next to the sulfone moiety or the 3-d orbitals of the sulfur of the sulfone. Thus the thioether sulfur of the immobilized ligand may interact with hydrogens on the edge of an aromatic ring.

Such an interaction appears to be quite general and widespread in proteins. The electron acceptor of the ligand may interact with the π-electrons of the aromatic ring of a protein. A similar cooperative and multi-point interactions between the immobilized ligands and the protein molecule can also be envisaged for the interaction of proteins with IP-gels. The pyridinium ring of the ligand is a good electron acceptor in the formation of a charged transfer complex and the "hetero-atom," such as O, N, or S in the substituent, can serve as an electron donor. The formation of charge transfer complexes between pyridinium rings and aromatic rings of aromatic amino acids is a well-known phenomena.[19] The aza-arenophilic interaction described here is a form of charge transfer adsorption discussed by Porath.[32,33]

The proposed protein adsorption mechanism presented here is no doubt a simplification of a complicated adsorption process. It nevertheless has served as a useful guide and helped us to design further experiments. The development of a neutral pH elution buffer using electron-rich compounds to compete with the electron donors of IgG is the consequence of such an adsorption model.

The successful development of IP-gels as an effective charge-transfer adsorbent has been possible because of: (1) the excellent electron accepting property of the pyridinium ring; (2) the extensive delocalized-electron system; and (3) the presence of a hetero-atom, such as O, N or S, in the substituents to the pyridine ring.

REFERENCES

1. R. Lindmark, K. Thoren-Tolling and J. Sjoquist, J. Immunol. Methods, 62, 1-13 (1983). "Binding of immunoglobulins to Protein A and immunoglobulin levels in mammalian sera."
2. J. Porath, F. Maisano and M. Belew, FEBS Lett., 185, 306-310 (1985). "Thiophilic adsorption - a new method for protein fractionation."
3. M. Belew, N. Juntti, A. Larsson and J. Porath, J. Immunol. Methods, 102, 173-182 (1987). "A one-step purification method for monoclonal antibodies based on salt-promoted adsorption chromatography on a "thiophilic" adsorbent."
4. T. W. Hutchens and J. Porath, Biochem., 26, 7199-7204 (1987). "Thiophilic adsorption, "A comparison of model protein behavior."
5. T. T. Ngo, U.S.Patent 4,981,961, "Synthetic affinity ligand compositions and methods for purification and recovery of organic molecules."
6. T. T. Ngo and N. Khatter, J. Chrom., 510, 281-291 (1990). "Chemistry and preparation of affinity ligands useful in immunoglobulin isolation and serum protein separation."
7. T. T. Ngo and N. Khatter, Appl. Biochem. Biotech., 30, 111-119 (1991). "Rapid and simple isolation of multigram goat IgG from serum using Avid AL^{TM} and radical flow column."
8. T. T. Ngo and N. Khatter, J. Chrom., 597, 101-109 (1992). "Avid AL, a synthetic ligand affinity gel mimicking immobilized bacterial antibody receptor for purification of immunoglobulin G."
9. A. R. Katritzky and J. M. Lagowski, "The principles of heterocyclic chemistry," Methuen & Co. Ltd., London, 1967, pp. 41-76.
10. T. L. Gilchrist, Heterocyclic Chemistry, Pitman, London, 1985, pp. 247-269.

11. R. D. Chamber, J. Hutchinson and W. K. R. Musgrave, J. Chem. Soc.,
    3736-3739 (1964). "Polyfluoroheterocyclic compounds. Part II.
    Nucleophilic substitution in pentafluopyridine."
12. R. E. Banks, J. E. Burgess, W. M. Cheng and R. N. Haszeldine, J.
    Chem. Soc., 575-581 (1965). "Heterocyclic polyfluoro-compounds. Part
    IV. Nucleophilic substitution in pentafluoropyridine : The
    preparation and properties of some 4-subsituted 2,3,5,6-
    tetrafluoropyridines."
13. R. D. Chamber and C. R. Sargent, in: "*Polyfluoroaromatic compounds
    in: Advances in Heterocyclic Chemistry*," Volume 28, A. R. Katritzky
    and A. J. Boulton, Eds., Academic Press, New York, 1982, pp. 47-59.
14. R. D. Chamber, W. Kenneth, R. Musgrave and P. G. Urben, Chem. Ind.,
    89 (1975). "Pyridium salts of halogenated heterocyclic compounds."
15. J. Porath, Biopolymers, **26**, S193-S204 (1987). "Salting-out adsorption
    techniques for protein purification."
16. G. Cilento and P. Giusti, J. Am. Chem. Soc., **81**, 3801-3802 (1959).
    "Electron transfer from the indole nucleus to the pyridine
    coenzyme."
17. G. Cilento and P. Tedeschi, J. Biol. Chem., **236**, 907-910 (1961).
    "Pyridine coenzymes. IV. Charge transfer interaction with the indole
    nucleus."
18. S. G. A. Alivisatos, G. A. Mourkides and A. Jibril, Nature, **186**, 718-
    719 (1960). "Non-enzymic reactions of indoles with coenzyme I."
19. E. M. Kosower, in: "*The Enzymes, Volume 3, "Charge-transfer
    complexing of pyridium rings*," P. D. Boyer, H. Lardy and K. Myrback,
    Eds., Academic Press, New York, 1960, pp. 171-194.
20. F. Ungar and S. G. A. Alivisatos, Biochim. Biophys. Acta, **46**, 406-408
    (1961). "Spectrophotometric evidence of certain anion-DNP$^+$
    interactions (including orthophosphates)."
21. S. G. A. Alivisatos, F. Ungar, U. A. Jibril and G. A. Mourkides,
    Biochim. Biophys. Acta, **51**, 361-372 (1961). "Non-enzymic reaction of
    indole with pyridine coenzymes and related structures."
22. J. Porath, in: "*Protein Recognition of Immobilized Ligands*," T. W.
    Hutchens, Ed., Alan R. Liss Inc., New York, pp. 101-122, 1989.
    "Electron-donor-acceptor chromatography (EDAC) for biomolecules in
    aqueous solutions."
23. M. A. Petti, T. J. Shepodd, R. E. Barrans, Jr., and D. A. Dougherty,
    J. Am. Chem. Soc., **110**, 6825-6840 (1988). "Hydrophobic" binding of
    water-soluble guests by high-symmetry, chiral hosts. An electron-
    rich receptor site with a general affinity for quaternary ammonium
    compounds and electron-deficient systems."
24. D. A. Dougherty and D. A. Stauffer, Science, **250**, 1558-1560 (1990).
    "Acetylcholine binding be a synthetic receptor: implications for
    biological recognition."
25. Y. Satow, G. H. Cohen, E. A. Padlan and D. R. Davies, J. Mol. Biol.,
    190, 593-604 (1986). "Phosphocholine binding immunoglobulin Fab
    McPC603, An X-ray difraction study at 2.7."
26. E. D. Getzoff, J. A. Tainer, R. A. Lerner and H. M. Geysen, in:
    "*Advances in Immunology*," Volume 43, F. J. Dixon, Ed., Academic
    Press, San Diego, pp. 1-98, 1988. "The chemistry and mechanism of
    antibody binding to protein antigens."
27. J. L. Sussman, M. Harel, F. Frolow, C. Oefner, A. Goldman, L. Toker
    and I. Silman, Science, **253**, 872-879 (1991). "Atomic structure of
    acetylcholinesterase from Torpedo californica: A prototypic
    acetylcholine-binding protein."
28. D. R. Davies and H. Metzger, Ann. Rev. Immunol., **1**, 87-117 (1983).
    "Structural basis of antibody function."
29. S. K. Burley and Petsko, FEBS Lett., **203**, 139-143 (1986). "Amino-
    aromatic interactions in proteins."
30. K. C. S. Reid, P. F. Lindley and J. M. Thornton, FEBS Lett., **190**,
    209-213 (1985). "Sulphur-aromatic interactions in proteins."

31. J. Porath and M. Belew, Trends in Biotech., **5**, 225-229 (1987). "Thiophilic" interaction and the selective adsorption of proteins."
32. J. Porath, J. Chrom., **159**, 13-24 (1978). "Explorations into the field od charge-transfer adsorption."
33. J. Porath, Pure & Applied Chemistry, **51**, 1549-1559 (1979). "Charge-transfer adsorption in aqueous media."
34. D. E. Atkinson, in *"Current Topics in Cellular Regulation,"* Volume 1, B. L. Horecker and E. R. Stadtman, Eds., Academic Press, New York, pp. 29-43, 1969. "Limitation of metabolite concentrations and the conservation of solvent capacity in the living cell.,
35. T. W. Hutchens and J. Porath, Anal. Biochem., **159**, 217-226 (1986). "Thiophilic adsorption of immunoglobulins - anaylsis of conditions optimal for selective immobilization and purification."

MOLDABLE DRY CHEMISTRY

J.E. Kennamer, A.D. Burke, and A.M. Usmani

Boehringer Mannheim Corporation
9115 Hague Road
Indianapolis, IN 46250

Results of thermal analyses of select diagnostic enzymes and their utility in determining stability of dry chemistries have been presented in this paper. An enzymatic compound consisting of glucose oxidase, peroxidase, tetramethyl benzidine indicator, alkylbenzene sulfonate and polyhydroxyethyl methacrylate was prepared and compression molded. Strips molded up to 150°C gave good glucose response. During molding the redox centers of the enzyme are not disturbed whereas the bulk of the enzyme, polymeric in composition, blends with PHEMA producing the unusual stability. There is a distinct possibility that a reaction injection molding (RIM) composition and processing can be developed for making dry chemistries.

INTRODUCTION

The chemistry of dry reagents useful in analysis of blood glucose, blood cholesterol and other analytes is now well established.[1-8] In a dry reagent chemistry, suitable reactants specifically enzymes and indicators, are coated as thin films, employing polymeric binders, onto a plastic base. A clinical sample, e.g., blood, 10-50 μL in volume, is then placed on the mini-size, disposable dry chemistry element. Several rapid reactions result in formation of color proportional to the concentration of the analyte. Quantification is usually done by reflectance spectroscopy, but more importantly with hand-held pocket size meters for "lay-use" monitoring at home.

Recently, our laboratory has researched non-aqueous polymers and coatings useful for dry chemistries.[9] In this method, enzymes are dispersed in organic solvents. A hydrophilic hydroxylated acrylic resin has been used to make non-aqueous diagnostic coatings. These coating-films gave excellent dynamic range, dose response and superior thermostability.

Because of deactivation of enzymes at elevated temperatures, molding compounds and moldings therefrom have not been researched earlier. We, therefore, proceeded to do thermal analysis of important diagnostic enzymes. Dry chemistries based on moldings of poly(hydroxyethyl methacrylate) (PHEMA) were explored and are discussed in this report. Mechanism of unusually high thermostability has also been presented in this paper.

ENZYMES

Enzymes are marvelously designed to be very specific catalysts. They are derived from plant and animal tissues; however, fermentation is currently the popular production method. Enzymes are extensively used in diagnostics, immunodiagnostics, and biosensors to measure or amplify signals of many, but specific, metabolites. Select diagnostic enzymes and their properties are presented below.

GLUCOSE OXIDASE (GOD)

GOD, a flavoenzyme, catalyzes the reaction sequence shown below. This enzyme contains two identical polypeptide chain subunits, each about 80,000 daltons, covalently linked by disulfide bonds. Each subunit contains one mole Fe and one mole of flavin-adenine dinucleotide (FAD). The molecule is about 74% protein, 16% neutral sugar and 2% amino sugars.

$$\text{ß-D-glucose} + \text{Enzyme-FAD} \longrightarrow \text{Enzyme-FADH}_2 + \text{δ-D-gluconolactone}$$

$$\text{Enzyme-FADH}_2 + O_2 \longrightarrow \text{Enzyme-FAD} + H_2O_2$$

PEROXIDASE (POD)

POD is a hemoprotein catalyzing the oxidation by hydrogen peroxide of a number of substrates, e.g., ascorbate, ferrocyanide and the leuco form of many dyes. The reaction sequence is shown below.

$$\text{POD} + H_2O_2 \longrightarrow \text{Compound I}$$

$$\text{Compound I} + AH_2 \text{ (oxidizable substrate)} \longrightarrow \text{Compound II} + AH$$

$$\text{Compound II} + AH_2 \longrightarrow \text{POD} + AH$$

$$2AH \longrightarrow \text{Oxidized product}$$

POD is suitable for the preparation of enzyme conjugated antibodies, because of its ability to yield chromogenic products, and also its relatively good stability characteristics. Seven isoenzymes have been described in literature. All contain photohemin IX as prosthetic group. Neutral and amino sugars account for about 18% of the enzyme. The "active site" involves apoprotein as well as the heme group.

CHOLESTEROL ESTERASE (CE) AND CHOLESTEROL OXIDASE (CO)

CE is sterol-ester hydrolase whereas CO is oxygen oxidoreductase for cholesterol, as shown in the reaction sequence below. Typical properties of select diagnostic enzymes are shown in Table 1.

Table 1. Typical properties of select diagnostic enzymes.

	Cholesterol Oxidase (CO)	Cholesterol Esterase (CE)	Glucose Oxidase (GOD)	Peroxidase (POD)
Source	Streptomyces	Pseudomonas	Aspergillus	Horseradish
EC	1.1.3.6	1.1.13	1.1.3.4	1.11.1.7
Molecular Weight	34,000	300,000	153,000	40,000
Isoelectric Point	5.1±0.1 and 5.4±0.1	5.95±0.05	4.2±0.1	
Michaelis Constant	4.3×10^{-5}M	2.3×10^{-5}M	3.3×10^{-2}M	
Inhibitors	Hg++,Ag+	Hg++,Ag+	Hg+,Ag+,Cu++	CN-, S-
Optimum pH	6.5-7.0	7.0-9.0	5.0	6.0-6.5
Optimum Temp.	45-50°C	40°C	30-40°C	45°C
pH Stability	5.0-10.0 (25°C,20h)	5.0-9.0 (25°C,20h)	4.0-6.0 (40°C,1h)	5.0-10.0 (25°C,20h)

$$\text{Cholesterol Ester} + H_2O \xrightarrow{\text{CE}} \text{Cholesterol} + \text{Fatty Acid}$$

$$\text{Cholesterol} + O_2 \xrightarrow{\text{CO}} \text{Cholest-4-en-3-one} + H_2O_2$$

In general, most enzymes are very fragile and sensitive to pH, solvent, and elevated temperatures. The catalytic activity of most enzymes is reduced dramatically as the temperature is increased. Typical enzymes used in diagnostics, e.g., GOD and POD, are almost completely deactivated around 65°C in solid form or aqueous solution. Despite wide and continued use of such enzymes in diagnostics for more than 30 years, limited or no thermal analysis work on these biopolymers has been reported until recently.

Our DSC analysis results indicating glass transition temperature (T_g), melting temperature (T_m), and decomposition temperature (T_d), are shown in Table 2.

A knowledge of the stability of enzymes in solid phase dry chemistries is very important (shelf-life, heat excursions encountered during transportation and storage). Stability of dry chemistries and how

Table 2. Thermal analysis of enzymes.

Enzyme	Source	T_g (°C)	T_M (°C)	T_d (°C)
Cholesterol Oxidase	Nocardia	50	98	210
Cholesterol Oxidase	Streptomyces	51	102	250
Cholesterol Esterase	Pseudomonas	43	88	162
Glucose Oxidase	Aspergillus	50	105	220
Peroxidase	Horseradish	50	100	225

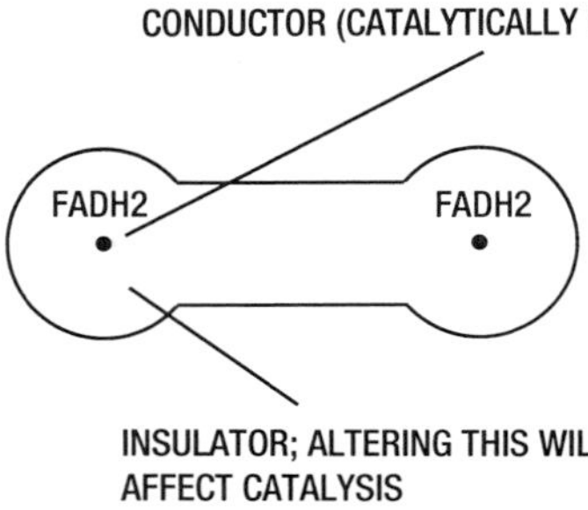

CONDUCTOR (CATALYTICALLY RELEVANT)

FADH2   FADH2

INSULATOR; ALTERING THIS WILL NOT
AFFECT CATALYSIS

Figure 1. Depiction of catalytic and electrical features of
glucose oxidase.

to predict lifetime have recently been discussed by Azhar et al.[10] Below
the glass transition temperature (T_g), the enzymes are in a glassy state
and should be thermally stable. Around T_g, onset of the rubbery state
begins, and the enzyme becomes prone to thermal instability. When the
enzymes melt around T_m, all the tertiary structures are destroyed, thus
making the enzyme completely inactive. The presence of chemicals can
considerably influence enzyme stability.

The electrical and catalytic features of GOD enzyme are depicted in
Figure 1. The redox center (FAD/FADH2) that can conduct electrons, is
catalytically relevant. To keep or sustain the enzyme's activity, the
redox centers must remain intact. The bulk of the enzyme, polymeric in
composition, is an insulator - altering it will not affect the enzyme's
catalytic activity.

ENZYMATIC COMPOUND

An enzymatic compound containing GOD, POD, tetramethyl benzidine
(TMB) indicator, linear alkylbenzene sulfonate, and PHEMA with weight
composition similar to a typical water-borne or non-aqueous coating
composition was ball-milled for 48 h. PHEMA was prepared by mass
polymerization using 1% benzoyl peroxide at 125°C/16h. PHEMA was crushed,
pulverized and finally ground in a ball mill to -80 mesh (<177 μm). Using
the molding compound, strips were made by compression molding. The molded
strips showed much improved thermal stability. This is schematically
shown as follows, with molding details in Table 3.

Table 3. Molding schedules.

Temp °C	Time, Min.	Activity (Response to Glucose)
105	1	Yes
105	5	Yes
125	1	Yes
150	1	Yes
200	1	No

$$(GOD, POD) \text{ Coating, Dry Film} \xrightarrow{65°C} \text{Rapid Inactivation}$$

$$(GOD, POD) \text{ Molding Compound} \xrightarrow[\text{Mold}]{\text{Heat/Pressure}} \text{Molded Strip}$$

THERMAL STABILITY

Why should the enzymes be stable in polymer melt whereas they quickly deactivate in water or dry form? We postulate a proposed mechanism, shown in Figure 2, for the excellent thermal stability of molded strips. We speculate the enzyme is surrounded very tightly by coils of PHEMA at room and elevated temperatures. Temporary melting of the crystalline insulator regions of the protein up to 150°C results in polyblending with PHEMA, the polyblending does not disturb the redox centers. When the molding is cooled, the enzyme/PHEMA blend crystallizes without destroying the redox centers. The catalytic activity of GOD is thus preserved during and after molding, and therefore, the molded strip responds to glucose. At 200°C the melt viscosity of the blend is lower

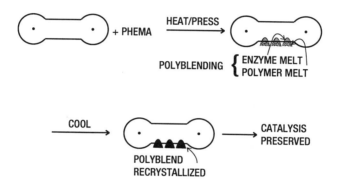

Figure 2. Preservation of catalyst in molded strips.

with increased mobility. The redox center is disturbed at 200°C and the cooled strip, therefore, becomes inactive. Certainly, the stability of the enzymes will depend on the concentration of PHEMA and the melt viscosity of the enzymatic compound.

CONCLUSIONS

DSC can be used for thermal characterization of enzymes and other biological polymers. Molding of strips using RIM may lead to useful chemistries. Modification of enzymes for added thermal stability by our technique is a distinct possibility. Extension of our concept in coating-films for biosensors is a distinct possibility.

REFERENCES

1. B. Walter, Anal. Chem., **55**, 449A (1983).
2. A. H. Free and H. M. Free, Lab. Med., **15**, 1595 (1984).
3. T. K. Mayer and N. P. Kubasik, Lab. Mgmt., 43 (April 1986).
4. J. A. Jackson and M. E. Conrad, Am. Clin. Products Rev., **6**, 10 (1987)
5. A. F. Azhar, A. D. Burke, J. E. DuBois and A. M. Usmani, Polymer Mat. Sci. Eng., **59**, 1539 (1988).
6. E. Diebold, M. Rapkin and A. Usmani, ChemTech, **21**, 462 (1991).
7. M. T. Skarstedt and A. M. Usmani, Polymer News, **14**, 38 (1989).
8. R. S. Campbell and C. P. Price, J. Intern Fed. Clin. Chem., **3**, 204 (1991).
9. A. Azhar, A. Burke, J. DuBois and A. Usmani, Polymer Mat. Sci. Eng., 66 (1992).
10. A. Azhar, A. Burke, J. DuBois and A. Usmani, in: "*Progress in Biomedical Polymers,*" C. G. Gebelein, Ed, Plenum, New York, 1990, pp. 149-156.

PREPARATION OF BACTERIAL ADSORPTION POLYMER AND ITS APPLICATION TO

BIOSENSORS

Rikio Tsushima, Akihiro Kondo, Masaru Sakata and Nariyoshi Kawabata[*]

Wakayama Res. Lab.
Kao Corp.
1334 Minato, Wakayama, 640 Japan
        and
*Department of Chemistry and Materials Technology
Faculty of Engineering and Design
Kyoto Institute of Technology
Matugasaki, Sakyo-ku, Kyoto, 606 Japan

Polymer particles immobilizing poly(N-benzyl-4-vinyl-pyridinium bromide) on their surfaces were prepared via reaction of a base polymer containing chloromethyl groups with poly(vinylpyridine), followed by quaternization with benzyl bromide. The polymer particles exhibited high adsorptive activity for microorganisms. A BOD sensor was prepared by a combination of a column packed with the polymer particles carrying activated sludge with an oxygen electrode. BOD values obtained using the BOD sensor agreed with those determined by the standard method.

INTRODUCTION

Crosslinked poly(N-benzyl-4-vinylpyridinium bromide) has been reported to capture many bacterial cells and viruses alive, on contact with them.[1,2] This insoluble pyridinium-type polymer is applicable to bacteria and virus removal from water, and as a carrier for immobilized microorganisms, which are useful for fermentation and as a microbial sensor. When an aerobic microorganism, adsorbed by the polymer, is located below a membrane oxygen electrode, a primary microbial sensor is formed.[3] The concentration of an organic material in sample solution can be determined by measuring the decrease in the dissolved oxygen concentration. By use of activated sludge as aerobic micro-organisms, the BOD (biochemical oxygen demand) of waste water can be determined within a short period.

From the practical point of view, it is preferable to provide a flow system made up by a column packed with a microbial adsorbent and oxygen electrode, etc. For the system, the microbial adsorbent is required to exhibit high adsorptive activity for microorganism and high strength. We now report the synthesis of a polymer particle, the surface of which is coated with poly(N-benzyl-4-vinylpyridinium bromide) and its application to a BOD sensor.

EXPERIMENTAL

1. MATERIALS

1A. Crosslinked poly(N-benzyl-4-vinylpyridinium bromide)

Suspension copolymerization of 4-vinylpyridine, styrene, and divinyl benzene was carried out using polyvinyl alcohol (PVA) and calcium carbonate as dispersing agent, respectively. The resulting cross-linked poly(4-vinylpyridine) was treated with benzyl bromide in refluxing ethanol to give cross-linked poly(N-benzyl-4-vinylpyridinium bromide).

1B. Polymer carrying poly(N-benzyl-4-vinylpyridinium bromide)

Suspension copolymerization of p-chloromethylstyrene , styrene, and divinylbenzene was carried out using PVA as dispersing agent. Then the resulting cross-linked polymer particles were treated with poly(4-vinylpyridine) in refluxing methanol. After filtration and washing with methanol, followed by reaction with benzyl bromide, polymer particles carrying poly(N-benzyl-4-vinylpyridinium bromide) were obtained. The amount of fixed pyridinium polymer was determined by the nitrogen content of the products.

2. MICROORGANISMS

Escherichia coli, strain B, was used as the microorganism for evaluating the microorganism adsorptive activity of the polymers. Activated sludge of the Wakayama plants of KAO Corp. was used as the microorganism for BOD sensor.

3. PROCEDURES

All procedures of adsorption experiments were done under aseptic conditions, in sterilized physiological saline, as was reported previously. The amount of adsorbed activated sludge was estimated by measuring oxygen consumption by its respiration.

RESULTS AND DISCUSSIONS

1. PREPARATION OF MICROORGANISM ADSORBENT POLYMERS

A terpolymer of 4-vinylpyridine, styrene and divinylbenzene was prepared by suspension polymerization using various dispersing agents. Because of the high water solubility of vinylpyridine, it was difficult to keep the suspension stable during the polymerization. However, addition of toluene to the monomers was found to stabilize the dispersed system.

The ability of the quaternized terpolymers to capture *Escherichia coli* was measured. As shown in Figure 1, the ability appeared to depend on the kind of dispersing agent used for suspension polymerization.

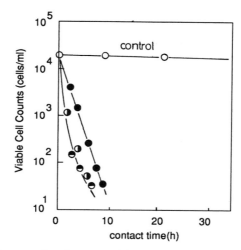

Figure 1. Removal of *E. coli* from water by crosslinked
poly(N-benzyl-4-vinylpyridinium bromide). (●) =
Polymer A; (◐) = Polymer B; (◖) = Polymer C;
(○) = the absence of resin. Copolymer
composition: vinylpyridine/styrene/divinylbenzene
= 8/1/1

Polymer A, prepared using PVA, exhibited lower adsorptive activity than
that of polymer B, prepared using calcium carbonate, which was similar to
that of polymer C, prepared by precipitation polymerization without a
dispersing agent. These results suggest that the PVA grafted unto polymer
surfaces prevents the adsorption of *Escherichia coli*. It is important to
make the surface structure of polymer identical to that of pure poly(N-
benzyl-4-vinylpyridinium bromide) in order to get an excellent microbial
adsorbent.

    Thus, immobilization of poly(N-benzyl-4-vinylpyridinium bromide) on
the surface of base polymer was examined according to Scheme 1, that is,
the reaction of base polymer containing chloromethyl group with polyvi-
nylpyridine, followed by quaternization with benzyl bromide. Table 1
shows the results of the immobilization, and Figure 2 shows the adsorption behavior of the resulting polymer carrying poly(N-benzyl-4-
vinylpyridinium bromide). The polymer E exhibited adsorptive activity as

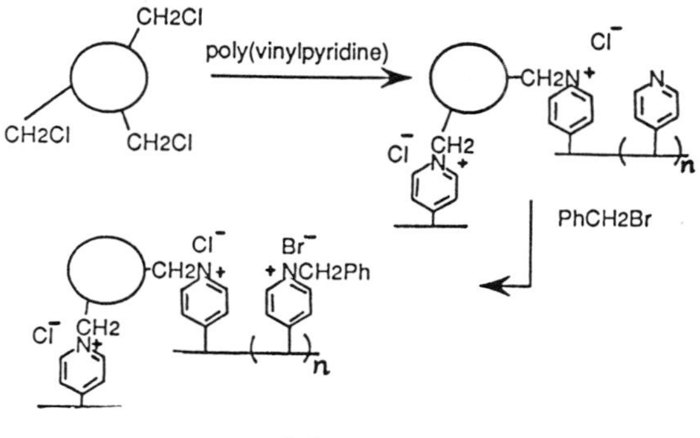

Scheme 1.

203

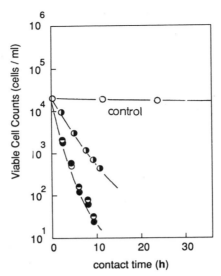

Figure 2. Comparison of adsorptive activity for *E. coli*
between cross-linked poly(N-benzyl-4-vinyl-pyri-
dinium bromide (C) and surface-modified polymers
D and E. (●) = polymer C; (◖) = polymer D; (●) =
polymer E; (o) = the absence of resin.

high as that of pure poly(N-benzyl-4-vinylpyridinium). This method can be
applied to prepare fibrous or filmy adsorbent.

2. BOD SENSOR

The adsorptive activity of the polymers A, B, C, and E was measured
for activated sludge. As shown in Table 2, the adsorptive activity of the
polymer C prepared by precipitation polymerization was higher than those
of the polymer A and B prepared by suspension polymerization, while the
adsorptive activity for of polymer D was similar to that of polymer C.
The adsorptive activity for activated sludge seems to be sensitive to the
surface structure of polymers.

Table 1. Immobilization of poly(N-benzyl-4-vinylpyridinium
bromide) on base polymer containing chloromethyl
group.

| Polymer | Base Polymer Chloromethylstyrene | | Product | |
	wt%	N%	PVP[a] mg/g-resin	BVP[b] mg/g-resin
D	9.0	0.24	18.3	48
E	21.0	0.42	32.5	86

(a) Poly(vinylpyridine), MW = 50,000
(b) Poly(N-benzyl-4-vinylpyridinium bromide).

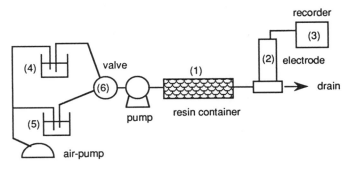

Figure 3. Flow-diagram of the BOD sensor.

A system illustrated in Figure 3 was prepared, in which (1) is a
resin container packed with polymer D carrying activated sludge, and (2)
is an oxygen electrode, the resulting measurement of which is registered
in a recorder (3). The containers for the sample (4) and the buffer
solution (5) were aerated, and the buffer solution was made to flow con-
stantly through valve (6). The measurement of the BOD was carried out by
feeding a definite amount of the sample by adjusting the valve (6). The
dissolved oxygen concentration loss, and the decreasing rate thereof,
were measured and the BOD value was determined from the calibration
curve. Figure 4 shows the relationship between the reciprocal of dissolved oxygen concentration loss (ΔOD) and the reciprocal of BOD. The
graph shows a linear relationship. Table 3 shows BOD values determined by
the BOD sensor and those according to a method specified in JIS K0102
(the term JIS means Japanese Industrial Standard). These results indicate
that the data obtained using the BOD sensor agree well with those determined by the JIS method. The determination of BOD by the JIS method
requires 5 days, while the application of the BOD sensor makes it possible to determine BOD in a few minutes.

Table 2. The adsorption activity of various poly(N-benzyl-4-
vinylpyridinium bromide for activated sludge.

Polymer	origin	Amount of adsorbed sludge mg/g-resin
A	PVA	2.6
B	$CaCO_3$	4.1
C	precipitation	6.5
E	immobilization	6.9

Table 3. Results of BOD measurements.

Sample	BOD by sensor mg/l	BOD by JIS method mg/l
1	371	360
2	182	180
3	147	150
4	72	75
5	548	560
6	86	80

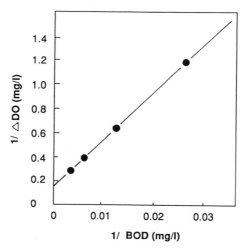

Figure 4. Relationship between 1/ΔDO vs 1/BOD.

REFERENCES

1. N. Kawabata, T. Hayashi and T. Matsumoto, Appl. Environ. Microbiol., **46**, 203 (1983).
2. N. Kawabata, T. Hayashi and T. Matsumoto, Agric. Biol. Chem., **50**(6), 1551 (1986).
3. N. Kawabata and N. Nakamura, Sen-i Gakkai Symp. Preprints, **21** (1986).

IMMOBILIZATION OF ADENOSINES ON SILICA GEL AND SPECIFIC SEPARATION OF

THYMIDINE OLIGOMERS

Yoshiaki Inaki, Kenji Matsukawa and Kiichi Takemoto

Department of Applied Fine Chemistry
Faculty of Engineering
Osaka University
Suita, Osaka 565, Japan

Adenosine and deoxyadenosine derivatives were immobilized on silica gel, which was able to be used as HPLC resins for the selective separation of oligothymidine, $pd(T)_n$, from the mixture of oligonucleotides. The chromatograms were obtained in 10% methanol-phosphate buffer aqueous mobile phase through 5°C to 30°C. The longest retention time was observed for the complementary $pd(T)_4$, and increased with decrease of temperature. The most hydrophobic $pd(A)_4$ did not show a long retention time and showed a small temperature dependency. This fact suggested that the hydrophobic interaction in this system was negligible, and the main separation factor was the base pairing between complementary nucleic acid bases. The retention times of $pd(T)_n$ became longer with a decrease of temperature. The $-\Delta H$ value for $pd(T)_4$ was 16.8 kcal/mol, which was approximately 4 kcal per one thymine base. An increase of the degree of polymerization caused retardation of the retention times, and gave higher $-\Delta H$ values. Interestingly, an increase of one thymidine unit {$pd(T)_2$ to $pd(T)_3$, and $pd(T)_3$ to $pd(T)_4$} caused an increasing $-\Delta H$ value of approximately 4 kcal, which is the same value obtained for $pd(T)_4$. These resins may be useful for separation of components of nucleic acids and polynucleotides as a specific separation system.

INTRODUCTION

DNA is known to have the double helical structure proposed by Watson and Crick, where the pair of bases through hydrogen bonding and the stacking of bases are significant interactions (Figure 1). The base pairing between complementary nucleic acid bases is a specific interaction, but the stacking of the bases is a nonspecific and hydrophobic one. The specific interaction of nucleic acids can be applied to high performance liquid chromatography (HPLC) for specific separation of nucleic acid fragments.

Biotechnology and Bioactive Polymers, Edited by C. Gebelein
and C. Carraher, Plenum Press, New York, 1994

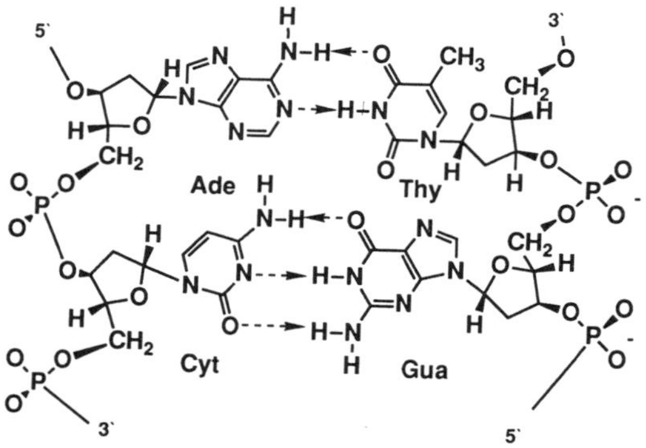

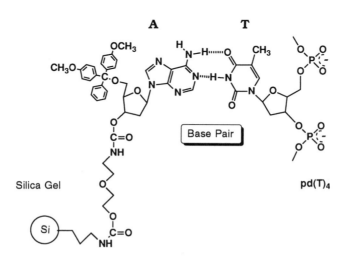

Figure 1. Base pairs in DNA

In biotechnology and genetic engineering, separations and purifica-
tions of the products from the mixtures are troublesome, but very important process. In this process of separation and identification of the
complex mixture, HPLC is a powerful tool. For the separation of nucleic
acid components by a HPLC system, commercially available ion-exchange and
reversed phase column have been used, although the separations in these
systems are not base specific separations. The specific interactions
between nucleic acid bases has not yet been applied for the HPLC system,
which should give a very useful HPLC system for biological, medical, and
biotechnological fields.

In the present paper, hydrophobic adenosine derivatives are bonded
on silica gel, and the resins obtained are applied to HPLC system for
base specific separation of oligonucleotides (Figure 2).

Figure 2. Illustration of the base specific separation on
adenosine immobilized silica gel

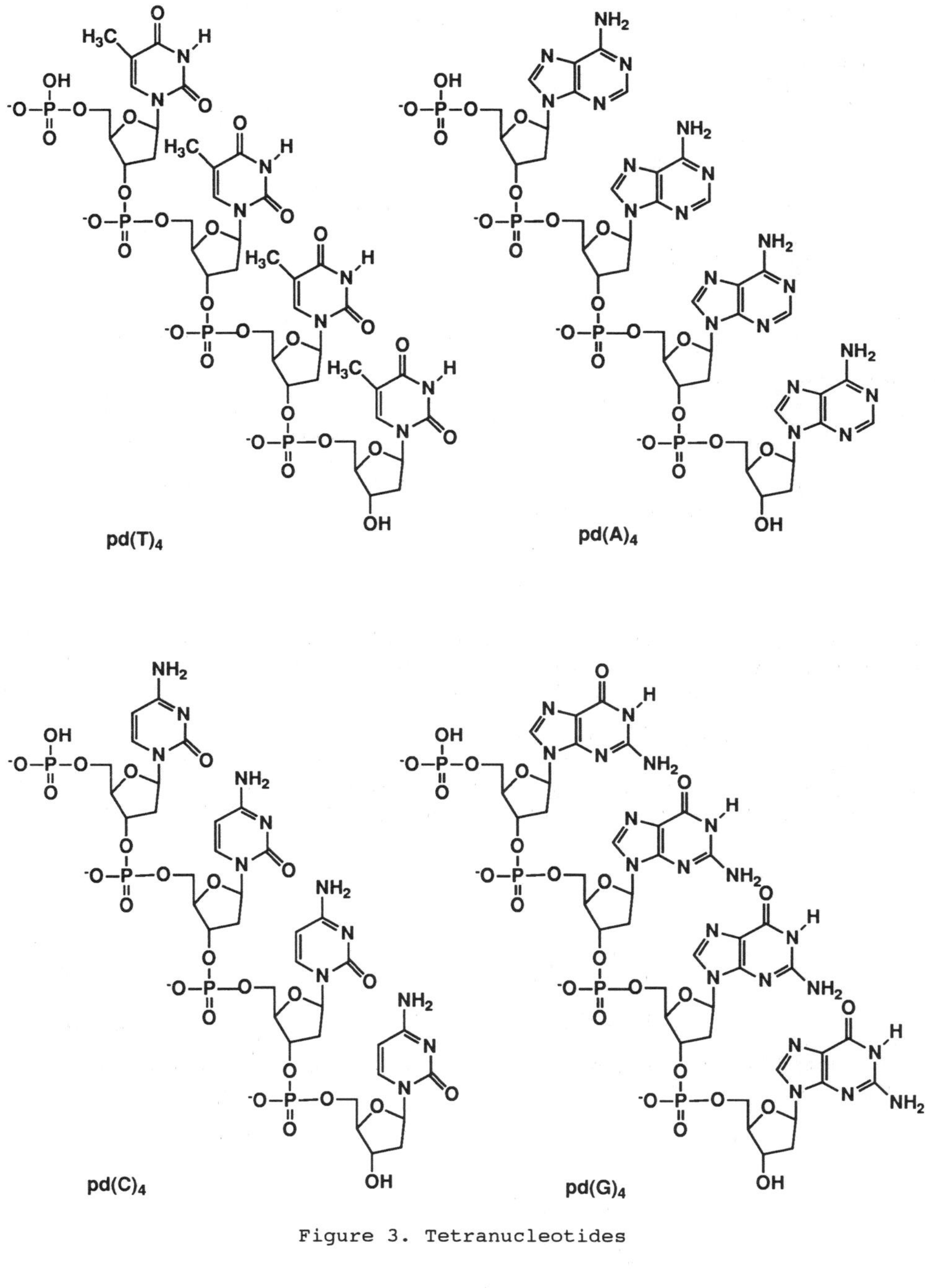

Figure 3. Tetranucleotides

For the separation of tetranucleotides, such as pd(T)$_4$, pd(A)$_4$, pd(C)$_4$, and pd(G)$_4$ (Figure 3), by ion-exchange HPLC, a textbook shows the order of elution to be pd(A)$_4$, pd(C)$_4$, pd(T)$_4$, and pd(G)$_4$ (Figure 4). When a reversed phase HPLC is used, where the hydrophobic long alkyl chain is immobilized on silica gel, the order of elution is pd(C)$_4$, pd(G)$_4$, pd(A)$_4$, and pd(T)$_4$ (Figure 4).[1,2] In these HPLC system, however,

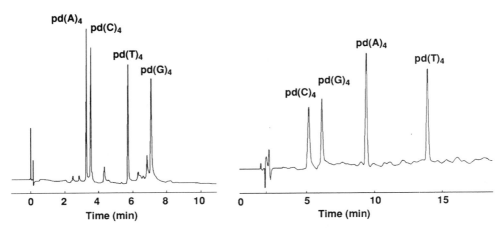

Ion Exchange

pd(A)₄
pd(C)₄
pd(T)₄
pd(G)₄

Time (min)

Reversed Phase

pd(A)₄
pd(T)₄
pd(G)₄
pd(C)₄

Time (min)

Figure 4. Typical separation of tetranucleotides by ion
exchange and reversed phase HPLC

it is difficult to change the order of elution. Therefore, the base
specific separation method may be convenient because the elution order
can be estimated from the base pairs in nucleic acids.

In our previous studies, nucleic acid bases were immobilized on
silica gel and applied to HPLC for the separation of nucleic acid fragments (Figure 5).[3-10] The complementary separations were observed in the
methanol mobile phase. In order to separate the nucleic acid fragments,

Si-Thy

Si-Ade

Si-PLL-T

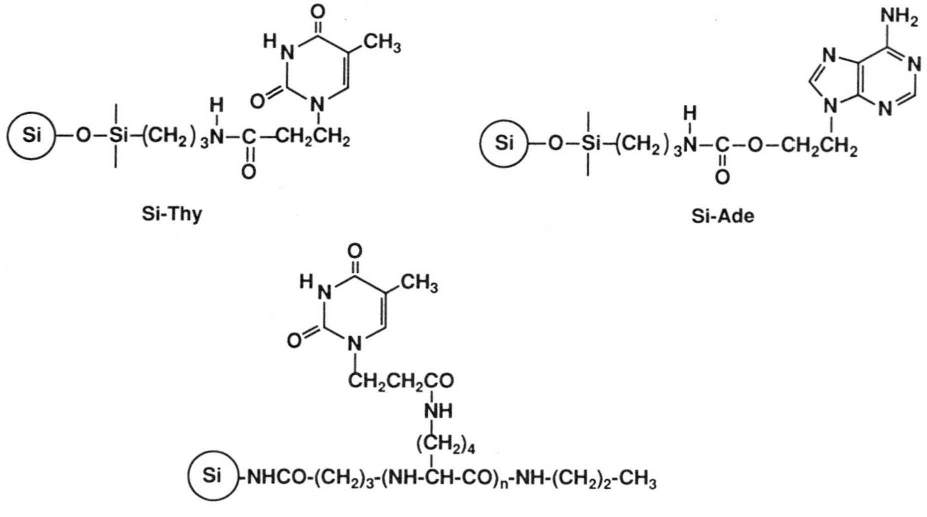

Figure 5. Nucleic acid base immobilized silica gels

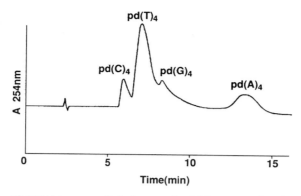

Figure 6. Chromatogram of tetranucleotides on Si-Ade. Mobile phase 0.1M triethylammonium acetate in 10% aqueous methanol. Flow Rate: 1 mL/min. Temperature: 35°C

however, water or aqueous methanol mobile phase should be used because the compounds are insoluble in organic solvents. When 10% methanol aqueous solution was used as a mobile phase, however, the separation was not complementary, and elution of the complementary thymidine oligomer was fast, and the most slow elution was the hydrophobic adenosine oligomer (Figure 6). This fact may explain by a hydrophobic interaction between the stationary phase and the solute.

To increase hydrophilicity of the resin, nucleosides are immobilized on silica gel in this paper. In the case using 10% methanol aqueous solution as a mobile phase, elution of the complementary thymidine oligomer is the last, thus the base specific separation is observed.

EXPERIMENTAL

1. SAMPLES

Nucleosides and nucleotides, and oligonucleotides were commercially available reagent grade.

2. DEOXYADENOSINE HAVING SPACER GROUP (2)

To the solution of the protected deoxyadenosine (1) (9.06 g, 14.9 mmol) in tetrahydrofuran (20 mL), 1,1'-carbonyldiimidazole (3.62 g, 22.3 mmol) was added at 0°C, and stirred for 2 hours.[11] The completion of the reaction was confirmed by appearance of the absorbance at 1760 cm^{-1} of activated carbonyl group in IR spectra. The solvent was removed by evaporation, and the residue was dissolved in chloroform and was washed with water to remove imidazole. After evaporation of chloroform, the imidazolide was obtained as solid (7.63 g, 67.2%).

To the solution of the 3'-O-imidazolide of the protected deoxyadenosine (7.63 g, 10 mmol) in tetrahydrofuran (100 mL), 2-aminoethoxyethanol (1.58 g, 15 mmol) was added at 0°C and the mixture was stirred for 3 hours. After the reaction, the solvent was removed under reduced pressure to give a light tan residue. The solid was dissolved in ethyl acetate, and insoluble part was removed by filtration. The solution was washed with water to remove unreacted 2-aminoethoxyethanol and imidazole. After evaporation of ethyl acetate, spacer bound deoxyadenosine (2) was obtained.

3. SILANE DERIVATIVE OF DEOXYADENOSINE (3)

The compound obtained above (2) (1.48 g, 2 mmol) was dissolved in tetrahydrofuran (30 mL), and was added 1,1'-carbonyldiimidazole (0.48 g, 3 mmol) at room temperature and the mixture was stirred for 24 hours. After reaction, the solvent was removed by evaporation, and the obtained residue was dissolved in chloroform and washed with water. The chloroform solution was dried by magnesium sulfate, and the solvent was removed by evaporation to give the imidazole (1.34 g, 83%).

To the obtained imidazolide (1.34 g, 1.66 mmol) in N,N'-dimethylformamide (30 mL), 3-aminopropyltriethoxysilane (0.443 g, 2.00 mmol) was added at room temperature and was stirred for 2 hours. The solvent was removed under reduced pressure, the residue was washed with ether, and the excess 3-aminopropyl-triethoxysilane and imidazole were removed under reduced pressure at 80°C to give the triethoxysilane derivative (3).

4. PROTECTED DEOXYADENOSINE IMMOBILIZED SILICA GEL (4)

To the solution of the triethoxysilane compound (3) in toluene (30 mL), dry silica gel (5 g; Cosmosil 5SL, particle size: 5 μm, pore size: 120 Å, surface area: 330 m^2/g) was added, and was kept stand for 3 days at 80°C. The silica gel was filtered and washed with acetone, methanol, ethanol, acetone, and diethyl ether in this order. Endcapping of the unreacted silanol group was carried out with hexamethyldisilazane (1 mL) for 48 hours at 60°C. The content of the deoxyadenosine derivative in the obtained silica gel derivative (4) was 0.3 mmol/g silica gel.

5. REMOVAL OF THE PROTECTING GROUP

Acetic acid aqueous solution (0.5%) was added to the silica gel (4), and was kept standing for 8 hours to give the deprotected deoxyadenosine immobilized silica gel (5) (Si-DMTr-O-dAdo). A similar reaction with 80% acetic acid aqueous solution gave the fully deprotected deoxyadenosine immobilized silica gel (6) (Si-dAdo).

6. CHROMATOGRAPHIC PROCEDURE

HPLC resins thus obtained were packed into stainless steel tube (150 x 4.6 mm i.d.). The experiments were performed with a Toyo Soda HPLC: CCP & 8000 equipped with a thermostated water bath and a UV detector operating at 254 nm.

RESULTS AND DISCUSSION

1. PREPARATIONS OF HPLC RESINS

The nucleoside bonded silica gel (4) were prepared by the reactions of the protected nucleoside (2) with 3-aminopropyl triethoxysilane,

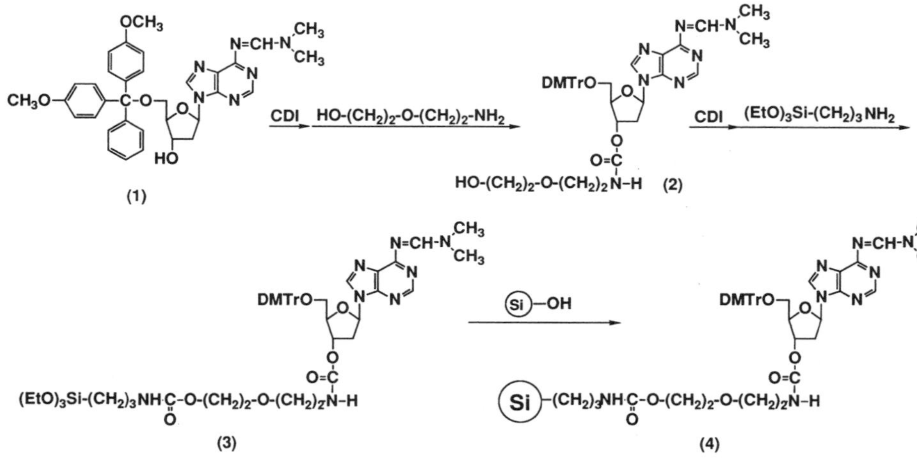

Scheme 1

followed by the direct reaction with silica gel, and free silanol groups were end-capped by trimethylsilyl groups (Scheme 1).

The amino blocking groups of Si–DMTrO–dmm–dAdo (4) was removed by 0.5% acetic acid aqueous solution to give Si–DMTrO–dAdo (5). The free deoxyadenosine bonded silica gel (6) (Si–dAdo) was obtained by deprotection with 80% acetic acid aqueous solution of Si–DMTrO–dmm–dAdo (Scheme 2). Bonding of the adenosine derivatives on silica gel was confirmed by IR spectra and elementary analysis to be 0.3 mmol/g silica gel.

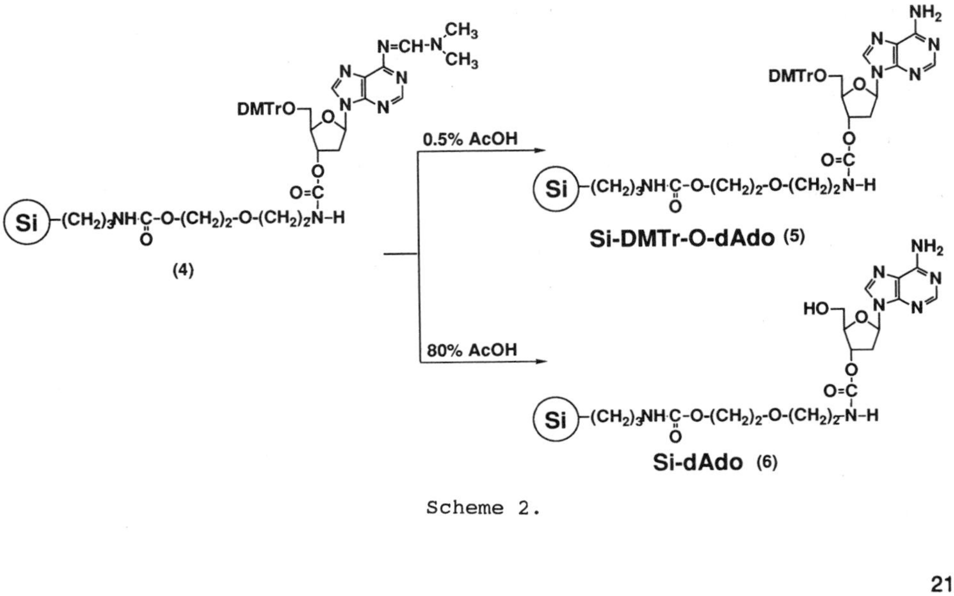

Scheme 2.

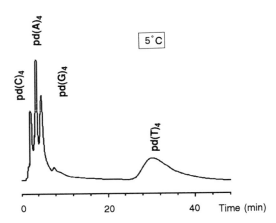

Figure 7. Chromatogram of pd(T)$_4$, pd(G)$_4$, pd(A)$_4$, and pd(C)$_4$ mixture. Column: Si-DMTrO-dAdo. Mobile phase: 10% methanol-phosphate buffer (pH 7), 1.0 mL/min. Temperature: 5°C. Detection: UV absorbance at 254 nm.

2. BASE SPECIFIC SEPARATION OF OLIGONUCLEOTIDES

Specific separation of pd(T)$_4$ was observed on Si-DMTrO-dAdo (5), as shown in Figure 7. The chromatogram was shown for the mixture of tetranucleotides with 10% methanol-phosphate buffer aqueous mobile phase (pH 7) at 5°C. The longest retention time was observed for the complementary pd(T)$_4$, and was separated completely from the other tetranucleotides.

Thymine base in nucleic acid is known to form base pair with the complementary adenine base by hydrogen bonds. The separation above, therefore, may be caused by the complementary base pair formation between thymine and adenine.

3. EFFECT OF TEMPERATURE

Figure 8 shows chromatogram at 30°C for 4 kinds of tetranucleotides using the same silica gel. At this temperature, the slowest elution was the complementary pd(T)$_4$, while the retention time decreased compared with the data at 5°C.

The retention times of the four tetranucleotides were plotted against the column temperature (Figure 9). The longest retention time was observed for the complementary pd(T)$_4$, and increased with decrease in temperature. At 5°C, separation of the thymidine oligomer was complete. The most hydrophobic pd(A)$_4$ did not show long retention time and showed small temperature dependency. These facts suggested that the hydrophobic interaction in this system was negligible, and the main separation factor was the base pairing between complementary nucleic acid bases.

The interactions in DNA are known to be temperature dependent. In the profile of UV absorbance against temperature, called the "melting curve," the UV absorbance increases when the temperature is increased, reflecting a lessened interaction between the bases in DNA. In the present paper, the temperature dependency of HPLC is observed. The complementary separation of oligothymidine on the adenosine bonded column

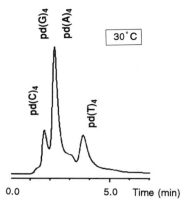

30°C

0.0   5.0   Time (min)

Figure 8. Chromatogram of pd(T)$_4$, pd(G)$_4$, pd(A)$_4$ and pd(C)$_4$
mixture. Column: Si-DMTrO-dAdo. Mobile phase: 10%
methanol-phosphate buffer (pH 7), 1.0 mL/min.
Temperature: 30°C. Detection: UV absorbance at 254
nm.

was excellent at low temperature, but the noncomplementary oligonucleotides had small temperature dependency. The significant temperature
effect for pd(T)$_4$ may also suggest that the separation of pd(T)$_4$ from
other tetranucleotides by the adenosine bonded silica gel was caused by
the specific base pairing between thymine and adenine.

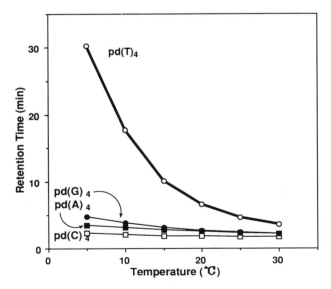

Figure 9. Temperature dependency of retention times for
tetranucleotides by Si-DMTr-O-dAdo.

4. THERMODYNAMIC CONSIDERATION (VAN'T HOFF PLOTS)

For the interaction between adenine in silica gel and thymine in solution (Equation 1), the following equilibrium constant can be assumed as Equation 2. where A is adenosine in the stationary phase (molar), T is thymine compound in solution, and A=T is the complex between adenine and thymine compound.

Silica gel   Solute   Complex

A   **T**   **A=T**

(1)

(Equation 1)

$$K = [A=T]/[A][T]$$ (Equation 2)

The temperature dependency of the equilibrium constant gives enthalpy (ΔH) and entropy (ΔS) differences according to the van't Hoff Equation 3.

$$R\ln K = -\Delta H/T + \Delta S$$ (Equation 3)

On the other hand, capacity factor k' for the silica gel in the HPLC system can be obtained from the HPLC data (Equation 4), where t_r is the retention time of the sample, and t_o is the unretained time.

$$k' = (t_r - t_o)/t_o$$ (Equation 4)

Since the capacity factor is the ratio of the solute in the stationary phase to the solute in mobile phase, the capacity factor relates to the equilibrium constant for base pairing, as shown in Equation 5, where V_s is the volume of stationary phase and V_m is the volume of the mobile phase.

$$k' = K(V_s/V_m)[A]$$ (Equation 5)

Therefore, Equations 3 and 5 combine to give Equation 6,[12,13] where ß is defined by Equation 7.

$$R\ln(k') = -\Delta H/T + \Delta S + ß$$ (Equation 6)

$$ß = R\ln\{(V_s/V_m)[A]\}$$ (Equation 7)

Figure 10 shows the van't Hoff plots for tetranucleotides on Si-

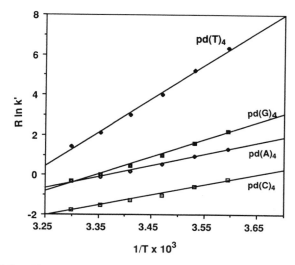

Figure 10. The van't Hoff plots of Rln(k') vs. 1/T for pd(T)$_4$, pd(G)$_4$, pd(A)$_4$ and pd(C)$_4$. Column: Si-DMTrO-dAdo.

DMTrO-dAdo using Equation 6. Straight lines were obtained for 4 kinds of tetranucleotides. The $-\Delta H$ value for pd(T)$_4$ was 16.8 kcal/mol, and the $-\Delta H$ values for other tetranucleotides were about 5 kcal/mol. The intercept of these lines gave the apparent ΔS values that contained constant value of ß for this column. The obtained values are tabulated in Table 1.

Figures 11 and 12 show the van't Hoff plots for nucleosides and nucleotides on Si-DMTrO-dAdo, respectively. Straight lines were obtained, and the $-\Delta H$ values were almost similar (2 - 5 kcal/mol) (Table 1). These data suggest that significant separation by base pairing was not observed

Table 1. Enthalpy and entropy differences of solutes determined from van't Hoff plots of ln(k') vs. 1/T.

Resins	Si-DMTrO-d-Ado		Si-d-Ado	
Sample	$-\Delta H$ kcal/mol	$\Delta S + ß$ cal/mol·deg	$-\Delta H$ kcal/mol	$\Delta S + ß$ cal/mol·deg
Cyd	2.1	$-$ 7.4	1.5	$-$ 5.0
Urd	3.0	$-$ 9.9	1.8	$-$ 5.7
Guo	4.7	-14.7	3.6	-11.0
Ado	5.3	-15.6	4.6	-13.5
5'-CMP	1.4	$-$ 6.2	1.5	$-$ 5.9
5'-UMP	1.8	$-$ 7.3	1.5	$-$ 5.8
5'-GMP	3.4	-12.2	2.9	-10.0
5'-AMP	4.2	-14.2	3.9	-12.8
pd(C)$_4$	5.2	-18.8	5.2	-18.1
pd(T)$_4$	16.8	-54.2	12.5	-41.1
pd(G)$_4$	8.5	-28.5	6.2	-20.9
pd(A)$_4$	5.7	-19.1	5.5	-18.2

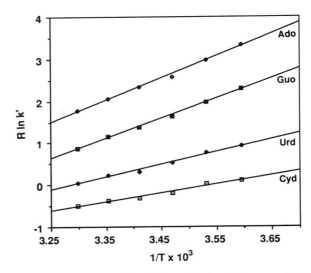

Figure 11. The van't Hoff plots of Rln(k') vs. 1/T for nucleosides. Column: Si-DMTrO-dAdo.

for nucleosides and nucleotides. Multiple interaction sites are necessary
for the base specific separation.

5. RELATIONSHIP BETWEEN $-\Delta H$ AND ΔS

The enthalpy values were plotted against the apparent entropy values
($\Delta S + \beta$) in Figure 13. Straight lines were obtained from all the data.
The observed isokinetic relationships suggest that the separation factor
is the same for all the data. It is difficult to obtain the absolute
value of entropy because the value of $(V_s/V_m)[A]$ is not known. However,
relative entropy value can be obtained for the same column under the same
chromatographic condition. Therefore, the enthalpy and entropy
differences from nucleotide (UMP) to tetranucleotide ($pd(T)_4$) were
obtained as 15.0 kcal/mol and -46.9 cal/mol·deg, respectively. These
values give the relative enthalpy and entropy values as 5.0 kcal/mol and
-15.6 cal/mol·deg for unit base pair.

6. EFFECT OF PROTECTING GROUP

An interesting fact was obtained for the effect of protecting group
at the 5' position of deoxyadenosine. Removal of the protecting group did
not cause increase in the retention time, but decrease in the retention
time. The retention time of $pd(T)_4$ on the deoxyadenosine immobilized
silica gel (Si-dAdo) was shorter than that on Si-DMTrO-dAdo (Figure 14).
In the case of Si-dAdo, however, the last retention was $pd(T)_4$, suggesting base specific separation. This fact may relate to the conformation of
the nucleoside where the *anti* conformation is predominant for the protected adenosine and is favorable to interaction with the thymine compound (Figure 15).

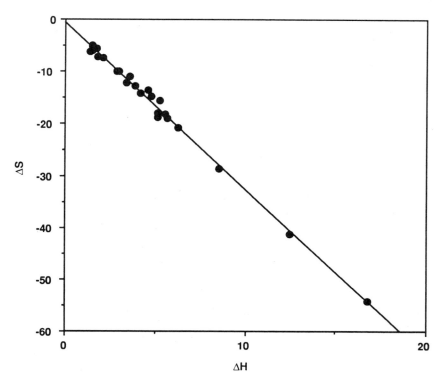

Figure 12. The van't Hoff plots of Rln(k') vs. 1/T for nucleotides. Column: Si-DMTrO-dAdo.

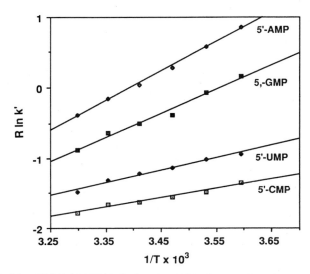

Figure 13. Relationship between enthalpy and entropy values.

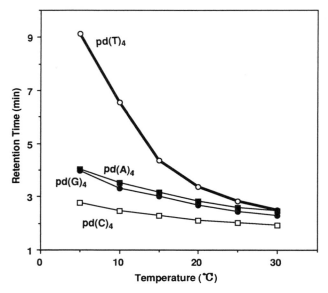

Figure 14. Temperature dependency of retention times for tetranucleotides by Si-dAdo.

Figure 16 shows the van't Hoff plots for the separation of tetranucleotides on Si-dAdo with 10% methanol-phosphate buffer mobile phase. Tetranucleotide containing complementary thymine base gave higher $-\Delta H$ values compared with other tetranucleotides. These results also indicate that the separations of pd(T)$_4$ on the adenosine bonded silica gels are caused by the base pairing between complementary nucleic acid bases.

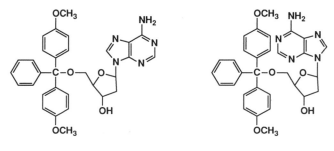

anti conformation   **syn conformation**

Figure 15. *Anti* and *syn* conformation of the protected deoxyadenosine.

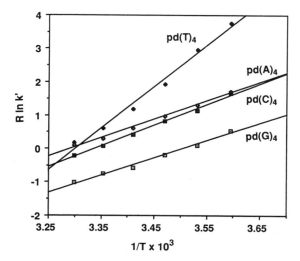

Figure 16. The van't Hoff plots of Rln(k') vs. 1/T for pd(T)$_4$, pd(G)$_4$, pd(A)$_4$, and pd(C)$_4$. Column : Si-dAdo. Mobile phase: 10% methanol-phosphate buffer (pH 7), 1.0 mL/min.

For nucleosides and nucleotides, base specific separation was not observed using Si-dAdo column (Figures 17 and 18).

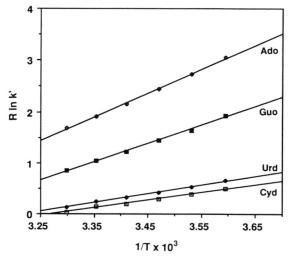

Figure 17. The van't Hoff plots of Rln(k') vs. 1/T for nucleosides. Column: Si-dAdo. Mobile phase: 10% methanol-phosphate buffer (pH 7), 1.0 mL/min.

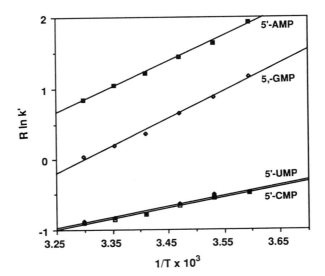

Figure 18. The van't Hoff plots of Rln(k') vs. 1/T for nucleotides. Column: Si-dAdo. Mobile phase: 10%
methanol-phosphate buffer (pH 7), 1.0 mL/min.

Base specific separation of oligonucleotides was also observed for
cytidine bonded silica gel, and guanosine bonded silica gel (Figure 19).
These data should also be reported in the proceeding paper.

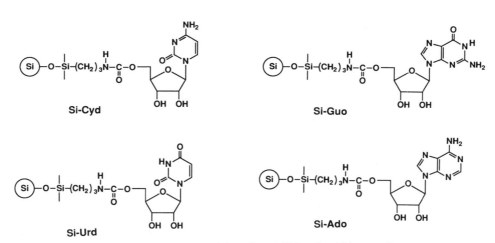

Si-Cyd

Si-Guo

Si-Urd

Si-Ado

Figure 19. Nucleosides immobilized silica gels.

CONCLUSION

The adenosine bonded silica gels were able to use as HPLC resins for specific separation of thymidine oligomer, $pd(T)_n$, in 10% methanol-phosphate buffer aqueous mobile phase. Using the deoxyadenosine bonded silica gel, complementary tetranucleotides of thymidine was completely separated from other tetranucleotides. This was caused by specific base pairing between adenine and thymine. The effect of temperature was significant on the base specific separation, and the blocking groups on sugar did not affect the separation. These resins may be useful for separation of components of nucleic acids and polynucleotides as a specific separation system.

REFERENCES

1. P. R. Brown, *"High Pressure Liquid Chromatography,"* Academic Press, 1973.
2. W. S. Hancock, Ed., *"High Performance Liquid Chromatography in Biotechnology,"* John Wiley & Sons, 1990.
3. Y. Inaki, S. Nagae, T. Miyamoto, Y. Sugiura, and K. Takamoto, Poly. Mater. Sci. Eng., **57**, 286 (1987).
4. Y. Inaki and K. Takemoto, Nucleic Acids Res., Sym. Ser., **19**, 45 (1988).
5. Y. Inaki, Y. Sugiura, T. Miyamoto, H. Hojho, S. Nagae, and K. Takemoto, Nucleic Acids Res., Sym. Ser., **20**, 59 (1988).
6. S. Nagae, T. Miyamoto, Y. Inaki, and K. Takemoto, Anal. Sci., **4**, 575 (1988).
7. Y. Inaki, S. Nagae, T. Miyamoto, Y. Sugiura, and K. Takemoto, in: *"Applied Bioactive Polymeric System,"* C. G. Gebelein, C. E. Carraher, Jr., and V. R. Foster, Eds., Plenum Publishing, p. 185 1988.
8. S. Nagae, T. Miyamoto, Y. Inaki, and K. Takemoto, Polymer J., **21**, 19 (1989).
9. S. Nagae, Y. Suda, Y. Inaki, and K. Takemoto, J. Polym. Sci. Part A, Polym. Chem., **27**, 2593 (1989).
10. S. Nagae, Y. Inaki, and K. Takemoto, Polymer J., **21**, 425 (1989).
11. A. Holy, *"Synthetic Procedures in Nucleic Acid Chemistry,"* W. W. Zorbach and R. S. Tipson, Eds., Interscience Publisher, p. 172 (1968).
12. Y. Baba, J. Chromatography, **485**, 143 (1989).
13. B. Feibush, A. Figueroa, R. Charles, K. D. Onan, P. Feibush, and B. L. Karger, J. Am. Chem. Soc., **108**, 3310 (1986).

ADVANCES IN ANTIMICROBIAL POLYMERS AND MATERIALS

Tyrone L. Vigo

USDA, ARS, SRRC
1100 R. E. Lee Blvd
New Orleans, LA 70124

Significant advances in the last three decades are reviewed for the synthesis and application of polymers and fibers to prevent microbial attack and material degradation. Diverse end uses for polymers and fibrous materials can be subdivided into aesthetic, biomedical and preservation of materials. During this period, the concepts and techniques of controlled release, permanent biobarrier and microencapsulation were formulated. This led to the discovery of several new polymers with bioactive groups and resultant materials with resistance to various microorganisms. Methods were developed for the early detection of microbial infiltration on surfaces. Factors that promote adhesion to and persistence of bacteria and fungi on surfaces and polymers are discussed. Such adhesion is based on fundamental interactions between microbes and polymer surfaces. Similar modes of attachment of viruses occur via adsorption on polymer surfaces. A prospective is also given on how related disciplines and newly mandated health regulations for protective clothing may lead to further advances in the understanding and development of new and effective antimicrobial polymers and fibers.

INTRODUCTION

Significant advances in the past three decades have been made in the synthesis and application of polymers to prevent microbial attack and degradation for diverse end uses (aesthetic, biomedical and preservation of materials). The concepts and techniques of controlled release, permanent biobarrier, microporous coatings and microencapsulation were discovered and initially developed in the 1960s and 1970s. This also led to the discovery of a variety of new polymers with bioactive functional groups. More current and novel approaches utilize hydrogels with active antimicrobial agents, fibers with different pore size to selectively diffuse agents, and crosslinked polymers with antimicrobial activity and several other improved functional properties. Methods of early detection of infiltration of surfaces and polymers by microbes are also discussed as well as the current knowledge on factors that contribute to the adhesion and persistence of microorganisms on various types of polymeric and fibrous surfaces. A prospective is also given on how other disciplines and application of fundamental scientific principles as well as new regulations on the use of protective clothing against biohazards, may lead to further advances in the understanding and development of antimicrobial polymers.

MODES OF ANTIMICROBIAL ACTIVITY AND PROTECTION

A landmark paper by Gagliardi described the principles and strategies for imparting antibacterial activity to fibrous materials.[1] Several important concepts were formalized and advanced in this paper. The first concept was the importance of measuring the microbiostatic and microbiocidal activity of the fiber directly rather than relying on tests (such as the agar plate) that are diffusion-controlled. The second important concept was the description of an effective mode of antimicrobial activity on the part of the agent, namely controlled release of the active agent migrating from the fiber to the microorganism. The third concept was the regeneration principle; this concept stated that the proper choice of a chemical species forming a metastable bond with a polymeric substrate could provide a continual reservoir of antimicrobial agents activated by hydrolysis, ultraviolet light or other methods. Such concepts led to the first quantitative test (Quinn test) that directly measured the antibacterial and antifungal activity of agents on a fiber,[2] and the first controlled-release, broad-spectrum antimicrobial fiber.[3] The latter was achieved by slow release of 5-nitrofurylacrolein from hydrolysis of its acetal with poly (vinyl alcohol) fiber (Figure 1). Microencapsulation of bactericides into fibrous substrates and materials approached the goals of the "regeneration principle" by employing a substantial reservoir of

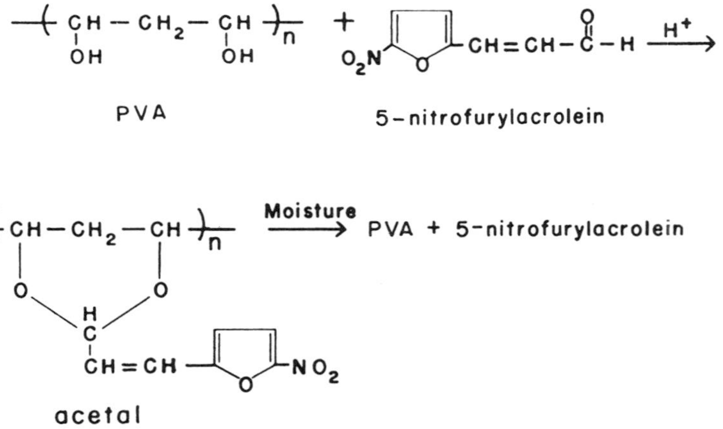

PVA   5-nitrofurylacrolein

acetal

Figure 1. Preparation of the broad-spectrum antimicrobial fiber Letilan. Poly (vinyl alcohol) fibers are converted to an acetal by their reaction with 5-nitrofurylacrolein in the presence of an acid catalyst. Antimicrobial activity is produced by slow release of the nitro compound in the presence of moisture. (From ref. 4).

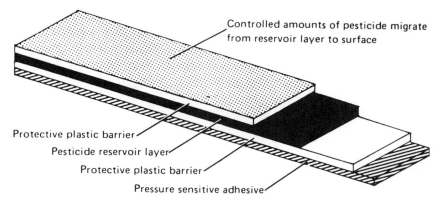

Figure 2. Microencapsulation technique for controlled release of pesticides from a polymeric surface. (From reference 4).

antimicrobial agent in a laminated structure such as that shown in Figure 2.[5] The concept of controlled release has been further refined by producing acrylic fibers of various pore sizes that can selectively release active antimicrobial agents.[6] This latter approach has several advantages: smaller amounts of active agent are required, localized application is possible, chemicals that are too toxic to humans may be applied in a discrete, controlled manner, and a wide variety of antimicrobial agents may be used that do not depend on a chemical reaction for their bonding or effectiveness.

There are two types of biobarriers that have been produced to protect materials against infiltration by microorganisms. The first is an inherently biostatic surface that does not support microbial growth; it is a coating comprised of a silicon polymer containing a pendant bioactive quaternary ammonium group.[7] However, other investigations have not been able to substantiate such claims for biocidal activity of this silicone coating, and the test methods for evaluating the antibacterial activity of the coating have also been called into question.[8] The second biobarrier is a physical barrier that was initially produced by waterproofing fabrics to make them resistant to bacteria contained in body fluids. This technology was further advanced with the advent of microporous breathable coatings (such as GoreTex^{TM}) that allow water vapor to be released but liquids not to penetrate. The concept of physical biobarriers against liquids containing microorganisms has been critically reviewed.[9] This approach has currently assumed great importance, since OSHA has recently mandated use of protective clothing for health care workers and many other personnel exposed to blood and other body fluids that contain the insidious hepatitis and HIV viruses.[10]

BIOACTIVE GROUPS IN ANTIMICROBIAL POLYMERS

Chemical agents that are effective microbiocides have been known since ancient times. Such agents include the halogens, sulfur, silver and mercury salts and more recently hypochlorite, peroxide and various phenols. However, it was only in the last three to four decades that these functional groups were incorporated into polymers and fibrous materials in such a manner as to cause the latter to function as antimicrobial agents. Table 1 shows representative examples of polymers with bioactive groups that have been discovered and used over the past 20 years.

Table 1. Examples of bioactive antimicrobial polymers.

Polymer	Bioactive group	Application	Ref.
Polyamide	Mercurated allyl triazines	Antimycotic	11
Biguanides	Biguanide	Biocides	12
Cellulose & polyester	Zinc peroxide polymer	Aesthetic & biomedical	13
Plastics & fibers	Fluorescein dyes with organotin groups	Mildew-proofing & antibacterial	14
Copolymer of salicylic acid/ thiourea/trioxane	Various metal complexes	Antibacterial & antifungal	15
Synthetic zeolites	Ag^+, Hg^{++}, or Zn^{++}	Aesthetic & biomedical	16
Polyester	Derivatives of nalidixic acid	Vascular grafts & catheters	17
Polyvinyl pyrrolidone	Gossypol complexes	Antiviral agents	18
Cotton cellulose	Phthalocyanine derivs.	Socks	19
Poly(vinyl) chloride	Imides and imidazole	Air filters	20
Aramids	Bisphenoxyarsines	Outdoor use	21
Synthetic hydrophobic polymers	Blends with hydrogels	Wound dressings & implants	22
Synthetic fibers	Phosphate esters	Carpets	23
Polyamide	Silver coating activated by direct current	Sutures	24
Microporous acrylic fibers	Chlorohexidine diacetate or PVP-iodine complex	Sanitary products Incontinence Hospital gowns	6
All types of natural and synthetic fibers	Crosslinked poly-ethylene glycols derived from a cyclic urea-formaldehyde resins	Biomedical Hosiery Sportswear	25

The first example of a bioactive group is that of mercurated allyltriazines; these compounds were incorporated into nylon socks and hosiery and were effective in preventing dermatophytic foot infections (such as athlete's foot).[11] Polymeric biguanides, prepared by reaction of diamines with diamine salts of dicyanamide, were effective biocides against gram-negative bacteria.[12] The *in situ* formation of a zinc peroxide polymer on cellulosic and polyester fibers produced antibacterial activity in fabrics against representative gram-positive (such as *Staphylococcus aureus*) and gram-negative bacteria (such as *Escherichia coli*) for up to 50 launderings.[13] Polymers obtained by reaction of diphenyltin dichloride with fluorescein dyes exhibited broad spectrum antimicrobial activity in paints, fibers and plastics.[14] Metal complexes of o- and p-hydroxybenzoic acid copolymers exhibited antibacterial and antifungal activity.[15] Zeolites, when treated with silver, copper or mercury salts, were effective bactericides and fungicides.[16] Polyester vascular grafts and catheters prevented *Pseudomonas* and *Staphylococcus* growth when they were complexed with derivatives of nalidixic acid.[17] When gossypol was complexed with poly (vinylpyrrolidone), it became less toxic and functioned as an effective agent against vesicular stomatitis virus.[18] Cotton socks containing phthalocyanine derivatives (such as octasodium cobalt phthalocyanine-octacarboxylate) were effective in controlling the growth of microorganisms and foot odor.[19] Woven and nonwoven air filters made of poly (vinyl chloride) reduced the growth of bacteria such as *Bacillus subtilis* when they contained a variety of imides or imidazoles such as N-(fluorodichloromethylthio) phthalimide.[20] Incorporation of bisphenoxy-arsines and phenarsazines into aramid fibers during melt spinning produce antimicrobial effects useful in fibers, films and carpets.[21]

The use of hydrogel composites in wound dressings and implants has been critically reviewed and represents a unique approach that has become increasingly important in the past decade.[22] The incorporation of an active antibacterial agent into a hydrophilic natural or synthetic polymer or blend provides a hydrogel structure and material that will prevent infection and promote wound healing. Representative hydrogel polymers used in this type of application are poly (2-hydroxyethyl methacrylate), block copolymers of polyethylene and polypropylene glycols, hydrocolloids of sodium carboxymethyl cellulose or gelatin dispersed in a elastomeric gum-like binder such as polyisobutylene and grafts of acrylamide onto polyurethane films. A variety of antibacterial agents can be incorporated into the polymer hydrogels. Some of the more commonly used agents are silver sulfadiazine, gentamicin, silver nitrate, antimicrobial nitrofurans and poly (vinyl pyrrolidone).[22]

When biocidal phosphate esters or their partially neutralized metal derivatives were combined with inert inorganic or organic particulate or polymer carriers, these compositions imparted broad spectrum anti-microbial activity to paints, plastics and fibrous materials. Modified surfaces with the desired antimicrobial properties may be achieved by appropriate types of microencapsulation; these antimicrobial surfaces are claimed to be effective even after prolonged washing and exposure to wear.[23]

The need for antimicrobial sutures has been demonstrated since bacterial growth is ten times faster than when such materials are not present in the body. Although sutures have been impregnated with a variety of antimicrobial agents (such as antibiotics, sulfa drugs, iodine and quaternary ammonium salts), these modified sutures normally do not exhibit broad spectrum activity against a diversity of infectious microorganisms and are effective for only several hours. In contrast, the antibacterial effectiveness of a silver-coated nylon suture has been demonstrated to have broad-spectrum antimicrobial activity that can be continually regenerated by application of a direct electric current.[24]

As mentioned earlier, the concept of controlled release of microbiocides has been refined by development of acrylic fibers that vary in pore size and thus vary in their rate of release of active antimicrobial agent from these pores.[6] Such fibers have been developed that contain antimicrobial agents such as ionic silver, chlorhexidine acetate and quaternary ammonium salts; applications are numerous and include incontinence and sanitary products, mattress covers, wound dressings, and hospital drapes and gowns.

The incorporation of crosslinked polyethylene glycols into a diverse group of fiber types and fabric constructions have produced materials with several improved functional properties. One of the more interesting new properties discovered is antibacterial activity against representative gram-positive and gram-negative bacteria. Because the modified fabrics or materials are thermally adaptable and have improved hydrophilicity, reduced particle release and resistance to static charge, they are currently being considered for clean room garments, hospital gowns, wound and burn dressings and other biomedical applications.[25]

PHYSICAL AND ANALYTICAL ASPECTS OF SURFACE/MICROBE INTERACTIONS

The detection of biodeterioration and infiltration by microorganisms on polymeric surfaces and fibers, their persistence or viable lifetime on these surfaces, and the manner in which they adhere to such surfaces are important considerations in producing the best antimicrobial materials used inside of and outside the body. While there are some differences in interactions of different classes of microorganisms (bacteria, fungi, algae, viruses) with polymeric surfaces, there are many similarities with regard to how microorganisms become attached to surfaces and persist on these surfaces.

Most of the earlier advances in detecting and classifying biodeterioration of polymeric materials were based on the numerous fundamental and tropical field studies conducted by the U. S. Army Quartermaster Corps during World War II.[26] In these studies, the morphological and taxonomic aspects of microorganisms were described that caused damage and deterioration to cellulosic fibers as well as the mechanisms of fiber degradation. From these studies, several relevant tests evolved, such as the current EPA humidity jar test that consists of inoculating a fabric with a test fungus, incubating the fabric, then inspecting it visually for fungal growth every week to a maximum of four weeks.[27]

Another conventional method for detecting biodeterioration and infiltration of all types of materials (natural and synthetic polymers and fibers) by microorganisms (bacteria as well as fungi) consists of the diffusion-controlled agar plate method in conjunction with scanning electron microscopy. The effectiveness of an antimicrobial treatment and/or the presence of microorganisms are determined by observing the growth around the sample in a nutrient such as agar; this test is rapid and qualitative but has the disadvantage that it is diffusion-controlled and measures only the ability of any agent to be released from the fiber. However, to directly measure the antibacterial and/or antifungal activity of a fiber, several quantitative tests have been developed.[28] These tests consist of sterilizing fabrics or fibrous materials in an autoclave or by chemical methods (e.g., ethylene oxide), inoculating the material with the appropriate test organism, incubating in an agar medium, then determining the antimicrobial activity of the fiber as % reduction of microbial colonies on the substrate using a low power microscope. While there are no tests specifically developed to determine virucidal activity of surfaces or detection of viruses, there have been several fundamental studies (described later) on the persistence of viruses on polymeric surfaces and factors that promote transfer of viruses from one surface to another.

For fabrics that must function as biobarriers to resist penetration of body fluids containing infectious microorganisms, several qualitative tests have been developed that measure the strike-through of microorganisms in solution by changes in hydrostatic pressure or by filtration efficiency of microbial aerosols through fabrics.[29] The protective clothing to prevent strike-through of body fluids containing these insidious viruses should lead to further refinements in these test methods.

A variety of rapid methods for early detection of microbial attack of substrates (particularly by fungi) have been recently described.[30] Most of these methods rely on chemical changes caused by enzymes released by the microbes. Such methods include: (a) determination of esterase activity with p-nitrophenyl acetate or by hydrolysis of fluorescein diacetate; (b) detection of ergosterol produced by certain classes of fungi; (c) production of ammonia by urealytic bacteria; and (d) bioluminescent assay of ATP with luciferin-luciferase mixtures.

Persistence of microorganisms on surfaces is an important aspect of microbial contamination and growth that has received some attention, but constitutes an area that requires further study. Most of the comprehensive studies on persistence time of disease-causing bacteria, fungi, viruses and other types of microorganisms on fibrous substrates were conducted over twenty years ago and have been critically reviewed.[28] The type of fibrous polymer has some influence on microbial persistence. A comparative study of the survival rate of *Streptococcus pyogenes* on swabs made from various unsterilized fibers indicated that only cotton and calcium alginate were effective in killing bacteria; all other major fiber types promoted growth of these bacteria.[31] It has also been determined that synthetic fibers generally retain odor-causing bacteria[32] (such as *Staphylococcus epidermidis*) and dermatophytic fungi[33] (such as *Trichophyton interdigitale*) to a greater extent than do natural fibers. More importantly, the persistence time of pathogenic bacteria (such as *Staphylococcus aureus*) and viruses (such as polio and vaccinia) on fibrous substrates are dependent on the relative humidity and the mode (solution or aerosol contact) by which the surfaces become contaminated. Aerosol contamination of fibrous surfaces that were held at low humidity (35%) was the most effective set of conditions conducive for microbial survival, while contamination by microorganisms in solutions of fabrics subsequently held at high humidity (78%) produced the shortest persistence time for microbes.[34,35] It was also shown that most types of fabrics were effective in transferring viruses under a variety of conditions unless they were laundered or treated with virucides. While there are no specific studies on the persistence time of the hepatitis and HIV viruses on fibrous substrates, current concerns dictate that reliable investigations should be conducted on this topic. However, it is known that inactivation of hepatitis viruses on any type of contaminated surface is much more difficult than corresponding inactivation of surfaces contaminated by the HIV virus.

As noted earlier, it has been demonstrated that sutures and other foreign objects in the body increase the tendency for undesirable microbial growth and infection.[24] However, the ability of microorganisms to adhere to polymeric surfaces inside the body and subsequently proliferate on that surface is the most crucial factor for the long term utility of *in vivo* biomedical devices and implants. Thus, the mechanisms and factors that promote attachment of microflora to fibrous and other types of polymeric substrates are of paramount importance in protection of personnel against biohazards and infectious diseases and effective use of implants and other biomedical devices inside the human body. If surfaces can be designed to minimize or alleviate adhesion of microbes, then the materials will be more effective in many situations and environments. Although there are several studies on how bacteria and other microorganisms adhere to polymeric (e.g., plastic) substrates, surprisingly little information and research have been conducted on this topic for fibrous substrates.

The mechanisms of bacterial adhesion and their physiological significance are discussed in a comprehensive monograph.[36] While there is no specific discussion of how bacteria and other deleterious microorganisms become attached to fibrous surfaces, there is a detailed discussion of how such attachment takes place on inert plastic prostheses placed in the human body (e.g., cardiac pacemakers and catheters)[37] and on the properties of nonbiological surfaces that are important for such adhesion.[38] Microorganisms such as bacteria and fungi adhere to all types of surfaces by a sequence of four processes shown in Figure 3.[39] Microorganisms approach surfaces by either active movement, diffusive transport, or convective transport. The initial steps of bacterial adhesion may be described by colloid chemical theories such as the DLVO theory depicted in Figure 4. Such theories describe changes in Gibbs energy as a function of the distances between two bodies (in this instance the microorganisms and the surface); the overall interaction is the sum of Van der Waals and electronic interactions. Ionic strength and distances of separation of the two bodies are the most important factors in predicting initial adhesion. At low ionic strength, the total interaction has a positive maximum that constitutes a barrier for adhesion in the primary minimum. Although DLVO theory is able to predict whether primary minimum adhesion can occur, it cannot quantify the depth of this minimum. Steric hindrance and/or polymer bridging may occur in addition to DLVO interactions in the presence of extended polymer molecules on the microbial cell or solid surface or both.[39] Such initial adhesion may be reversible in that when microorganisms exhibit Brownian motion, they can be readily removed by shear forces or adhesion may be irreversible when no Brownian motion occurs. The actual attachment step is achieved by the bacterium forming either fibrils or polymers to form strong links between the cell surface and the solid surface. The bacteria then form microcolonies or biofilms that adhere to the solid surface (Figure 3) forming a biofilm comprised of naturally-occurring polysaccharides and related structures. While there may be differences in the chemical nature of polymers of one bacterial species to another, biofilm formation is a general phenomenon whereby the microbes affix themselves to a surface. Adherence of the pathogenic bacterium *Staphylococcus aureus* to a cardiac pacemaker is shown in Figure 5. Once the microorganisms adhere to such a biomedical device, their removal normally proves to be quite difficult by injection of even the most potent antibiotics. Removal of the biomedical device is normally required to control microbial growth and infection. Thus, any design of the device and/or modification of its polymeric surface that would minimize bacterial adhesion is exceedingly desirable and clinically important.

The reason it is difficult to remove adhered microbes from a surface is their ability to form an embedded matrix with the substrate via a biofilm. This mode of adhesion gives then the capability of complexing or

232

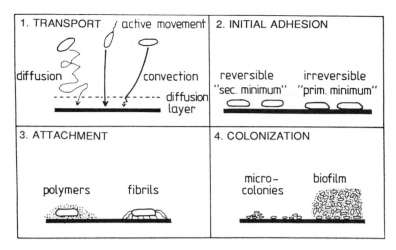

1. TRANSPORT / active movement	2. INITIAL ADHESION
diffusion	convection	diffusion layer	reversible "sec. minimum"	irreversible "prim. minimum"
3. ATTACHMENT	4. COLONIZATION
polymers	fibrils	micro-colonies	biofilm

Figure 3. Schematic representation of the sequencing steps in the colonization of surfaces by microorganisms (from ref. 39).

rendering ineffective antibiotics and/or antimicrobial chelating agents. The only effective methods for removing biofilms from surfaces once they are formed are (a) the use of oxidizing agents (such as hypochlorite) and (b) physical removal at low temperature via ice formation to destroy the physical integrity of the biofilm. Oxidizing agents are effective because they not only kill the bacteria, but also because the biofilm is depolymerized and thus becomes physically detached from the surface. Obviously, these two methods would normally require removal of the biomedical device or prosthesis for safety of the patient and maximum effectiveness of biofilm removal.

A fundamental study on the surface thermodynamics of bacterial adhesion of five different bacterial strains with five different polymeric surfaces was conducted.[40] It was determined that the extent of bacterial adhesion was determined by the surface properties of all three phases: surface tensions of adhering particles, of the suspending liquid medium, and of the polymeric substrate. Adhesion to hydrophobic surfaces was more excessive when the surface tension of the bacteria was larger than that of the liquid medium; this trend was reversed when the surface tension of the liquid medium was larger than that of the bacteria. A later study demonstrated that the degree of adhesion of 16 different bacterial strains on a solid sulfated polystyrene surface was proportional to their contact angles; i.e., the higher the contact angle of the bacterium, the greater its adhesion to the surface.[41] Adhesion of the yeast-like fungi *Candida albicans* to plastic surfaces appears to be promoted by the same surface phenomena that occur with adhesion of bacteria to such surfaces.[42]

The adhesion of a representative gram-positive bacterium (*S. aureus*) to six films (cellulose acetate, nylon 66, polyester, polyethylene, polypropylene and Teflon) and four fabrics (cotton, nylon, polyester and polypropylene) was measured after an initial hour contact of these surfaces with aqueous suspensions of the bacteria.[43] Although there was a good correlation between bacterial adhesion and increasing hydrophobicity of the film surfaces, no clear correlation was observed between the

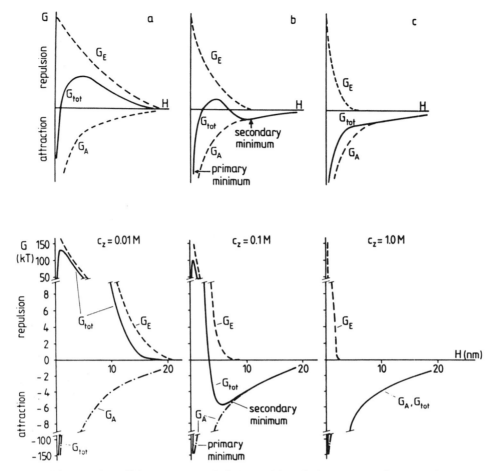

Figure 4. Gibbs energy of interaction between a sphere and a
flat surface having the same charge sign,
according to DLVO theory. (Upper) Schematic
representation for (a) low, (b) intermediate, and
(c) high ionic strength. (Lower) Calculated graphs
for a bacterium and a flat surface. Values taken:
Hamaker constant, 2×10^{-21} J; bacterial cell
radius, 0.5 μm; electrokinetic potential of
bacterium and surface, 15 and -20 mv,
respectively; ionic strength (C_z for 1-1
electrolyte) as indicated. GE, Electrostatic
interaction; GA, Van der Waals interaction; G_{tot},
total interaction; H, shortest separation distance
between the two surfaces. (From ref. 39).

surface wettability of fabrics and bacterial adhesion. Thus, further
studies are warranted to elucidate and characterize factors of fibrous
surfaces that promote or minimize adhesion by bacteria and other
microorganisms.

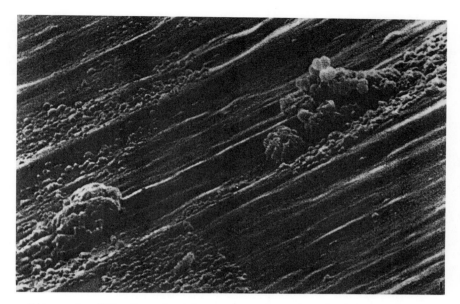

Figure 5. SEM of a plastic surface associated with a cardiac pacemaker that had been colonized by a bacterimic route by cells of *Staphylococcus aureus*. These coccoid bacteria can be seen within one of the microcolonies (lower) because their amorphous glycocalyx exopolymers had condensed during dehydration to reveal the enclosed bacteria cells. (From ref. 37).

Attachment of viruses to solid surfaces has been critically reviewed and occurs by adsorption rather than adhesion.[44] This mode of attachment occurs due to the small size of a representative virus relative to bacteria and other microorganisms. However, the DLVO theory advanced for initial adhesion of bacteria is also applicable to adsorption of viruses, since the latter exhibit colloidal behavior. The two most important factors that influence virus adsorption to a surface are isoelectric point and hydrophobicity; since viruses and surfaces differ widely in these parameters, specific studies are warranted to determine affinity of viruses for various fibrous and polymeric surfaces.

PROSPECTIVE

Advances in the discovery of new antimicrobial polymers and materials may be galvanized by current concerns about the spread of insidious viruses such as HIV and hepatitis and by contact with contaminated surfaces that lead to more stringent health and medical regulations to treat most materials with microbiocides. Modification of polymer and fiber surfaces by plasma and other high-energy sources should lead to the bonding of dissimilar surfaces and enhance control of diffusible agents to produce sophisticated and new antimicrobial polymers. Modification of polymeric and fibrous surfaces, and changing the porosity, wettability and other characteristics of the polymeric substrates, should produce implants and biomedical devices with greater resistance to microbial adhesion and biofilm formation. The greater acceptance and need for synthetic biomedical devices and artificial replacement parts for our bodies will also give incentive for the discovery and use of materials that are biocompatible and fight infection and antibody rejection. Colonization of space will require that materials be resistant to terrestrial and even extraterrestrial microorganisms.

REFERENCES

1. D. D. Gagliardi, Am. Dyest. Reptr., **51**, 49 (1962).
2. H. Quinn, Appl. Microbiol., **10**, 74 (1962).
3. L. S. Vol'f, A. I. Meos, V. V. Koteskii & S. Hillers, Khim. Volonka, **6**, 16 (1963).
4. T. L. Vigo in: *"Chemical Processing of Fibers and Fabrics, Functional Finishes, Pt. A,"* M. Lewin & S. B. Sello, Eds., Marcel Dekker, New York, 1983, Chapter 4, p. 367.
5. Anon., Chem. Week, **115**, 43 (Aug. 21, 1974).
6. C. D. Potter, J. Coated Fabrics, **18 (2)**, 259 (1989).
7. A. J. Isquith, E. A. Abbott & P. A. Walters, Appl. Microbiol., **24 (6)**, 859 (1972).
8. A. F. Turbak, private communication, Jan. 10, 1992.
9. C. E. Wood, J. Coated Fabrics, **19 (1)**, 143 (1990).
10. Occupational Safety and Health Administration, Fed. Regist., **56 (235)**, 64004 (Dec. 6, 1991).
11. L. Gip & B. Magnusson, Nord. Med., **82 (44)**, 1375 (1969).
12. A. J. Buckley, E. G. Gazzard, J. N. Greenshields, D. Pemberton, J. H. Wild & D. S. Leitch, Brit. Pat. 1,531,717 (Nov. 12, 1974).
13. G. F. Danna, T. L. Vigo & C. M. Welch, U. S. Pat. 4,199,322 (Apr. 22, 1980).
14. C. E. Carraher, Jr., R. S. Venkatchalam, T. O. Tiernan & M. L. Taylor, Org. Coat. Appl. Poly. Sci. Proc., **47**, 119 (1982).
15. M. M. Patel, R. Manavalan & K. Prabhakaran, Proc. Indian Acad. Sci. Chem. Sci., **92 (3)**, 271 (1983).
16. H. Zenji, O. Hideo, H. Shigetaka, N. Saburo, I. Shunya & T. Kenichi, Eur. Pat. Appl. EP 103,214 (Mar. 21, 1984).
17. C. J. Fox, Jr., S. Modak & K. Reemtama, U. S. Pat. 4,563,485 (Jan. 7, 1986).
18. M. G. Mukhamediev, S. A. Auelbekov, Z. T. Sharipova, T. M. Babaev, U. N. Musaev & K. A. Aslanov, Khim-Farm. Zh., **20 (4)**, 450 (1986).
19. E. Takahasi, T. Aihane, T. Shibagaki & H. Satomi. Jpn. Kokai Tokkyo Koho JP 62,247,816 (Oct. 28, 1987).
20. M. Kamimura, Jpn. Kokai Tokkyo Koho JP 63,112,701 (May 17, 1988).
21. W. L. Burton, U. S. Pat. 4,769,268 (Sept. 6, 1988).
22. P. H. Corkhill, C. J. Hamilton & B. J. Tighe, Biomaterials, **10 (1)**, 3 (1989).
23. R. H. McIntosh, Jr., A. F. Turbak & R. H. McIntosh, Sr., U. S. Pat. 4.908,209 (Mar. 13, 1990).
24. C. C. Chu in: *"High-Tech Fibrous Materials,"* T. L. Vigo and A. F. Turbak, Eds., Am. Chem. Soc., Washington, DC, ACS Symp. Ser. 457, 1991, Chapter 12, p. 167.
25. T. L. Vigo & J. S. Bruno, *"Proc. 26th Intersoc. Energy Conversion and Eng. Conf. Vol. 4,"* Boston, MA, 1991, p. 161.
26. R. H. Siu, "Microbial Decomposition of Cellulose", Reinhold, New York, 1951.
27. EPA, Pesticide Program: Guidelines for Registering Pesticides in the United States, Fed. Reg., **40 (123)**, 26802 (June 25, 1975).
28. T. L. Vigo & M. A. Benjaminson, Text. Res. J., **51**, 454 (1981).
29. R. E. Seaman, "6th INDA Tech. Symp. Proc.", INDA, Atlanta, GA, 1978, p. 66.

30. B. J. McCarthy, Int. Biodeterior. Bull., **19 (2)**, 53 (1983).
31. A. H. Dadd, V. P. Dagnall, P. H. Everall & A. C. Jones, J. Med. Microbiol., **3 (4)**, 561 (1970).
32. K. Gutfrend & A. Dravnicks, "*Odor Retention in Cotton and Synthetic Fabrics,*" Ill. Tech. Res. Inst. Rept. No. C 8224-6, Chicago IL, Jan. 1973.
33. L. Chalmers, Dragoco Rep., **11**, 215 (1972).
34. L. J. Wilkhoff, L. Westbrook & G. J. Dixon, Appl. Microbiol., **17 (2)**, 268 (1969).
35. G. J. Dixon, R. W. Sidwell & E. McNeil, Appl. Microbiol., **14 (2)**, 183 (1966).
36. D. C. Savage & M. Fletcher, Eds., "*Bacterial Adhesion: Mechanisms and Physiological Significance,*" Plenum Press, New York, 1985.
37. J. W. Costerton, T. J. Marrie, K.-J. Cheng in: "*Bacterial Adhesion: Mechanisms and Physiological Significance,*" D. C. Savage & M. Fletcher, Eds., Plenum Press, New York, 1985, Chapter 1, p. 3.
38. G. Loeb in: "*Bacterial Adhesion: Mechanisms and Physiological Significance,*" D. C. Savage & M. Fletcher, Eds., Plenum Press, New York, 1985, Chapter 5, p. 111.
39. M. C. M. van Loosdrect, J. Lyklema, W. Norde & A. J. B. Zehnder, Microbiol. Rev., **54 (1)**, 75 (1990).
40. D. R. Absolom, F. V. Lamberti, Z. Policova, W. Zingg, C. J. van Oss & W. Neumann, Appl. Envir. Microbiol., **46 (1)**, 90 (1983).
41. M. C. M. van Loosdrecht, J. Lyklema, W. Norde, G. Schraa & A. J. B. Zehnder, Appl. Envir. Microbiol., **53 (8)**, 1893 (1987).
42. R. A. Calderone & P. C. Braun, Microbiol. Rev., **55 (1)**, 1 (1991).
43. Y. -L. Hsieh & D. Timm, J. Colloid Interface Sci., **123 (1)**, 275 (1988).
44. C. P. Gerba in: "*Adv. Appl. Microbiol. 30,*" A. I. Laskin, Ed., Academic Press, New York, 1984, p. 133.

DEVELOPMENT OF POLYMERIC BETA-LACTAM SYSTEMS - SYNTHETIC STUDY AND

CHARACTERIZATION

Malay Ghosh*

Schering-Plough Research Institute
2000 Galloping Hill Road
Kenilworth, NJ 07033

Over the past 40 years beta-lactam derivatives have been extensively used by the physicians as one of the best antimicrobial agents. Beta-lactam chemistry is well developed, however, reports on polymers containing beta-lactam system is very rare in literature. This paper will describe our study on synthesis of new polymeric systems containing a variety of substituted beta-lactam moieties. Penicillin based polymeric systems are also prepared and will be reported. The characterization of the new monomers, polymers and the result of the *in vitro* release study will be described.

INTRODUCTION

The observation of Alexander Flemming,[1] in 1929, that a Penicillium mold has the ability to inhibit bacterial growth has resulted in intensive research activity among various groups of scientists all over the world. Chain, Florey and coworkers were successful in isolating the active compound from the mold as an impure brown powder in 1940.[2] The elucidation of chemical structure of the compound, however, took a longer time and finally was established by Robert Woodward.[3] The compound was named Penicillin [1] and is the first compound found in nature that contains a beta-lactam ring [2]. Beta-lactam systems are also known as 2-azetidinone. Penicillin and its derivatives show a remarkable difference between toxicity to bacterial cell wall and mammalian host and for that reason are much safer antibiotics.[4] In the late 1960s, another beta-lactam derivative was isolated from a marine source which was named cephalosporin [3].[5] Since then literally hundreds of new penicillin and cephalosporin derivatives were made and were tested for their biological activity. Another breakthrough came in late 1970 when several other structurally novel beta-lactam compounds were isolated from natural sources, and they are generally known as carbapenam [4].[6] Together with cephalosporin and carbapenams, the penicillin occupy the primary position

---

* Current address: Alcon Laboratories, Preformulation: Product Design
Group, 6201 South Freeway, Ft. Worth, TX 76134

Biotechnology and Bioactive Polymers, Edited by C. Gebelein
and C. Carraher, Plenum Press, New York, 1994

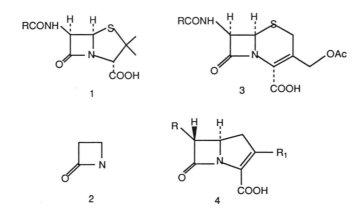

Figure 1. Chemical structures of penicillin, cephalosporin,
beta-lactam functionality and carbapenams.

in the field of antibacterial chemotheraphy. The structures are shown in
Figure 1.

     There have been very few investigations carried out on the
development of polymeric systems containing beta-lactam moieties. The
chemistry of beta-lactam is rich yet the unstable nature of this four
membered ring system is considered to be the main barrier to synthesizing
polymeric beta-lactam systems.[7-9] Azetidinone systems, in general, are
sensitive towards any nucleophile and/or electrophile.   Very often
chemical interaction with nucleophile/electrophile leads to the cleavage
of beta-lactam ring.   This is particularly true for bicyclic systems
containing beta-lactam moiety where opening of four membered ring will
reduce steric interactions.   One case in point is sodium salt of
ampicillin.   The stability of ampicillin are found to depend on pH,
temperature as well as interactions with other ingredients present in the

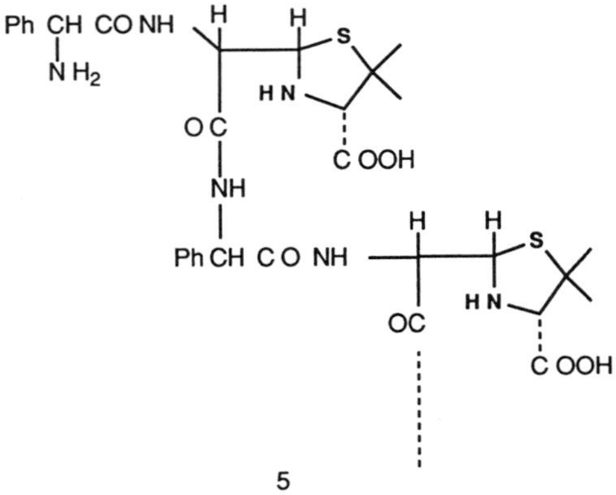

Figure 2. Polymer of ampicillin prepared by ring opening of
beta-lactam group.

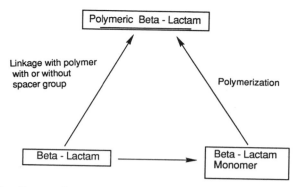

Scheme 1. General synthetic protocol to prepare polymeric
beta-lactam derivatives.

formulations. Polymer [5] (Figure 2) is formed from ampicillin when a
solution of ampicillin in water is kept at room temperature for a few
days.[10] The product is formed by opening of beta-lactam system followed
by condensation reaction with an amine, and thus devoid of any beta-
lactam system. As a part of our investigation in the area of beta-
lactams,[11,12] we have developed new polymers containing azetidinone
functionality. The present paper will describe the synthesis,
characterization and *in vitro* release of polymeric beta-lactams.

RESULTS AND DISCUSSION

Polymeric beta-lactams could be prepared either by step growth
polymerization reaction or by polycondensation reaction from appropriate
monomers, and the general synthetic protocol is shown in Scheme 1. Our
experience in the area of beta-lactam chemistry has been helpful to
develop novel azetidinones which are used as pivotal synthetic
intermediates to prepare polymeric azetidinones.[11-13]

The synthetic methodology adopted to prepare monomers are given in
Schemes 2 and 3 respectively. Thus the reaction of aromatic amines [6a]
or [6b] with methyl glyoxalate [7],[14] in presence of sodium sulfate or
molecular seive in methylene chloride at room temperature gave the imine
[8a] and [8b] as a mixture of syn and anti isomers. IR spectra of the
imines exhibit two sharp bands at 1635 and 1730 cm^{-1} which are
characteristics of imine and ester functionality respectively. Proton
NMR spectra shows the presence of two peak at 8.0 and 8.1 ppm indicating
the presence of syn and anti isomer. The imines are not very stable and
should be kept under nitrogen atmosphere. The appropriate beta-lactam
system was then constructed by using an acid chloride-imine cycloaddition
reaction.[7] Treatment of benzyloxyacetyl chloride with the imines
(mixture of syn and anti) in methylene chloride, in presence of
triethylamine at room temperature, gave the cis beta-lactam [9a] and [9b]
in approximately 75 % yield. The compounds were purified by column
chromatography.

The stereochemistry of the beta-lactam system has been established
from proton NMR data which shows the coupling constant of the C3 and C4
protons are 5 Hz ($J_{3,4}$ = 5 Hz). For a beta-lactam having a cis

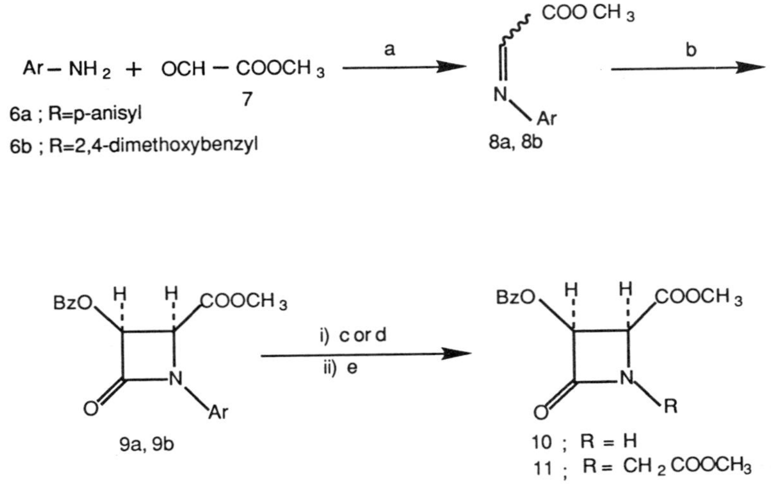

Ar−NH₂ is: Ar − NH$_2$ + OCH − COOCH$_3$

6a ; R=p-anisyl
6b ; R=2,4-dimethoxybenzyl

7

COO CH$_3$

N—Ar

8a, 8b

BzO H H COOCH$_3$

O N Ar

9a, 9b

i) c or d
ii) e

BzO H H COOCH$_3$

O N R

10 ; R = H
11 ; R = CH$_2$COOCH$_3$

a : Sodium sulfate or molecular seive in methylene chloride
b : Benzyloxyacetyl chloride, triethylamine at room temperature
c : Cerric (IV) ammonium nitrate
d : Potassium persulfate
e : Methyl bromoacetate, sodium hydride

Scheme 2. Synthetic route to prepare beta-lactam intermediates.

stereochemistry the coupling constant of the C3 and C4 protons ranges from 4 to 6 Hz whereas a trans substituted beta-lactam shows a coupling constant of 1-3 Hz. The IR spectra of the compounds showed the presence of sharp band in the region of 1760-1775 cm^{-1} which is clear indicative of the formation of beta-lactam ring. The chemical structure is also confirmed by CIMS and elemental analysis. Although the reason is not clear, it is important to note that the beta-lactam ring formation reaction is stereospecific in nature. No trace of trans beta-lactam was observed from proton NMR study of the crude reaction mixture. Reaction of [9a] with cerric (IV) ammonium nitrate in acetonitrile-water solvent system at room temperature resulted in the removal of para methoxy phenyl group on nitrogen, and the compound [10] is obtained in 58% yield. The progress of the reaction is monitored by thin layer chromatography. Purification of the compound is done by recrystallization from ethyl acetate-hexane solvent system.

Similarly, reaction of potassium persulfate with [9b] in buffered tetrahydrofuran-water gave the same product [10]. The yield in this case is 42%. N-alkylation of [10] was performed by reacting [10] with methyl bromoacetate in presence of suitable base for example NaH or Triton B. This is not a straight forward reaction and the yield of the alkylated product [11] ranges from 24 to 46% depending on reagent, solvent and the temperature of the reaction. We found that a combination of NaH, DMSO and 40°C gave the highest yield. The proton NMR data showed the presence of two distinct doublets corresponding to C3 and C4 protons of azetidinone system with a coupling constant of 5 Hz. This indicates the fact that no isomerization took place during the course of the reactions.

Although the overall yield of [11] is good via Scheme 2, the Scheme 3 involves lesser number of synthetic transformations to prepare the same

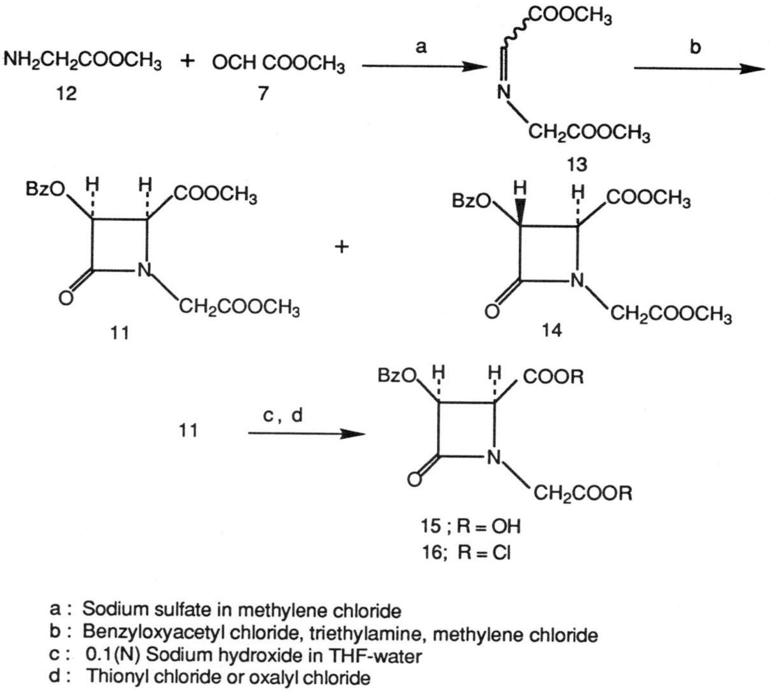

Scheme 3. Alternate synthetic pathway to prepare beta-lactam
intermediates.

a : Sodium sulfate in methylene chloride
b : Benzyloxyacetyl chloride, triethylamine, methylene chloride
c : 0.1(N) Sodium hydroxide in THF-water
d : Thionyl chloride or oxalyl chloride

intermediate [11]. In this route, commercially available glycine methyl
ester hydrochloride is neutralized with triethylamine to give glycine
methyl ester [12] which without separation was immediately reacted with
[7] to get the unstable imine [13] in 72% yield. Here also the imine was
obtained as a mixture of syn and anti isomers. The IR of the compound
exhibited bands corresponding to imine functionality at 1650 cm^{-1}.
Proton NMR of [13] shows peak at 7.8 and 8.2 ppm corresponding to the
proton attached to imine functional group. Reaction of benzyloxyacetyl
chloride with imine [13] gave the beta-lactam in 40% yield as a mixture
of cis [11] and trans [14] isomers. The relative ratio of cis and trans
compound is 4 : 1 which is established from the proton NMR data of the
crude reaction product. Cis-beta-lactam [11] was treated with decinormal
sodium hydroxide in either tetrahydrofuran-water or acetone-water solvent
system at room temperature to give beta-lactam diacid [15] in 76% yield.
The diacid is stable at room temperature and IR shows sharp bands at
1765 and 1700 cm^{-1}, characteristics of beta-lactam and acid
functionalities respectively. Addition of triethylamine causes the
disappearance of 1700 cm^{-1} band which confirms the presence of carboxyl
functional group. It is interesting to note that no epimerization of cis
ring system was observed during hydrolysis step. This is confirmed by
the coupling constant of C3-C4 protons (J = 5.2 Hz). Treatment of diacid
[15] with thionyl chloride or oxalyl chloride gave the desired monomer
[16] in very good yield. The monomer is crystallized from chloroform-
hexane solvent system under nitrogen atmosphere. IR spectra of the
compound shows two very sharp band at 1800 and 1760 cm^{-1} indicating the
presence of acid chloride and beta-lactam group. Finally the structure
is confirmed by proton NMR and CIMS results.

Table 1 lists yield and some of the important spectroscopic data of the intermediates synthesized. The monomer [16] was then reacted with aliphatic diamines [17] and [18] in DMF or NMP in presence of triethylamine or dimethylaminopyridine as acid acceptor (Scheme 4). The polymers [19] and [20] are obtained as grey solid in 79-88% yield. The purification of the polymers have been performed by reprecipitation from DMF-methanol. Attempted polymerization of [15] with the diamines by using reagents like dicyclohexyl carbodiimide, triphenylphosphine/ carbontetrabromide, posphorous oxytrichloride, however, was not successful. In all cases starting material was isolated without any degradation product. The prepared polymers are soluble in most of the organic solvents. Some of the spectroscopic characteristics of the polymers will be found in Table 2.

In general, penicillin and related systems are not stable in nature. Our initial attempt to polymerize 6-aminopenicillanic acid in presence of various coupling reagent was not successful and in all the cases cleavage of beta-lactam ring was found. We then focused our effort to develop monomers from penicillin. The synthetic pathway adopted for this purpose is given in Scheme 5. Penicillin G potassium salt [21] was transformed to thermodynamically more stable acid sulfoxide [22] which in turn was reacted with allyl bromide in presence of sodium hydride and 18-crown-6 to give the allyl ester of penicillin [23] in overall 40% yield.

The presence of 18-crown-6 is very important in this ester forming reaction. Proton NMR study shows the stereochemistry of C5-C6 protons (penicillin system) is cis since it has a coupling constant of 5.3 Hz. CIMS and elemental analysis also confirmed the structure. The polymerization reaction of the monomer [23] was then conducted by a free radical pathway using AIBN as free radical initiator and Dioxane or THF as solvent (Scheme 6). The polymer was obtained as light gray powder. The polymer was purified by reprecipitation technique. The IR spectra of the prepared polymer exhibits bands at 1785, 1730 and 1660 cm^{-1} which are due to the presence of beta-lactam, ester and amide functional groups respectively. Proton NMR study shows the presence of two doublets at 4.9 and 5.4 ppm having a coupling constant of 5.1 Hz. These signals are due to the presence of proton at 5 and 6 positions in the penicillin system.

Table 1. Yield and spectroscopic data of beta-lactam intermediates.

Compound	Yield (%)	IR (cm^{-1})	^{1}HNMR* (ppm)	Stereochemistry
8a	86	1730, 1635	8.0, 8.1	Syn, Anti (1:2)
8b	89	1740, 1635	7.9, 8.05	Syn, Anti (1:2)
9a	76	1770, 1740	4.3, 4.7	Cis only
9b	72	1775, 1735	4.5, 4.8	Cis only
10	69	1775, 1730	4.7, 5.0	Cis only
11	46	1770, 1740	4.6, 5.1	Cis only
13	70	1725, 1650	7.9, 8.2	Syn, Anti (1:3)
14	40	1765, 1730	4.6, 4.9	Trans
15	65	1775, 1705	4.8, 5.2	Cis only
16	62	1800, 1760	4.7, 5.1	Cis only

* The data in ^{1}H-NMR column are for C-3 and C-4 protons except for compounds 8 and 9. The stereochemistry is based on coupling constant of C3 and C4 protons of beta-lactam.

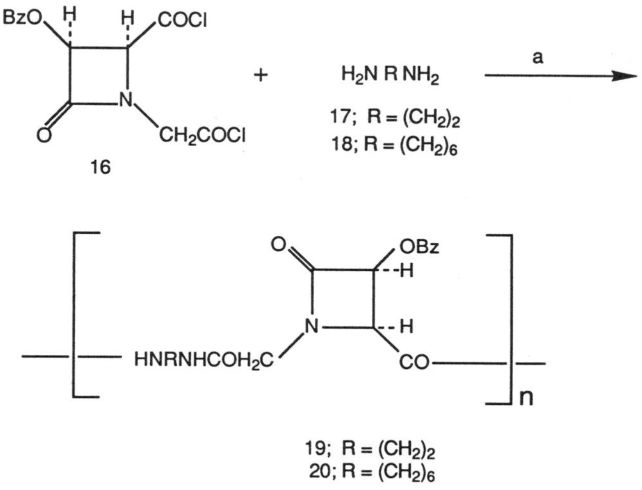

16

17; R = (CH$_2$)$_2$
18; R = (CH$_2$)$_6$

19; R = (CH$_2$)$_2$
20; R = (CH$_2$)$_6$

a : Triethylamine or Dimethylaminopyridine , DMF or NMP

Scheme 4. Synthesis of polymeric beta-lactam derivatives.

A sharp singlet at 4.5 ppm is indicative of the presence of C3 proton of penicillin. All the data suggested the chemical structure of polymer as [24] . This is also confirmed by elemental analysis.

The *in vitro* release of penicillin sulfoxide from monomer [23] and polymer [24] in deionized water have been performed at 37°C, and the result is shown in Figure 3. As expected the releases the drug is much faster from monomer than the polymer.

Table 2. Yield and spectroscopic data of polymeric beta-lactams.

Compound	Yield[1] (%)	IR[2] (cm^{-1})	^{1}H-NMR[3] (ppm)	Solubility
19	52	1755, 1650	4.2, 4.9	DMF, THF, DMSO NMP, Dioxane, EA
20	40	1765, 1645	4.4, 4.85	DMF, NMP, DMSO EMK, EA
24	31	1785, 1720	4.9, 5.4	DMF, NMP, DMSO

(1) Yield of purified polymer.
(2) IR was recorded by using KBr pellet.
(3) Protons attached to beta-lactam group. No trans product was noticed in crude reaction mixture.

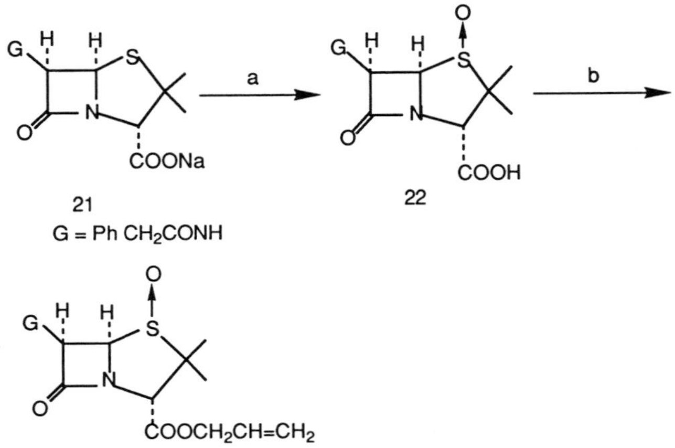

21
G = Ph CH₂CONH

22

23

a : p-TSOH, AcOOH, Water
b : Br CH₂ CH=CH₂ , Sodium hydride , 18-Crown -6

Scheme 5. Monomer synthesis from penicillin G.

CONCLUSION

In summary, by using step growth polymerization reaction new polymers containing monocyclic or bicyclic azetidinone system have been prepared. The overall yield of the polymer prepared from penicillin derived system is good. The synthetic route has enough versatility to develop new polymeric beta-lactam systems.

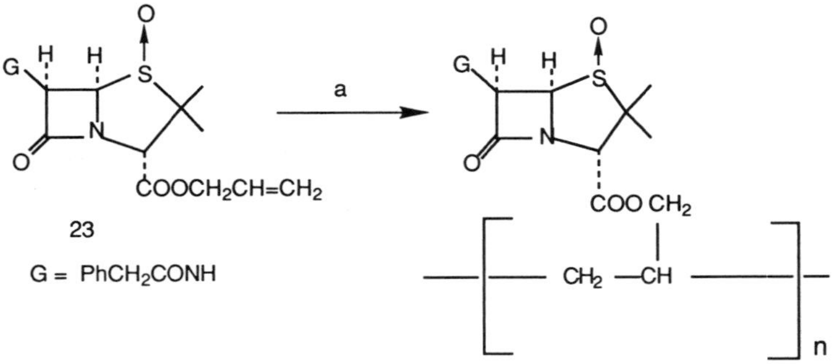

23

G = PhCH₂CONH

24

a : AIBN, Dioxane

Scheme 6. Synthesis of penicillin based polymer.

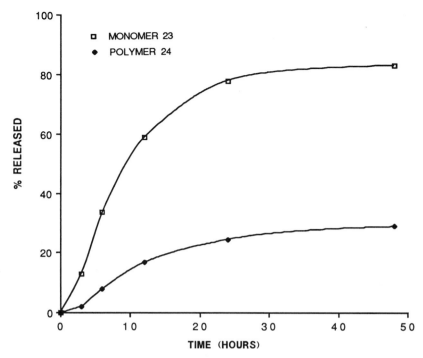

Figure 3. *In vitro* release of penicillin sulfoxide in deionized water from monomer 23 and 24.

ACKNOWLEDGMENT

        The author wishes to thank Dr. Carey Bottom and Dr. Francisco
Alvarez of Schering-Plough Research Institute for their help.

REFERENCES

 1. A. Fleming, Brit. J. Exp. Pathol., **10**, 226 (1929).
 2. H.W. Florey, E. Chain, N. G. Heatley, M. A. Jennings, A. G. Sanders,
    E. P. Abraham and M. E. Florey, "*Antibiotics*," Vol. 2, Oxford
    University Press, London, 1949.
 3. J. Elks, Ed., "*Recent Advances in the Chemistry of Beta-lactam
    Antibiotics*," The Chemical Society, London, 1977.
 4. F. E. Hahn, in: "*Progress in Molecular and Subcellular Biology*,"
    Vol.8, F. E. Hahn, Ed., Springer-Verlag, Berlin, 1983, p. 1-21.
 5. E. P. Abraham and G. G. F. Newton, Biochem. J., **79**, 377 (1961).
 6. R. B. Morin and M. Gorman, Eds., "*Chemistry and Biology of Beta-
    Lactam Antibiotics*," Vol. 1 and 2, Academic Press, New York, 1982.
 7. G. A. Koppel in: "*Small Ring Heterocycles-Part 2*," A. Hassner, Ed.,
    John Wiley, New York, 1983.
 8. J. E. Baldwin, Ed., Tetrahedron, **39**, Symposia-in-Print #10, 1983.
 9. M. S. Manhas, M. Ghosh and A. K. Bose, J. Org. Chem., **55**, 575 (1990).
10. H. Aki, N. Sawai, K. Yamamoto and M. Yamamoto, Pharm. Res., **8**, 119
    (1991).

11. A. K. Bose, M. S. Manhas, M. Ghosh, M. Shah, V. S. Raju, S. S. Bari, S. N. Newaz, B. K. Banik, A. G. Chaudhury and K. J. Barakat, J. Org. Chem., **56**, 6968 (1991).

12. M. Ghosh, J. K. Roy and B. G. Chatterjee, Ind. J. Chem., **24B**, 144 (1985).

13. M. Ghosh and B. G. Chatterjee, J. Ind. Chem. Soc., **LXII**, 457 (1985).

14. Methyl glyoxalate was prepared by ozonolysis of dimethyl malate followed by reduction with dimethyl disulfide. The compound is purified by vacuum distillation and is kept under nitrogen in refrigerator.

BIODEGRADABLE POLYMERS FOR THE CONTROLLED DELIVERY OF VACCINES

E. Schmitt, D.R. Flanagan and R.J. Linhardt

Division of Pharmaceutics
and
Division of Medicinal and Natural Products Chemistry
College of Pharmacy
University of Iowa
Iowa City, Iowa 52242

New methods for the efficient and effective administration of vaccines are required particularly with the advent of new synthetic subunit vaccines. This chapter describes the approaches of several research groups to administer vaccines using biodegradable polymer carriers. Delivery systems composed of synthetic poly[esters] and poly[iminocarbonates] and natural (crosslinked serum albumin) biodegradable polymers are described. The antibody levels in response to these systems are presented, and possible mechanisms responsible for the observed effects are discussed. The important questions that need to be answered before this technology can be successfully applied are also discussed.

INTRODUCTION

Biodegradable polymers have been used for a variety of medical applications including the controlled release of drugs and biologicals.[1] One potential application of biodegradable polymers is the controlled delivery of vaccines. Worldwide, vaccinations represent hundreds of millions of human treatments annually.[2] One problem with traditional vaccinations is the need for repeated booster regimens. Controlled release may obviate the need for such regimens. This could be a substantial advantage in situations where compliance makes it difficult to administer repeated injections. Additionally, many new synthetic subunit vaccines are insufficiently immunogenic alone to be successful and thus require the use of adjuvants and or new delivery techniques.[3] Current adjuvants such as aluminum hydroxide and Freund's complete and incomplete adjuvants are believed to exert as least part of their effect by prolonging the release of antigen from the injection site into the surrounding tissues. Our laboratory has shown that a wide range of potential *in vitro* release profiles may be achieved by careful selection of the polymer and formulation method.[4,5] Biodegradable and biocompatible microspheres or pellets seem well suited to provide slow or delayed release of antigen without the undesirable side effects of other adjuvants. Therefore, the modification of release kinetics or incorporation of both vaccine and adjuvant into biodegradable polymers may substantially improve the immunogenicity of many new vaccines.

INITIAL STUDIES

As early as 1976, Chang described the use of poly(lactic acid) for the microencapsulation of vaccines and antigens but did not study the release properties or immunogenicity of these formulations.[6] Preis and Langer later showed that the sustained release of bovine serum albumin, ribonuclease-A and bovine γ-globulin from ethylene-vinyl acetate pellets (0.3 mm^3) resulted in sustained antibody levels in mice for 25 weeks.[7] The antibody (IgG) levels in mice (n = 8) implanted with a single BSA/polymer pellet (50μg BSA) were comparable to those in mice (n = 4)

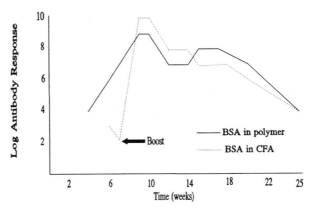

Figure 1. Antibody response to BSA in CFA given in two doses compared to sustained release from a polymeric matrix. (Prepared from data in reference 7.)

given a primary injection of BSA (50 μg) in complete Freund's adjuvant (CFA) followed by an identical booster regimen seven days later (100 μg total BSA) (Figure 1). These results gave the impetus for further study since they showed that sustained release of antigens from polymeric matrices could indeed elicit a substantial and prolonged immune response with a single treatment. In a second set of experiments these authors studied the ability of 3 different antigens of varying molecular weight to induce an immune response when given as antigen/polymer pellets. Bovine γ-globulin, bovine serum albumin and ribonuclease-A having molecular weights of 158,000, 68,000 and 14,000, respectively, all elicited a prolonged immune response. The ethylene-vinyl acetate pellets used in this study, however, do not degrade *in vivo* and thus would require surgical removal of the residual polymer making it less convenient than traditional inoculations. These experiments did however show that the sustained release of vaccines and adjuvants from polymeric matrices was a viable route for single-step immunization.

Kohn, et al. studied an antigen delivery system based on poly(CTTH-iminocarbonate) (IUPAC nomenclature: poly-[oxyimidocarbonyloxy-p-phenyl-ene-[2-(hexyloxycarbonyl)-ethylene]imino-[2-[1-(benzyloxy)-formamido]-1-oxotrimethylene]-p-phenylene]).[8] This polymer was designed with the aim that its primary degradation product, N-benzyloxycarbonyl-L-tyrosyl-L-tyrosine hexyl ester (CTTH) would show adjuvant properties as was shown for other L-tyrosine derivatives by Miller and Tees.[9] Indeed, the authors found CTTH to be as potent an adjuvant as Freund's complete and muramyl dipeptide (MDP).[8] In vitro release profiles appeared to follow a square-root of time dependence with most of the antigen released within 25 days (Figure 2). The square-root of time dependence and the observed effect of loading on release kinetics suggested that the release of antigen was controlled by a matrix diffusion mechanism and not determined by the degradation of the polymer. In vivo studies were carried out in mice using eight treatment groups of ten mice each. The treatment groups were as follows:

GROUP TREATMENT

A   25 μg BSA in 1 mL physiological saline solution (PSS), boosted at 4 weeks

B   25 μg BSA in 1 mL CFA (1:1 emulsion), boosted at 4 weeks

C   25 μg BSA + 100 mg MDP in 1 mL PSS, boosted at 4 weeks

D   25 μg BSA adsorbed onto 40 mg of solid tyrosine in 1 mL PSS, boosted at 4 weeks

E   25 μg BSA adsorbed onto 40 mg of solid dityrosine in 1mL PSS, boosted at 4 weeks

F   25 μg BSA adsorbed onto 40 mg of solid CTTH in 1 mL PSS, boosted at 4 weeks

G   50 μg BSA incorporated in a poly(BPA-iminocarbonate) device

H   50 μg BSA incorporated in a poly(CTTH-iminobonate) device

The devices implanted in groups G and H were 0.5 mg films cast from CH_2Cl_2 solutions containing suspended BSA. The anti-BSA antibody titers after 56 weeks were over twice as high in animals given a single injection of the poly(CTTH-iminocarbonate) formulation (Group H) compared to animals given BSA at 0 and 4 weeks. (Group A) (Figure 3)

Within the last few years, many researchers have been evaluating the polyesters- and copoly(esters) of lactic acid and glycolic acid as matrices for the delivery of peptides and proteins including vaccines and antigens. These polymers have already been used in humans for many years as surgical sutures and have well established biocompatibility and biodegradability. Thus, their regulatory status is well established. Also, a wide range of potential degradation times from a few weeks to over 1 year are attainable by selecting the proper copolymer and molecular weight. Miller, et al. have shown the influence of copolymer composition on the degradation time for copolymers of L-lactic and glycolic acids (Figure 4).[10] These polyesters have comparatively well characterized degradation profiles and are known to degrade by bulk hydrolysis. This means that if release is degradation-controlled, a biphasic release profile is expected for molecules, such as proteins, unable to diffuse through the intact polymer. A typical in vitro release profile for a model compound in shown in Figure 5. The model compound, amaranth, is a hydrophilic dye with three sulfonate groups which has been shown not to diffuse through the poly(D,L-lactide-co-glycolide, 50:50) polymer used.

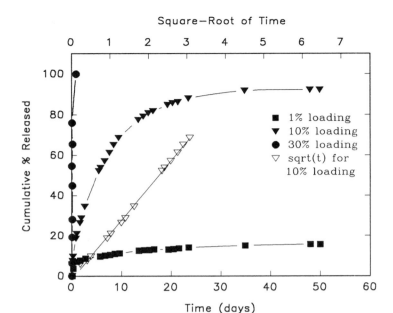

Figure 2. *In vitro* release profiles of Eosin Y from poly-
(CTTH-iminocarbonate) devices containing different
loadings. (Prepared from data in reference 8.)

This type of release profile may prove useful since it closely approxi-
mates the traditional release kinetics seen with an initial dose followed
by a booster regimen.

O'Hagan, et al. studied the immunogenic response of mice to ovalbumin (OVA) entrapped in poly(D,L-lactide-co-glycolide, 50:50) microparticles by two routes of immunization, subcutaneous (s.c.) and intraperitoneal injection (i.p.).[11] The microparticles were formed from the copolymer (MW = 9000) by an oil-in-water emulsion method. Characterization
showed the microparticles to have an average loading of 1% w/w OVA and a
volume-mean diameter of 5.34 μm. Antibody (IgG) levels in response to
injections of soluble ovalbumin, ovalbumin emulsified in CFA and ovalbumin entrapped in poly(ester) microcapsules were measured. Both primary
and secondary responses were measured. Ovalbumin is a rather poor immuno-
gen thus the results of this study were useful in assessing the adjuvant
effect of the microparticulate delivery system. The results for the
primary immune response showed that after single i.p. injections (100 μg
OVA), the mice given encapsulated OVA had a significantly greater IgG
titer than both the primary and secondary responses to soluble OVA
throughout the 12 week study. The responses to the encapsulated OVA were
also significantly greater than the OVA/CFA preparation for 10 weeks
following injection. At 12 weeks, however, the two treatment groups were
not significantly different. The adjuvant effect of the microparticulate
formulation was proposed to be due to the uptake of small particles by
macrophages which then present the antigen. Thus particle size would be
very important for this mechanism. The previous results of Langer, et
al., which used 0.3 mm³ pellets however, do not seem to show this same
importance of particle size since their delivery system was found to be
nearly as effective as a CFA emulsion even though the pellets were much
too large to be phagocytized by macrophages.[7] The secondary response was

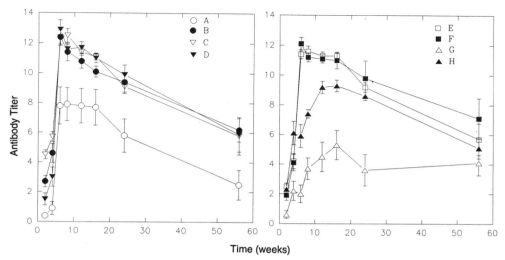

Figure 3. Mean hemagglutination BSA antibody titers as a
function of time for the eight treatment groups
used by Kohn et al. Data represent the mean of ten
animals ± SEM. (Prepared from data in reference
8.)

evaluated using three groups of ten mice each. The treatment groups were
given primary s.c. injections of (1) soluble OVA, (2) encapsulated OVA,
or (3) OVA emulsified in CFA followed by booster injections at six weeks.
The booster injections were identical to the primary injections except
for group (3) where the booster was OVA dispersed in Freund's incomplete
adjuvant (FIA). The difference between the encapsulated OVA and OVA/CFA
formulations were not significant for the secondary response but both
showed a greater response than was achieved with soluble OVA. In a subsequent publication the authors reported that OVA entrapped in poly(lac-
tide-co-glycolide) microspheres gave a significantly lower response than
OVA/CFA when studied in rats.[12] This was true for both the primary and
secondary responses. The microsphere formulations used in these two
studies were from the same batch. The only difference between the two
experiments was the animal model. The first study used the mouse and the
second used the rat. The authors addressed this disparity but were unable
to offer any explanation for the observed results.

Eldridge, et al. found that poly(D,L-lactide-co-glycolide) microspheres represented a potent adjuvant system for Staphylococcal enterotoxin B (SEB) capable of inducing both circulating and mucosal immunity
when given orally.[13] Eldridge, et al. also found that orally administered
poly(D,L-lactide-co-glycolide) 1-10 μm microspheres containing SEB were
specifically taken up into the Peyer's patch lymphoid tissue of the
gut.[14] Furthermore, SEB- containing microspheres induced circulating
anti-SEB antibodies and a secretory IgA anti-SEB response in saliva and
gut fluid, while soluble SEB did not. These results suggest that the
polymer was able to protect the antigen from the gastrointestinal
environment. In another study, Eldridge, et al. evaluated the immune
response of mice given SEB-containing poly(D,L-lactide-co-glycolide)
microspheres subcutaneously.[15] The results showed that the kinetics,
magnitude and duration of the immune response to encapsulated SEB were
similar to those obtained after an equivalent dose emulsified in CFA

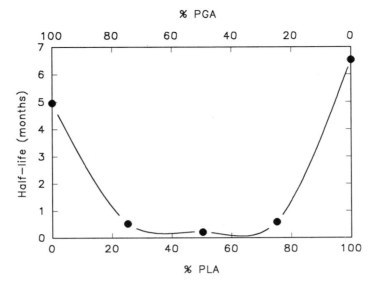

Figure 4. Degradation half-life for copolymers of L-lactic acid and glycolic acid as a function of copolymer composition. (Prepared from data presented in reference 10.)

(Figure 6). The microspheres had the additional benefit that they did not induce the inflammation and granulomata observed when CFA was present. Antigen-containing microspheres 1 to 10 μm in diameter exhibited stronger adjuvant activity than those >10 μm suggesting that uptake of the microspheres by macrophages is an important factor in the adjuvant activity thus supporting the conclusions of O'Hagan, et al.[11]

Beck, et al. have described a system for delivering antigens and antibodies to the female reproductive tract.[16,17] In these studies Pneumococcus bacteria, herpes simplex viral antigens and bovine chorionic gonadotrophin hormone (HCG) were microencapsulated with lactide/glycolide polymers. The *in vivo* effects of vaginally administered microcapsules were followed in rabbits. The results of vaginal washes showed that immunization was successful at two weeks post treatment. This showed the primary response to be intact however no data was given for the secondary response. Workers at Stolle International have described a delivery system designed to release antigen to dairy cattle over 6 to 12 months.[18] Animals were injected intramuscularly with antigen-containing microparticles suspended in an aqueous vehicle. The antibody titers in milk of animals given a single dose of microparticles were as high as those found in animals after conventional immunizations which had the inconvenience of requiring several booster injections.

Our laboratory has recently formulated a biodegradable microparticulate vaccine preparation by a coacervation process using poly(D,L-lactide-co-glycolide, 50:50) which is currently undergoing *in vivo* evaluation. (Figure 7)

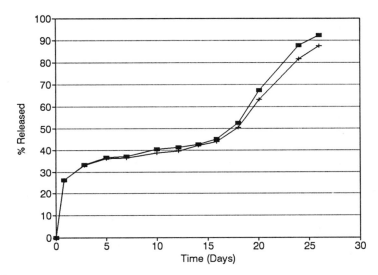

Figure 5. Release of amaranth in phosphate buffered saline at
37°C from poly(D,Llactide-co-glycolide, 50:50)
microparticles prepared by a coacervation method.

NATURAL BIODEGRADABLE POLYMERS

Lee, et al. first reported the preparation of solid albumin microspheres produced by mild chemical crosslinking of serum albumin with
glutaraldehyde.[19] They found that the *in vitro* release profiles of steroids from these systems could be varied by controlling the cross-link

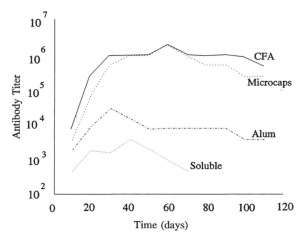

Figure 6. Antibody response to SEB toxin formulations.
(Prepared from data presented in reference 15.)

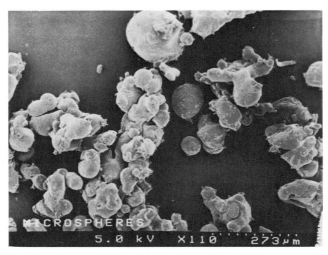

Figure 7. Scanning electron micrograph of a poly(D,L-lactide-co-glycolide, 50:50) vaccine formulation prepared by coacervation.

density. The *in vivo* degradation of these microspheres would proceed via endogenous proteases. The method of Lee et al. was subsequently used by Dewar, et al. to entrap virus particles.[20] The model virus employed, Nodamura virus, a small, nonpathogenic, RNA virus was entrapped in rabbit serum albumin (RSA) beads and later studied *in vivo* in rabbits. It is important to mention that proteins entrapped into albumin beads in this way would be covalently bound to the albumin and potentially to other protein molecules. The purpose of the study was to determine whether the formulation procedure or the release characteristics had any deleterious effects on the antigenicity of the virus. The results showed that the viral antibody titers in rabbits given a single injection of the RSA bead formulation paralleled those given a single injection of the virus in CFA for 60 days, the duration of the study. This suggested that the RSA bead formulation possessed an adjuvant action approximating CFA, presumably due to slow release of the virus from the protein matrix since the beads were too large (100-200 μm) to be phagocytized by macrophages. Continuing this work, Martin, et al. studied the effectiveness of serum albumin beads as a delivery system for subunit vaccines.[21] The 40 kdalton Nodamura capsid protein was used as a model. The antibody responses to the capsid protein and the whole virus given in RSA beads were slow compared to soluble vaccines. The entrapped vaccines showed a continuous increase in the levels of circulating antibodies which, at 60 days post-injection, was approximately equal to the peak responses obtained with soluble vaccines. The authors were unable to show any antigenic proper-ties of the empty beads in homotypic animals.

Langhein, et al. studied the effectiveness of different antigenic materials entrapped in the same way in albumin beads.[22] Model vaccines containing *Clostridium botulinum* type D toxin and *Klebsiella pneumoniae* capsular polysaccharide antigen were prepared. The *in vivo* evaluation of this system employed four treatment groups: (1) *Cl. botulinum* toxoid; (2) *Cl. botulinum* toxoid emulsified in Freud's incomplete adjuvant; (3) *Cl. botulinum* toxin covalently bound into RSA beads; and (4) Cl. botulinum toxin bound into RSA beads and stored for 4 months at room temperature.

The stored formulation retained nearly all of its immunogenic activity. This finding could be of practical importance since the formulation of a dry, stable vaccine could reduce storage and transportation costs.

The synthesis of cross-linked starch, a second type of natural biodegradable polymer, has been described by Artursson, et al.[23] To date, this polymer has been applied to the delivery of enzymes to intracellular compartments of the reticuloendothelial system after intravenous injection.[24] The authors found the poly(acrylstarch) microparticles to be potent adjuvants in testing the tendency of the particles to induce an autoantibody response to autologous proteins.[25,26] These polymers, however, have not yet been applied to vaccine delivery systems.

There seems to be a potential problem with these naturally derived biodegradable polymers which is related to their enzymatic degradation mechanism. Since the presence or concentration of some enzymes can vary substantially among individuals, in addition to interspecies variability, the rate of matrix degradation and thus the rate of vaccine release could vary considerably among different patient or animal populations. Synthetic biodegradable polymers such as the poly(esters), poly(iminocarbonates), poly(anhydrides)[27] and poly(orthoesters)[28] were designed to degrade through chemical hydrolysis in water and should show less biologic variability.

FUTURE APPROACHES

In the future, the mechanism by which vaccines encapsulated in biodegradable polymers induce immunization requires study. The size of the delivery system may be of paramount importance as suggested by O'Hagan, et al.,[11] or it may only play a small role as would be suggested by the data of Langer and Preis.[7] Alternatively, the release kinetics may determine the magnitude of the effect. Ideal release kinetics may not be zero-order or matrix-controlled but instead may be pulsatile mimicking standard immunization protocols. The answer to these questions will largely determine the direction taken to formulate these systems. The optimization of vaccine delivery using biodegradable polymers will require that these fundamental questions be answered. This may suggest the use of alternate biodegradable polymers having adjuvant activity or the novel use of existing materials such as the lactide/glycolide polymers. Continued research should address these important questions.

ACKNOWLEDGEMENT

Mr. Eric A. Schmitt gratefully acknowledges the financial support of Bristol-Myers Squibb and the American Foundation For Pharmaceutical Education.

REFERENCES

1. R. J. Linhardt, in: "*Controlled Release of Drugs: Polymers and Aggregate Systems*," M. Rosoff, Ed., VCH Publishers, New York, 1989, p. 53.
2. D. H. Lewis, in: "*Biodegradable Polymers as Drug Delivery Systems*," M. Chasin & R. Langer, Eds., Marcel Dekker, Inc., New York, 1990, Chapter 1, p. 27.

3. D. W. Barry, E. Stanton, & R. E. Maynor, Infect. Immun., **10**, 1329 (1974).
4. H. T. Wang, E. Schmitt, D. R. Flanagan & R. J. Linhardt, J. Controlled Release, **17**, 23 (1991).
5. H. T. Wang, H. Palmer, D. R. Flanagan, and E. Schmitt, Biomaterials, **11**, 679 (1990).
6. T. M. S. Chang, J. Bioeng., **1**, 25 (1976).
7. R. S. Langer & I. Preis, J. Immunol. Methods, **28**, 193 (1979).
8. J. Kohn, S. M. Niemi, E. C. Albert, J. C. Murphy, R. Langer & J. G. Fox, J. Immunol. Methods, **95**, 31 (1986).
9. A. C. Miller & E. C. Tees, Clin. Allergy, **4**, 49 (1974).
10. R. A. Miller, J. M. Brady & D. E. Cutright, J. Biomed. Mater. Res., **11**, 711 (1977).
11. D. T. O'Hagan, D. Rahman, J. P. McGee, H. Jeffery, M. C. Davies, P. Williams, S. S. Davis & S. J. Challacombe, Immunology, **73**, 239 (1991).
12. D. T. O'Hagan, H. Jeffery, M. J. J. Roberts, J. P. McGee & S. S. Davis, Vaccine, **9**, 768 (1991).
13. J. H. Eldridge, J. K. Staas, J. A. Meulbroek, J. R. McGhee, T. R. Tice & R. M. Gilley, Molecular Immunol., **28**, 287 (1991).
14. J. H. Eldridge, R. M. Gilley, J. K. Staas. Z. Moldoveanu, J. A. Meulbroek & T. R. Tice, Curr. Top. Microbiol. Immun., **146**, 59 (1989).
15. J. H. Eldridge, J. K. Staas, J. A. Meulbroek, T. R. Tice & R. M. Gilley. Infect. Immun., **59**, 2978 (1991).
16. L. R. Beck, C. F. Flowers, Jr., D. R. Cowsar & A. C. Tanquary, U.S. Patent 4,756,907 (1988).
17. L. R. Beck, C. F. Flowers, Jr., D. R. Cowsar & A. C. Tanquary, U.S. Patent 4,732,763 (1988).
18. Stolle Milk Biologics International, Cincinnati, Ohio, Product Bulletin (1988).
19. T. K. Lee, T. D. Sokoloski & G. P. Royer, Science, **213**, 233 (1981).
20. J. B. Dewar, D. A. Hendry & J. F. E. Newman. S. Afr. Med. J., **65**, 564 (1984).
21. M. E. D. Martin, J. B. Dewar & J. F. E. Newman, Vaccine, **6**, 33 (1988).
22. C. Langhein & J. F. E. Newman, J. Applied Bacteriology, **63**, 443 (1987).
23. P. Artursson, P. Edman, T. Laakso & I. Sjöholm, J. Pharm. Sci., **73**, 1507 (1984).
24. P. Artursson, P. Edman & I. Sjöholm, J. Pharmacol. Exp. Ther., **231**, 705 (1984).
25. P. Artursson, P. Edman & I. Sjöholm, J. Pharmacol. Exp. Ther., **234**, 255 (1985).
26. P. Artursson, I. L. Martensson & I. Sjöholm, J. Pharm. Sci., **75**, 697 (1986).
27. H. B. Rosen, J. Chang, G. E. Wnek, R. J. Linhardt & R. Langer. Biomaterials, **4**, 131 (1983).
28. J. Heller, J. Controlled Release, **2**, 167 (1985).

STRATEGIES FOR TREATING ARTERIAL RESTENOSIS USING POLYMERIC CONTROLLED RELEASE IMPLANTS

Robert J. Levy[1]*, Gershon Golomb[2], Joseph Trachy[1], Vinod Labhasetwar[1], David Muller[1], and Eric Topol[3]

1. The University of Michigan Medical School
   Ann Arbor, Michigan
2. School of Pharmacy
   The Hebrew University of Jerusalem
   Jerusalem, Israel
3. The Cleveland Clinic Foundation
   Cleveland, Ohio

Coronary artery obstruction is currently being treated with a number of invasive approaches involving catheter based angioplasty procedures. These have included most recently balloon angioplasty combined with expansion of obstructed coronary arteries using balloon expandable stainless steel stents. However, angioplasty itself, especially with stenting, leads to an accelerated reobstruction process, known as restenosis. Research reported in this paper has investigated an approach to preventing restenosis using controlled release drug-polymer implants for local inhibition of the pathophysiologic events of restenosis. Model therapeutic compounds were chosen including aspirin, as an antiplatelet agent, hirulog, as an antithrombin, and colchicine as an antiproliferative. Controlled release polymer matrices were successfully formulated and characterized. Retention of anticoagulant activity for the peptide, hirulog, was demonstrated *in vitro*. These polymers are suitable for investigations in periadventitial implants and animal models of restenosis. Eventually, controlled release strategies for preventing restenosis will involve integrating of ideal agents including gene therapy, with stents and related devices in order to develop a drug delivery systems approach.

INTRODUCTION

Current invasive approaches for treating coronary obstruction include balloon angioplasty, and most recently balloon angioplasty

---

* Address for Reprints and Correspondence: Robert J. Levy, M.D. Kresge II, Room 5014, The University of Michigan, Ann Arbor, MI 48109-0576, Phone: 313-936-2850, Fax: 313-747-3270

Biotechnology and Bioactive Polymers, Edited by C. Gebelein and C. Carraher, Plenum Press, New York, 1994

combined with expansion of obstructed coronary segments using balloon expendable stainless steel stents[1] to compensate for reobstruction due to elastic recoil of the arterial wall. However, the arterial stenting procedure itself induces a pronounced pathophysiologic response compara- ble to balloon dilatation of a diseased artery, also leading to reob- struction. This overall process has been termed restenosis. Restenosis has been the subject of a number of recent reviews.[1-3] It is known to affect up to approximately 50% of stent angioplasty procedures, within six months after stenting.[2]

The pathophysiologic mechanisms of restenosis are at present incompletely understood. Several specific events seem to predominate in the early development of the restenosis process, and these are the basis of the therapeutic strategy discussed in this paper (Table 1). Platelet thrombi form initially both along the surface of the stent itself, and the injured artery beneath it. Furthermore, without stenting, there is also subintimal exposure by balloon trauma of the vessel wall. Presumably, adherent platelets in addition to inducing subsequent fibrin thrombosis, also release important growth factors such as the "platelet derived growth factor" (PDGF), and various other active compounds such as prostaglandins, into the arterial wall, thereby influencing the subsequent proliferative response.[4] Fibrin thrombosis also contributes to acute arterial obstruction, and organization and remodeling of the fibrin thrombus can further complicate the restenosis process. A variety of growth factors and cytokines derived not only from platelets but also

Table 1. Therapeutic strategies in restenosis.

PATHOLOGIC MECHANISM	AGENT	MODE OF ACTION
Platelet Binding:		
	(a) Aspirin	(a) Prostaglandin synthesis inhibitor
	(b) Antibody to IIb/IIIa glycoprotein	(b) Blocks aggregation
	(c) Platelet derived growth factor receptor	(c) Inhibits platelet contribution to proliferation
Fibrin Clot Formation:		
	(a) Heparin*	(a) Bind to antithromb. III
	(b) Hirulog/hirudin**	(b) Binding to thrombin
Cellular Proliferation:		
	(a) Angiopeptin	(a) Inhibit smooth muscle cell proliferation
	(b) Colchicine	(b) Mitotic inhibitor
	(c) Dexamethasone	(c) Antiinflammatory steroidal effects

   * Also has activity against antiplatelet Factor 4 and has
     antiproliferative activity.
   ** Inhibits thrombin binding to platelets.

endothelial cells, smooth muscle cells, and macrophages are released at the time of vascular injury.[5] These growth factors include such well known proteins as the fibroblast growth factor, and, interleukin 1, to name a few. Proliferative events begin within the first several days after the stent-induced injury and are clearly evident experimentally within a week after stenting.

A number of strategies for ameliorating restenosis have been investigated thus far both clinically and experimentally. Clinical trials and animal protocols with various systemic drug administrations to prevent restenosis recently have been reviewed.[6] These have included investigations of aspirin (as an antiplatelet agent), heparin (an antithrombotic), colchicine (an antiproliferative), as well as a host of other agents.[6] Some of the various drugs studied experimentally thus far have included calcium channel blockers, steroids, antimetabolites such as methotrexate, and recently discovered growth antagonists such as angiopeptin.[6] In addition, coated stents have also been proposed for clinical trials.[7] The principal coating agent studied to this date has been heparin.[7] The results of all the systemic administration clinical trials, and the limited coated stent work in the clinical arena, have been thus far disappointing. No positive effects have been discovered with respect to controlling restenosis. A number of new strategies have emerged for clinical use including biodegradable stent designs, which may also include drug loadings.[8] In addition, balloon angioplasty prior to stenting with the so-called microporous or "sweating" balloon, through which agents can hopefully be seeped into the arterial wall, has also been investigated in preliminary studies clinically.[9] Spears and others have proposed combining the microporous balloon strategy with photoactivated drug use, initializing drug activity with laser energy administered through a catheter.[10]

All of the above clinical strategies have also been investigated in various animal studies, which have indicated some preliminary benefit. Most recently, studies by Edelman and his colleagues, have demonstrated that periarterial drug administration using heparin-ethylenevinyl acetate composites significantly inhibited restenosis in a rat arterial injury model.[11,12] This initial success of a controlled release drug delivery approach to restenosis has stimulated interest in the field. Controlled release drug implants have been used by our group and others to treat a variety of cardiovascular diseases, and this approach is uniquely suited for this general group of disorders.[13] Controlled release may be defined as formulations of drug polymer composites, either as monolithic matrices or reservoirs with rate limiting membrane configurations, in which drug administration can be sustained through the use of polymeric materials. Implantation of controlled release polymer systems at the site of a cardiovascular disease process offers the advantages of regional high levels of drug, with optimal drug action, as well as lowering systemic drug exposure, and thereby minimizing the possibility of side effects.

Controlled release drug administration is being used in one cardiovascular clinical application thus far, the dexamethasone releasing cardiac pacing catheter.[14] This unique application of a silicone rubber controlled release system, placed at the tip of a cardiac pacing catheter, results in inhibition of scar tissue forming near the myocardial-electrode contact site, which would otherwise raise electrical resistance, and increase pacing energy requirements. Experimental studies, by our group and others, have shown that site-specific controlled release drug implants can inhibit cardiovascular calcification and cardiac arrhythmias, delay cardiac transplant rejection, and prevent bacterial endocarditis.[15]

The goals of the present paper are as follows:

1. We will present a strategy for a mechanism-based approach to cardiovascular drug delivery for restenosis based on drugs specifically administered to selectively inhibit either the platelet component of the restenosis process, or fibrin thrombus formation, or the arterial wall proliferative response.

2. We will also describe our efforts to formulate prototypical controlled release matrices to administer regional therapy targeted at each of the above mechanisms.

3. We will present data characterizing our prototype formulations in terms of their *in vitro* bulk drug delivery and maintenance of drug activity following incorporation and release. We will also describe an animal model approach for investigating these controlled release strategies.

EXPERIMENTAL

The silicone rubber used in these studies was Dow Corning Silastic Q7-4840 (Midland, MI). The drugs used included colchicine as both a nonradioactive preparation (Sigma, St. Louis, MO), and radioactive (Tritium Labeled, New England Nuclear, Billerica, MA). Aspirin (Sigma, St. Louis, MO), hirulog as both nonradioactive and tritium labeled (Biogen, Cambridge, MA), and dexamethasone, also nonlabeled, and tritium labeled (Amersham, Arlington Heights, IL). Kits for assaying prothrombin time were obtained from Sigma, St. Louis, MO. Controlled release matrices, in general, were formulated by sieving the desired agent as a dry powder to 90 - 120 mesh particle size, and levigating it with silicone rubber prepolymer and exposing to vacuum for 30 minutes. Polymerization techniques, included casting the drug-polymer composites (20% drug, 80% polymer) into thin slabs in aluminum molds, under 20,000 psi in a Carver Press (Fisher, Chicago, IL), polymerizing at 80°C for 50 minutes. In addition, surfaces of slab matrices were sealed using the same silicone polymer in order to permit unidirectional drug release and limit swelling.

Characterization of drug release from the matrices (1X1 cm), included *in vitro* release studies in a physiologic buffer HEPES, 0.5M, pH 7.4 under perfect sink conditions for the durations required for each individual study. Retained biologic activity for the peptide anticoagulant, hirulog, was assessed using prothrombin time assays on aliquots of *in vitro* releasing buffer from the hirulog matrix studies.

RESULTS AND DISCUSSION

The agents chosen for controlled release incorporations were selected based on the mechanistic rationale described above (Table 1). Aspirin is a well known platelet antagonist, and has advantages for local release because of its acetylization of the various enzymes required for prostaglandin biosynthesis.[16] In the course of gastrointestinal administration of aspirin, hydrolysis of the molecule results in its partial deacetylization. Therefore, regional aspirin administration would theoretically have advantages for inhibiting prostaglandin synthesis in the arterial wall. Hirulog is a potent direct antithrombin, which binds to thrombin's substrate recognition site, and catalytic binding site. Hirulog[17] is a peptide containing the partial amino acid sequence of hirudin, the leech derived antithrombin protein. Hirulog is a potent anticoagulant, and its local administration was hypothesized to result in local anticoagulation, and thus avoid systemic anticoagulant exposure. In addition, hirulog can normally be only administered by the intravenous route, since gastrointestinal hydrolysis and absorption of the peptide would result in its deactivation. Colchicine was chosen as a mitotic inhibitor, which would hypothetically inhibit the arterial wall proliferative response.[18] Colchicine is a well known agent, which acts to inhibit cell proliferation via inhibition of microtubule formation. Dexamethasone was selected as an established antiproliferative as well as an antiinflammatory agent, with immunosuppressive properties. This agents general effects were thought to hypothetically be of benefit for many of the various components leading to the proliferative response in the arterial wall as well.

In vitro drug release results for the matrices demonstrated similar release kinetic profiles. As can be seen in Figure 1, cumulative drug release for the various compounds studied was achieved without an initial burst effect, with a nearly constant release rate, and near depletion of the matrix by 30 days of incubation.

The controlled release hirulog studies provided an opportunity to assess retention of anticoagulant activity following incorporation and release from silicone rubber polymers. A simple prothrombin time assay was used to evaluate hirulog inhibition on the prothrombin activated clot formation in this assay system. As can be seen in Table 2, throughout the duration of the *in vitro* release studies, hirulog continued to exert a significant anticoagulant activity, at activity levels comparable to standard solutions of hirulog containing the same amounts of this agent. Hirulog is also of interest, since it has antiplatelet effects. Albeit indirect, the anti-platelet effects are based on hirulog's inhibition of thrombin binding to platelets, thereby inhibiting platelet activation.[19] However, hirulog has some important limitations which are noteworthy. Hirulog is a nematode-derived peptide, and thus, sensitization with subsequent humoral immune response to this compound could potentially limit its usefulness to several weeks, and probably a one-time usage as well. In addition, controlled release investigators are just beginning to consider protein denaturation issues after incorporation, at the liquid solid interface in monolithic matrices. Hirulog, like many other peptides and proteins may be denatured due to aggregation or conformational changes or both during the time course of implantation and the progression of the drug releasing front through the monolithic matrix.

Controlled release implant strategies for preventing restenosis will have to be studied in animal models of this disorder. At present, a number of animal models are available all of which have limitations when compared to restenosis as it occurs in the human arterial wall.[6] A number of rat and rabbit models of restenosis exist, and these are based on arterial injury caused either by balloon catheter or air desiccation. The end points in these model systems are pathologic assessments after several weeks of the morphology of the arterial wall. The validity of the rat models has been questioned by many, since a number of agents which appeared to be effective in rat arterial injury models, have failed to have efficacy in larger animal models or in clinical studies.[6] Pig coronary artery stenting, followed by restenosis, is another useful model

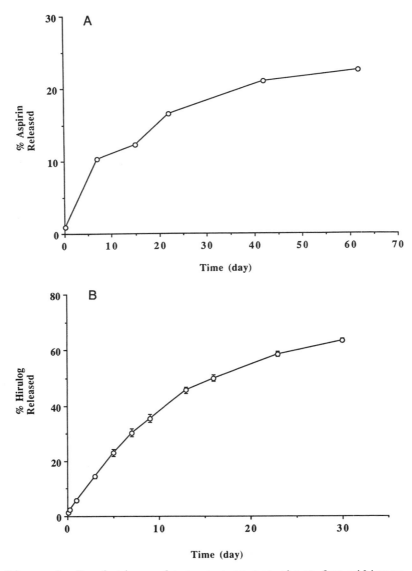

Figure 1. Cumulative release curves are shown for silicone
rubber controlled release polymers for the
various agents of interest. For all compounds
studied, no burst phase of release was noted, and
nearly constant release rates were observed for
the first 20 days of drug delivery. Aspirin
matrices were studied for the longest durations,
and demonstrated an exponentially declining
release rate with time (see A). In addition,
water solubility also governed net cumulative
release, as is evident for the less soluble
colchicine (See 1C, opposite page, compared to A
& B).

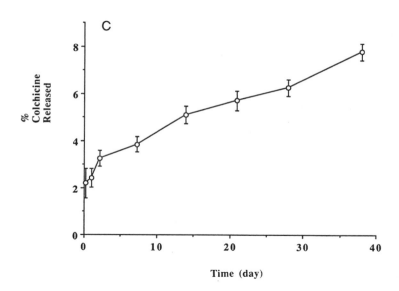

Time (day)

of this disorder investigated by Schwartz and others.[6,20] In these studies, pigs are subjected to coronary angioplasty, with balloon angioplasty of a coronary artery and expandable oversized stent placement. While angiography can document progression of restenosis *in vivo*, pathologic assessment of the coronary arteries is required to establish the extent of the chiefly proliferative response. In general, these studies have shown that the pig seems to have a more hyperplastic response to stenting than noted clinically. Nevertheless, the pig

Table 2. The effects of compounding and sterilization on the biological activity of Hirulog matrices

Prothrombin time test (sec); control = 8.7 ± 0.5 (mean ± SD)

	Sterilized		Non-sterilized
	γ-irradiated[*]	Heat[**]	
Hirulog powder[a]	–	21.1 ± 2.4	21.0 ± 3.5
Hirulog[b]	12.3 ± 1.3	–	12.3 ± 1.7

Hirulog matrices released *in vitro*[c]	>45 sec (24 hr release); >115 sec (48 hr release)
Explanted hirulog matrices after 5 days[d]	>23 (24 hr release)

* 2 Mrads for 2 hr.
** 110°C for 30 min.
(a) 100 μg/mL.
(b) 1 cm^2; sealed, 12 hr release.
(c) unsealed.
(d) explants from pigs' advential release and released *in vitro*. 2 < n < 7.

coronary model is probably the most comparable to the human disorder at this time. Histologically, the hyperplastic lesions produced are remarkably similar to those noted clinically following balloon angioplasty.

Another approach for studying restenosis in a larger animal model, is stent placement in the pig carotid artery.[6] This model system has been pioneered by Muller and others, and offers a number of advantages.[6] The pig carotid artery is a large and easily accessible vessel, which can be instrumented and accessed with a periadventitial controlled release matrix. Periadventitial polymer matrices are also of interest since their placement enhances intimal proliferation.[21] In addition, a number of studies have taken advantage of the fact that the contralateral carotid can be used as a control in the same animal if local strategies are to be considered.[6] Once again, the end point of this model system is typically the pathological evaluation of restenosis after a period of time following stenting, typically 30 days. However, shorter term studies can also examine the initial events in restenosis, including platelet and fibrin thrombus formation.

All of these animal model systems can be used to investigate controlled release polymer implants for restenosis. Optimal polymer configurations remain as yet to be determined. Drug delivery systems bonded to stents would be ideal, but are limited by the finite mass of the expandable stent in terms of the potential amounts of drug that could be incorporated. Periadvential drug delivery has been useful experimentally in providing adequate amounts of drug for local therapy investigations. However, periadvential controlled release would require an invasive surgical approach for clinical use, and thus alternatives to this will be needed. Polymers to be used might include biodegradable matrices, such as polylactic-polyglycolic acid, which have the advantage of disappearing without leaving a residual implant. Nondegradable matrices, such as silicone rubber or polyurethane, however have the advantage of material strength, and could in fact become part of a vascular prosthesis, if necessary, in order to maintain structural integrity in the region of the diseased blood vessel. Furthermore, nondegradable matrices could be configured as refillable drug delivery reservoirs. This design would have the advantage of replenishing drug, should long term therapy be required. The ability to use alternative agents is also a useful option in the case of reservoir implants.

Another important area of investigation for restenosis is the use of gene therapy. Pioneering work by the Nabels,[22] and by Dichek and Anderson,[23] has demonstrated effective gene transfer to the cells of the arterial wall. This has been achieved either by the use of various transection techniques to arterial wall segments, or by directly injecting genetically modified cells into an isolated arterial segment. Coating of stents and vascular grafts with genetically modified cells is also another strategy which has been successful in experimental studies. These approaches have the advantage of providing genetic coding for useful proteins, which could limit restenosis. A number of possibilities for genes encoding various proteins and peptides agents have been considered. Tissue plasminogen activator (TPA) has been successfully incorporated into endothelial cells, which have been bonded to an expandable stent and implanted into an artery in animal model studies by Anderson and Dichek.[23] TPA would presumably facilitate the clot lysis in the proximity of a stent. At this time gene therapy has a number of important limitations. Retroviral vectors for transfection of cells of the arterial wall have raised a number of concerns related to vector specificity as well as safety. Other DNA transfection techniques have focused chiefly on liposome formulations. Liposomes are limited by short term stability, and lack of specificity *in vivo*. In addition, all of the attempts at transfer of DNA directly to the cells of the arterial wall have been relatively inefficient, with the best results thus far achieving a transection frequency of 1% or less. Anti-sense mRNAs to proto-oncogenes have been introduced via the microporous balloon to inhibit the proliferative response in the rabbit and rat atherosclerosis models. Furthermore, gene therapy strategies can only address relatively late events in restenosis, since the time required for transcription precludes potential gene therapy of acute thrombotic events. Nevertheless, the gene therapy approach offers great promise.

Thus it can be seen that controlled release strategies for preventing restenosis will involve the integrating of ideal agents, including useful genes, with stents and related devices in order to develop a drug delivery systems approach. Ideally, if catheter based interventions are to continue to be increasingly used, drug implant interventions should be achieved by this route as well. Therefore, increasing investigative use of drug or gene loaded stents will become a priority, and their development will become an important research area.

ACKNOWLEDGEMENT

The authors thank Mrs. Catherine Wongstrom and Ms. Jill Van Cise for preparing the manuscript. This work was supported in part by NIH Grants HL38118 and HL416643, and American Association Grants in Aid, 890654 and 911538.

REFERENCES

1. D. W. M. Muller and S. G. Ellis. Coron. Art. Dis., **1**, 438 (1990).
2. D. R. Holmes, R. E. Vliestra, H. C. Smith, et al., Am. J. Cardiol., **53**, 77C (1984).
3. M. W. Ela, G. S. Roubin and S. B. King., Circulation, **79**, 1374 (1989).
4. R. Ross. N. Engl. J. Med., **314**, .488 (1986).
5. J. H. Ip, V. F. Fuser, L. Badimon, J. Badimon, M. B. Taubman and J. H. Chesebro., J. Am. Coll. Cardiol., **15**, 1667 (1990).
6. D. W. Muller, S. G. Ellis and E. J. Topol., J. Am. Coll. Cardiol., **19**, 418 (1992).
7. D. A. Cox, P. G. Anderson, G. S. Roubin, C. Y. Chow, S. K. Agrawal and J. B. Cavender, Circulation, Suppl. II, **84**, 71 (1991).
8. M. Ebecke, A. Buchwald, H. Stricker and V. Wiegand, Circulation, Suppl. II, **84**, 72 (1991).
9. H. Wolinsky and C. S. Lin, J. Am. Coll. Cardiol., **17**, 174B (1991).
10. J. R. Spears, K. K. Sourav and L. P. McNath, J. Am. Coll. Cardiol., **17**, 179B (1991).
11. E. R. Edelman, D. H. Adams and M. J. Karnovsky, Proc. Natl. Acad. Sci. U.S.A., **87**, 3773 (1990).
12. E. R. Edelman, M. A. Nugent, L. T. Smith and M. J. Karnovsky, J. Clin. Invest., **39**, 65 (1992).
13. R. J. Levy, T. P. Johnston, A. Sintov and G. Golomb, J. Cont. Rel, **11**, 245 (1990).
14. H. Mond, K. Stokes, J. Holland, L. Griggs, et al., PACE, **11**, 214 (1988).
15. R. J.Levy, S. F.Bolling, R. Siden, A. Kadish, Y. Pathak, et al, in: *"Cosmetic and Pharmaceutical Applications of Polymers,"* C. G.

Gebelein, T. Cheng & V. Yang, Eds., Plenum Publ. Corp., New York, 1991, p. 231.

16. L. Schwartz, M. G. Bourassa, J. Lesperance, et al. N. Engl. J. Med., **318**, 1714 (1988).

17. J. M. Maraganore, P. Bourdon, J. Jablonski, K. L. Ramachandran and J. W. Fenton, Biochemistry, **29**, 7095 (1990).

18. D. W. M. Muller, S. G. Ellis and E. J. Topol, J. Am. Coll. Cardiol., **17**, 126B (1991).

19. S. W. Tam, D. W. Fenton and T. C. Detwiler, J. Biol. Chem., **254**, 8723 (1979).

20. R. Schwartz, J. G.Murphy, W. D.Edwards, et al, Circulation, **82**, 2190 (1990).

21. J. Gebrane, J. Roland and L. Orcel, Virchow's Arch. Pathol. Anat., **396**, 41 (1982).

22. E. G. Nabel, G. Plautz and G. J. Nabel., Science, **294**, 1285 (1990).

23. D. A. Dichek, R. F. Neville, J. A. Zwibel, et al. Circulation, **80**, 1347 (1989).

THE USE OF POLYICLC IN THE TREATMENT OF AIDS

Hilton Levy[a], Andres Salazar[b], Javier Morales[c] and
Owen St. Clair Morgan[d]

(a) NIAIAD
    Bethesda, MD
(b) Walter Reed Army Hospital
    Washington, DC
(c) Ashford Hospital
    Puerto Rico
(d) University of West Indies
    Kingston, Jamacia

The use of a dsRNA, stabilized against hydrolysis by
enzymes in primate serum, in the treatment of AIDS and
tropical spastic paraparesis (TSP), is described. This
compound, poly inosinic.polycytidylic acid, stabilized with a
complex of polylysine and carboxymethylcellulose, (PolyICLC),
was given i.m. to AIDS patients with and without Zidovudine
(AZT). In TSP it was given alone, i.m. The drug was well-
tolerated in both groups of patients. Side effects were mild,
consisting of a mild myalgia and very slight fever. Some
beneficial effects are described.

INTRODUCTION

Shortly after Isaacs and Lindenmann described Interferon (IFN) in
1957,[1] it was realized that IFN was potentially a broad spectrum
antiviral agent that might be useful in the clinic. This potential was
very slow to be realized, because both mouse and human IFNs were
available only in very small quantities, not enough even to do in mice,
let alone humans. The situation has changed in the past few years, but
early on investigators started to look for non-replicating entities that
could cause the host to turn on the production of his own interferon in
large quantities. A number of compounds were found, but they either
produced only small quantities of IFN, or they were too toxic. Later it
was shown that several natural and synthetic double-stranded RNAs (dsRNA)
were effective inducers in rodents. The most effective was the dsRNA
containing one strand of poly-inosinic acid linked to a strand of polycytidylic acid, PolyI·PolyC. This compound could lead to the formation of
large quantities of IFN in mice, and was a good antiviral agent in mice.[2]
We showed that it was fairly effective agent in mice vs. a variety of
transplanted, chemically induced, and spontaneous tumors.[3] However, when

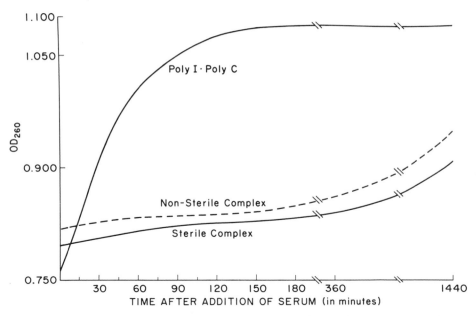

Figure 1. Comparison of rates of hydrolysis by human serum of
PolyI·PolyC and PolyICLC.

we tried it in monkeys, chimpanzees or people it had virtually no effect.
It was not toxic, but it did not induce IFN.

In collaborative studies with Nordlund and Wolfe,[4] we showed that in
primate serum there is a high concentration of hydrolytic enzymes that
hydrolyze and inactivate PolyI·PolyC. We prepared a derivative of
PolyI·PolyC by complexing PolyI·PolyC with Polylysine and carboxymethylcellulose, which we call Poly ICLC. The increased resistance to hydrolysis of PolyICLC as compared with PolyI·PolyC is shown in Figure 1.

A further demonstration of the increased stability of PolyICLC as
compared with plain PolyI·PolyC is shown in thermal denaturation studies.
When dsRNA is heated, the hydrogen bonds joining the two strands become
stressed, and at specific temperature (call the Tm), the two strands
separate. This separation is associated with a sharp increase in the
optical density at 260nm. The higher the Tm, the more stable is the
dsRNA. Figure 2 shows that the Tm of PolyICLC is 87°, while that of
PolyI·PolyC is 49°C.

When PolyICLC is given i.v. to primates, including humans, rather
high concentrations of IFN can be achieved, higher than is readily
achieved by administration of exogenous interferon. These IFN
concentrations are associated with significant dose related toxicity,
including fever and myalgia.

However, when the drug is given intramuscularly, adverse effects are
very much milder. Early studies were designed around the oncologists
maxim: give the maximum tolerated dose (MTD). However, it has been shown
repeatedly that with Biological Response Modifiers, like PolyICLC, the
MTD is associated with increased toxicity, and much less augmentation of
immune activities. Optimum effects are obtained at lower doses. We are
currently doing studies with low doses of PolyICLC, given i.m.

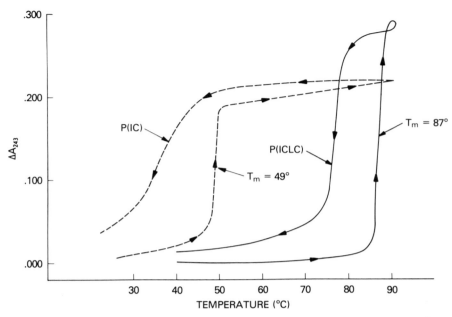

Figure 2. Comparison of thermal denaturation curves of
PolyI·PolyC and PolyICLC.

RESULTS

In a continuing open pilot study in patients with AIDS, low doses
(0.2–2 mg/patient) with and without oral zidovudine (AZT) were given,
i.m. to 22 patients. A profile of a representative group of patients is
shown in Table 1. These are seriously ill patients with Walter Reed
Classification mostly at 5 or 6, on a scale where 1 is minimal disease
and 10 is death.

PolyICLC was well-tolerated, with no significant clinical or
laboratory toxicity. Side effects consisted of a mild 12–24 flu-like
syndrome. There was low grade fever and malaise at the higher doses.

It is also apparent that, as used in these studies, PolyICLC is not
a cure. A number of these severely ill patients have died over the 24
months, several from non-AIDS causes, such as alcohol chirrosis. There
are some positive effects of the drug that should be mentioned.

There were, with a number of patients, a rise in T4 (helper
lymphocytes). These are the cells that are one of the primary sites of
attack by the virus. The T4 cells are important cells in developing
immune responses to infections, and their decrease in AIDS is the primary
cause of defective immune reactions against diseases.

The usual course for AIDS patients is for there to be a steady
decline in T4s, finally culminating in infection leading to death. In
most of our patients, after each injection of PolyICLC there was a
transient rise in T4 counts for 1–2 days, with a subsequent return to
pre-injection levels. A comparable transient rise is seen in the amount
of 2'5'A made by peripheral blood cells, Figure 3.

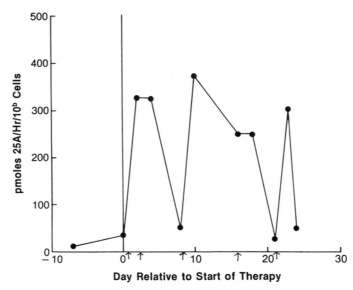

Day Relative to Start of Therapy

Figure 3. Transient changes in 2'5' oligo A levels after treatment with PolyICLC. (Arrows indicate days of injection).

In some of the patients there was seen a rise in T4s which persisted for a number of months. This stability was, not surprisingly, more evident in patients whose T4s were less severely reduced, than in the very advanced cases, Figure 4.

Table 1. Patient profile of AIDS patients receiving polyICLC plus AZT.

Group I: PolyICLC plus AZT			
Patient Number	Sex	Age	WR Stage
PR-01	F	41	6
PR-02	M	32	6
PR-03	M	25	6
PR-04	F	31	5
PR-05	M	25	6
PR-06	M	40	6
PR-07	M	40	5
PR-08	M	48	5
04-056	F	30	6
PR-10	M	35	2
PR-11	F	41	5
PR-12	M	40	3
PR-13	M	45	5
PR-16	M	43	6
PR-24	M	38	5
Mean		37	5
Std. Dev.		7	1

272

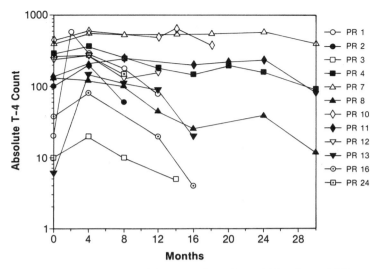

Figure 4. Long term effects of Poly ICLC plus AZT on T4 counts.

Another point of interest is the following: the administration of AZT frequently is associated with the development of anemia. None of the patients receiving the combination of Poly ICLC and AZT developed anemia. This is illustrated in Table 2 where the mean hemoglobin values of 16 patients, along with the standard deviations are listed. It can be seen that there was no decrease in hemoglobulin values in this group of patients. This possible protective effect is consistent with observations that Poly ICLC leads to an increase of bone marrow production of pluripotential stem cells, which can give rise to red cells.

In a number of patients there was a conversion from positive to negative for the presence of the P-24 antigen, a viral antigen. A few cases are shown in Table 3.

In a separate dosing study of PolyICLC in 8 AIDS patients, neuropsychological testing has shown a marked improvement in choice reaction time and the Purdue pegboard test at 16 weeks of treatment, with a heavy deterioration back to baseline when PolyICLC was discontinued (Salazar, Martin, Levy, et al.; unpublished) (Figure 5).

Table 2. Possible protective effect of PolyICLC on AZT induced anemia.

	Group I: PolyICLC plus AZT Hemoglobin				
	Base	4 mo	8 mo	12 mo	16 mo
Mean	12.5	12.3	11.8	12.0	12.1
Std. Dev.	1.9	1.7	2.3	2.0	2.7

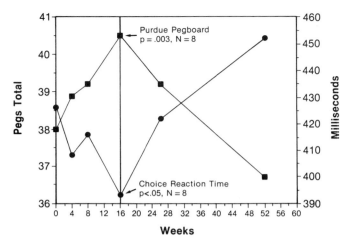

Figure 5. Effect of PolyICLC on neurologic performance in AIDS patients. (Injections were given weekly until week 16).

This contrasts with a gradual, statistically significant deterioration in choice reaction time seen in an untreated HIV + cohort (N = 41) over six months.

In concurrent clinical studies we have demonstrated a significant correlation (r -0.85, p < 0.0001) between deterioration in choice reaction time and a rise in cerebrospinal quinolinic acid (QUIN), an endogenous neurotoxin which shows marked elevation in AIDS dementia and mostly elevated in early HIV infection. QUIN formation from tryptophan is induced by interferon, and cerebral spinal fluid-QUIN levels correlate highly with beta-2-microglobulin in HIV-infected patients, suggesting a possible link between the interferon system defects and the mental symptoms in AIDS. It is unknown whether the improved performance was mediated through a decrease in QUIN.

There is another retroviral disease in humans, tropical spastic paraparesis, caused by the HTLV-1 virus. It resembles multiple sclerosis to some extent, but is not so severe. In studies done in Jamaica, we treated 20 patients with this disease, with low doses of PolyICLC, i.m. They were examined serially for several neurologic functions listed in Table 4.

Table 3. Effect of PolyICLC plus AZT on P-24 antigen levels.

	Virology, PolyICLC plus AZT				
		P-24 (pcg/mL)			
Patient number	Base	4 mo	8 mo	12 mo	16 mo
PR-01	89	15	21	305	–
PR-04	0	–	78	0	0
PR-11	80	37	–	0	0
PR-12	76	37	–	0	0
PR-13	85	–	16	11	–

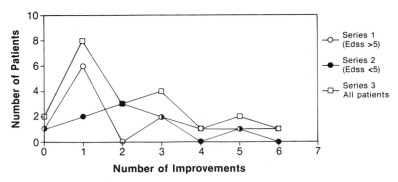

Figure 6. Number of improvements in tropical spastic paraparesis patients treated with PolyICLC.

The effect of PolyICLC was summarized as follows. The number of
neurological parameters that showed an improvement was determined for
each patient. The number of patients showing 1, 2 etc. improvements was
tabulated. The data was plotted in three different ways: (1) patients
less severely ill (disability score <5), (2) those with more severe
illness (disability ≤5), and (3) the total population. Figure 6 shows
this summary.

Table 4. List of tests done on tropical spastic paraparesis
patients.

CLINICAL HISTORY

Body weight (lbs)
Temperature
Fatigue (0, +1, +2, +3)
Bowel Movements/wk

NEUROLOGICAL EXAM

Kurtzke EDSS	(0-10)
Pyramidal	(0-6)
Cerebellar	(0-5)
Brain Stem	(0-5)
Sensory	(0-6)
Bowel & Bladder	(0-6)
Visual	(0-6)
Mental	(0-5)
Ambulation Index	(0-9)
Karnofsky Score	(0-100)

N/CLINICIAN MOTOR SCORE

Timed 20 ft walk (seconds)
20 ft Tamdem walk (seconds)
Nine Hole Peg Test:
Right Hand (seconds)
Left Hand (seconds)

Most of the patients showed improvements in one or more of the measured parameters. These are chronic patients who are not likely to show any improvement spontaneously. There was only one measurement of a decrease in a patient while on PolyICLC. The results are suggestive of at least a statistically significant, if not clinical significant effect.

The first phase of the study has been completed, and patients taken off the drug. Most of the patients are requesting that the treatment be reinstated because they felt better while on treatment. That could be placebo effect, but perhaps not.

CONCLUSION

In conclusion, the preliminary results of studies of the use of PolyICLC in treatment of two neurologic diseases show some signs of effectiveness. More detailed studies of doses and schedules would be required to optimize this therapy.

REFERENCES

1. A. Isaacs & J. Lindenmann, Proc. Roy. Soc., Sec. B, **147**, 258 (1957). "Virus Interferon, I - The Interferon."
2. A. K. Field, A. A. Tytell, G. P. Lampson, & E. M. R. Hilleman, Proc. Natl. Acad. Sci., USA, **58**, 1004 (1967). "Inducers of Interferon and Host Resistance, II - Multistranded synthetic Polynucleotides Complexes."
3. H. B. Levy, L. W. Law & A. S. Rabson, Proc. Natl. Acad. Sci., USA, **62**, 357 (1969). "Inhibition of Tumor Growth by Polyinosinic· Polycytidylic Acid."
4. J. S. Nordlund, S. M. Wolfe & H. B. Levy, Proc. Soc. Exp. Biol. Med., **133**, 439 (1970). "Inhibition of Biologic Activity by PolyI·PolyC by Human Plasma."

THE RELEASE OF 5-FLUOROURACIL FROM DEXTRAN AND XYLAN SYSTEMS

Charles G. Gebelein,[1,3,4]* Darrell Williams,[2] Kathy
Marshall[2] and Terri Slaven[2]

(1) Department of Chemistry
    Youngstown State University
    Youngstown, OH 44555
(2) Department of Chemical Engineering
    Youngstown State University
    Youngstown, OH 44555
(3) Department of Chemistry
    Florida Atlantic University
    Boca Raton, FL 33431
(4) LIONFIRE, Inc.
    Edgewater, FL 32132

ABSTRACT

    As part of a continuing study on the controlled release
of anticancer agents, 5-fluorouracil:poly(saccharide) adducts
were synthesized and their release profiles were determined.
The poly(saccharides) studied included dextran (15,000 and
60,000 molecular weight), xylan and chitosan. The 5-FU was
attached to the poly(saccharides) using 1,6-diisocyanato-
hexane as the coupling agent. In all cases, the release of
the 5-FU was rapid and was essentially complete within a few
days, when the adducts were prepared using a triethylamine
catalyst system. The kinetic patterns were neither true
Higuchi style plots, which are typical of monolithic systems,
nor were they zero-order profiles like previously described
methacrylate-type copolymers. When a tin catalyst was used,
however, the 5-FU was released over a much longer time
period, and the release profile was linear when plotted
against the square root of time. Only these tin catalyzed
adducts are potentially useful in controlled release of 5-FU.

INTRODUCTION

    Our laboratories have been pursuing a program aimed at developing
better controlled release systems involving 5-fluorouracil (5-FU), and

---

* To whom requests should be sent at the following address:
      Dr. Charles G. Gebelein, LIONFIRE, Inc., 1730 Umbrella
      Tree Drive, Edgewater, FL 32132-3111.

Biotechnology and Bioactive Polymers, Edited by C. Gebelein
and C. Carraher, Plenum Press, New York, 1994

other anticancer agents, for several years. Our previous research showed that certain copolymers which contain 5-FU, in the form of a monomer called [EMCF], release this anticancer drug in a zero-order release pattern.[1-5] Unfortunately, the polymeric systems were of a methacrylate-type and would not be biodegradable. The use of biodegradable polymers has been claimed, by some, to have importance in chemotherapy, although this could be debated when an affliction as severe as cancer is the topic of the contention. Nevertheless, our studies have aimed at evaluating the potential utility of different modes of controlled release therapy and the use of biodegradable materials is a currently active area of research in chemotherapy. Accordingly, we have been studying the use of biode-gradable matrices or supports for these anticancer agents. Some of our previous work using a biodegradable polyester in a monolithic system has been reported.[6,7] In those cases, the 5-FU was dispersed in the polymer as a monolithic system, and the release rates were not of the zero-order type. In this paper, we consider the use of various poly(saccharides) as support polymers. Related research along these lines has been reported by Ouchi.[8-13]

EXPERIMENTAL

1. MATERIALS USED

The dextran samples (15,000 and 60,000 MW) were obtained from ICN. The xylan used in most examples was isolated from oat spelts and was obtained from Aldrich Chemical. A second type of xylan, isolated from larchwood, was obtained from Sigma Chemical. The chitosan was obtained from Fluka. The 5-fluorouracil (5-FU) was obtained from PCR Research, and the 1,6-diisocyanatohexane (DIH) were obtained from Aldrich Chemical. The dimethyl sulfoxide (DMSO) was a spectroscopic, anhydrous grade. All other chemicals and solvents were of analytical grade, or better.

2. THE 1,6-DIISOCYANATOHEXANE DERIVATIVE OF 5-FU [DIFU]

In the general procedure, this reaction intermediate was prepared by reacting a slight excess of 5-fluorouracil [5-FU] (PCR, Inc.) with 1,6-diisocyanatohexane [DIH] (Aldrich), at room temperature in dimethyl sulfoxide (DMSO) solvent. A small quantity of triethylamine was used as the catalyst. An excess 5-FU was used to reduce the possibility of the reaction of two 5-FU molecules with the diisocyanate. (This excess was normally about 1 mole%.) The [DIFU] solutions were generally used without isolating the intermediate; the solutions always showed the expected strong N=C=O band in the IR spectra.

In a specific example, 12.04 g, (0.0926 m) 5-FU and 15.56 (0.0925 m) DIH were stirred in 100 mL DMSO, with 2 mL added triethylamine, for approximately one week. The initial N=C=O IR spectrum peak diminished to about half its intensity during this interval. The DIFU solution was used directly in the next step without any attempts at isolation.

3. PREPARATION OF POLY(SACCHARIDE) DERIVATIVES

In the general procedure, the poly(saccharide) was dissolved in DMSO and combined with the DIFU solution, prepared as described above, in the proportions necessary to prepare substitution levels ranging from 25% to about 100%. Note that these substitution levels are normally based on the repeat unit, and not on the number of hydroxyl groups (or amine groups) per repeat unit. These solutions were stirred magnetically, at room temperature, until IR monitoring showed no residual N=C=O peak. The 5-FU modified poly(saccharide) was then precipitated by the addition of excess methanol and was washed with this solvent until all traces of DMSO were removed. In some cases, the solid products were also washed repeatedly with ethyl acetate, to remove residual DMSO. The polymer samples were air and vacuum dried at room temperature. In general, the total amount of product was approximately the sums of the three reactants (5-FU, DIH and saccharide) and no attempts were made at further purification. The poly(saccharides) studied included dextran (15,000 and 60,000 MW)), xylan and chitosan.

In a specific example continued from the above section, 15.0 g (0.0925 m) dextran (15,000 MW) was dissolved in 100 mL DMSO and added to the DIFU solution described above. After stirring for one week, the isolated product showed no evidence of any remaining N=C=O band in the IR. The solution was then flooded with a large excess of methanol, to remove unreacted DIH, and the modified poly(saccharide) was separated by vacuum filtration. It was then air and vacuum dried. This reaction gave 100% substitution of the repeat units in the dextran. Presumably this reaction occurred at the single primary hydroxyl group, rather than the two secondary hydroxyl groups, but this was not studied in any detail.

4. PREPARATION OF THE XYLAN:5-FU ADDUCT USING A TIN CATALYST SYSTEM

Xylan, made from oat spelts, contains significant amounts of saccharic acids. Another type of xylan, isolated from larchwood, was also studied and this contained very low levels of free acid. In a second procedural modification, dibutyltindilaurate catalyst was used to make the DIFU solutions, and while adding it to the xylan. Dibutyltindiacetate was also tried but seemed less effective.

In a specific example, the DIFU solution was prepared by stirring a solution of 6.5 g (0.050 m) 5-FU and 8.1 g (0.048 m) DIH, plus 20 drops of dibutyltindilaurate in about 50 mL DMSO for one day. The viscous, paste-like, mixture showed about 50% of the initial level of the N=C=O peak in its IR spectrum. This was then mixed with 6.6 g (0.050 m) xylan, (from larchwood) previously dissolved in about 50 mL DMSO. After four days, there was no N=C=O peak was observed in the IR spectrum, and the mixture was poured into 1 L methanol, with stirring. The solid product was isolated by vacuum filtration, washed several times with methanol and then twice with ethyl acetate, and dried under vacuum at room temperature to give 16.7 g product (79% of theory). The sample did not melt, but did begin to decompose at about 240°C. This sample has 100% of the repeat units substituted, at either secondary hydroxyl group. It also has 50% of the hydroxyl groups reacted.

5. CONTROLLED RELEASE STUDIES

The poly(saccharide) derivatives were pulverized and sieved to obtain particles within the 60-100 mesh range. Exactly 0.500 grams of this powder was placed in special gas dispersion tubes which had a 13.0 mm cavity diameter, a 25 mm cavity height and a sinter glass thickness of 3.75 mm (tube porosity = "C"). These tubes were placed in one liter of distilled water, at 37°C, which was stirred mechanically in the manner previously described.[5] A various time intervals, an aliquot was removed, and the amount of 5-FU released, as a function of time, was determined using a Hewlett Packard Diode Array Spectrometer at 262 nm.

RESULTS AND DISCUSSION

Earlier research from our laboratories showed that the copolymers of a certain 5-FU containing monomer, called EMCF, released its 5-FU upon hydrolysis, in a zero-order kinetic pattern.[1-5] An example of these controlled release results is shown in Figure 1 for a series of EMCF copolymers with methyl methacrylate (MMA).

The reasons for this zero-order release profile have been postulated to arise because these copolymers are water insoluble, and, therefore, would not appear in the kinetic expression. In addition, the water used in the release studies was always in large excess, and thus also not in the kinetic expression. The net result of these features is that the copolymers always showed zero-order release of the 5-FU, at all levels of EMCF in the copolymer, and for MMA, MA (methyl acrylate), BA (butyl acrylate) and some other comonomers. While always linear, the exact rate of release could be varied widely by varying the percentage EMCF in the copolymer, and by varying the exact nature of the comonomer.

While this seems to be a desirable result, from the standpoint of cancer chemotherapy, some researchers suggested that a biodegradable carrier polymer would be preferred over the EMCF copolymers, since these methacrylate derivatives are not biodegradable. When the 5-FU was placed into a matrix of poly(caprolactone), however, the release rates were no longer zero-order, but instead followed the typical Higuchi pattern, wherein the release was linear when plotted against the square root of time.[14] Some typical results are shown in Figure 2.

Since the EMCF copolymers could be considered as prodrugs, some other studies from our laboratories tried to duplicate the zero-order kinetic pattern by using a monomeric 5-FU prodrug.[15] These attempts were unsuccessful, however. We decided to explore adducts of 5-FU to various natural, biodegradable polymers, and chose the poly(saccharides) as the

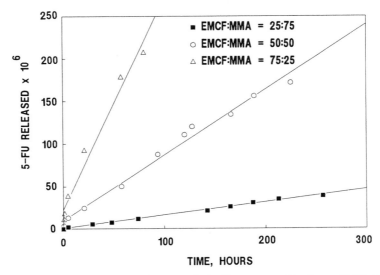

Figure 1. Zero order release of 5-fluorouracil from EMCF:MMA copolymers.[5]

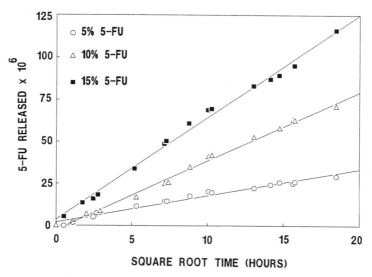

Figure 2. The release of 5-fluorouracil from poly(caprolactone) matrices.[7]

carrier molecule partly because Ouchi had reported some success with this kind of material.[8-13]

The 5-FU:poly(saccharide) adducts were prepared by the scheme outlined below in Equation 1, where xylan is used as the example of the poly(saccharide). This equation shows the use of dibutyltindilaurate as the catalyst although most of the polymers studied here were prepared using triethylamine as the catalyst. As will become apparent, however, the tin derivative is much more effective.

The poly(saccharides) used, and the number and type of groups available for possible reaction, per repeat unit, were: dextran [one primary –OH and two secondary –OH]; xylan [two secondary –OH]; chitosan [one primary –OH, one secondary –OH and one $-NH_2$]. The 5-FU adducts were prepared so that the maximum average substitution was only one 5-FU per repeat unit, and were prepared at levels of 25, 50, 75 and 100% based on this maximum substitution. Two different molecular weight ranges were studied for dextran (15,000 and 60,000). In addition, two different xylan samples were examined, one which was isolated from oat spelts and one isolated from larchwood. The latter contains very little amounts of free acid functionality, whereas the oat spelt material often has 10% or more free acid present. Less work was done with chitosan because this material is more difficult to use under our reaction conditions.

Ouchi has previously reported the synthesis of some chitin and chitosan derivatives of 5-FU.[8-13] More recently they have reported some derivatives of dextran with 5-FU.[12-13] The 5-FU:poly(saccharide) adducts were clearly shown to have antitumor activity in Ouchi's studies on animals, but little was published regarding the actual release rates of the 5-FU. While this seems almost trivial, there are several ways these adducts could function. They could, of course, serve as a source of bound 5-FU, which is released via hydrolysis or enzymatic cleavage. They could, however, be systems in which the 5-FU is entrapped within a crosslinked poly(saccharide) matrix. Since poly(saccharides) are relatively water

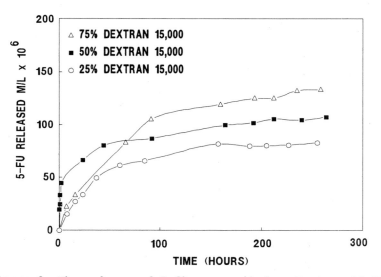

Figure 3. The release of 5-fluorouracil from Dextran 15,000 adducts.

sensitive, the 5-FU should tend to exit the matrix fairly rapidly, and then function as an antitumor agent. In either case, additional data was necessary to evaluate the actual release of the 5-FU from the adducts.

In this study we prepared 5-FU adducts of several poly(saccharides), of different structures, in order to access the actual release profiles and to determine which system, if any, has the most promise for a controlled release anticancer agent. Figure 3 shows the release profiles for three 5-FU adducts of dextran 15,000 molecular weight. Most of the 5-FU was released within 50 hours, and, except for the almost immediate burst effect, these release profiles were not zero-order.

Similar results are depicted in Figure 4 for the 60,000 molecular weight dextran. Likewise, the xylan adducts, prepared with triethylamine as the catalyst, also showed very rapid release of the 5-FU, with little indications of a zero-order release profile (Figure 5). The chitosan adduct, Figure 6, released its 5-FU almost immediately. All these adducts used the triethylamine catalyst system in the preparation of the polymers.

Figure 7 compares these release results for three different poly(saccharide) adducts; chitosan is omitted. It is readily apparent that all three systems release the 5-FU in less than 50 hours, although the total amount released varies. When this release data is plotted against the square root of time, Higuchi kinetics, Figure 8, there are some indications of linearity in the earliest time intervals, but, in general, the release does not seem to fit this pattern very well.

What is, perhaps, most surprising in this data is that the release occurs rapidly for all the 5-FU:poly(saccharide) adducts, regardless of the nature of the bonding to the repeat unit. For example, the most logical reaction site in the two dextrans is the primary hydroxyl, but xylan only has a pair of secondary hydroxyl groups. In spite of this, the initial release is faster from xylan than from either dextran. There does

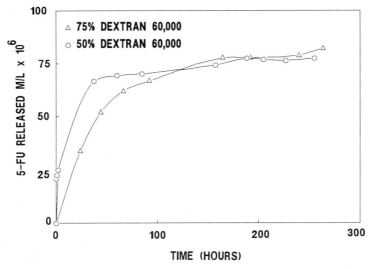

Figure 4. The release of 5-fluorouracil from Dextran 60,000 adducts.

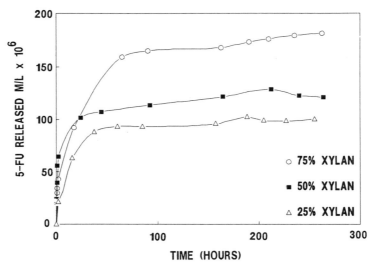

Figure 5. The release of 5-fluorouracil from xylan adducts.

not seem to be any major difference between the two dextrans, even though the molecular weight of the larger if four times great than for the smaller. In a similar manner, chitosan should react at the amine site rather than either hydroxyl group. Much prior data suggests that the resulting amide-type linkage should be less labile than the ester-type linkages in the other two poly(saccharides). In fact, chitosan releases the 5-FU even more rapidly than any of the others. Before drawing any firm conclusion on these matters, let's turn our attention to the other xylan adduct synthesis which used the tin catalysts.

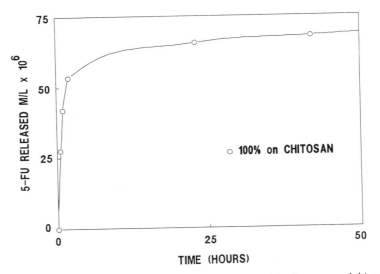

Figure 6. The release of 5-fluorouracil from a chitosan adduct.

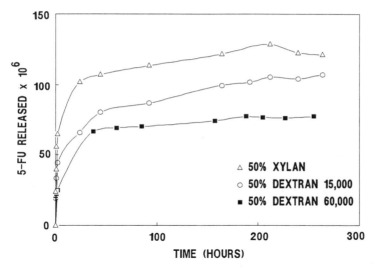

Figure 7. Comparison of the release of 5-fluorouracil from
xylan and Dextran adducts, at 50% loading.

Figure 9 shows the controlled release results for one of these xylan
adducts in which 50% of the hydroxyl groups have been reacted. Note that
this poly(saccharide adduct should have 100% of the repeat units reacted.
Again, we see that the release profile does not follow zero-order
kinetics, but when this same data is plotted against the square root of
time, as in Figure 10, we discover that it gives an excellent fit for the
Higuchi pattern. Whereas the triethylamine catalyzed adducts released all

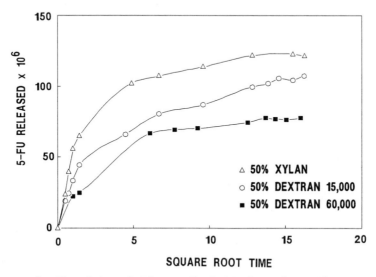

Figure 8. The data of Figure 7 plotted against the square
root of time.

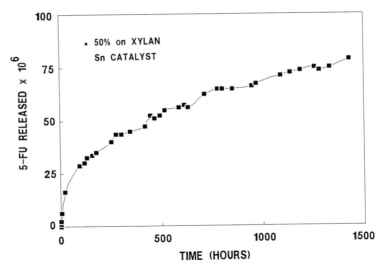

Figure 9. The release of 5-fluorouracil from a xylan adduct
prepared using a tin catalyst system.

their 5-FU in less than 50 hours, this sample is still releasing 5-FU
after nearly 1500 hours, some thirty times longer. Over essentially all
this time range, the 5-FU:xylan adduct obey the Higuchi kinetics typical
of a monolithic system.

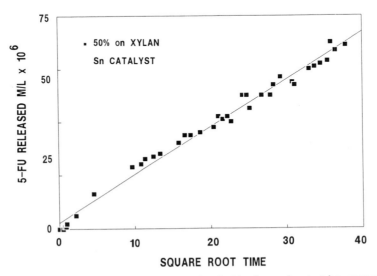

Figure 10. The data of Figure 9 plotted against the square
root of time.

The time lengths for the controlled release of 5-FU from this 5-FU: xylan adduct, prepared using a tin catalyst system, is comparable to the results obtained with the annealed 5-FU:poly(caprolactone) monolithic systems shown in Figure 2,[6,7] or the best EMCF copolymers, such as shown in Figure 1.[1-5] This adduct was the specific example described in the Experimental Section of this paper. All the examples in which the catalyst was triethylamine showed rapid release and relatively short times for total release. In short, the adducts formed via triethylamine catalysis do not work well for the controlled release of 5-FU, but the systems catalyzed with dibutyltindilaurate do function well. This result suggests a problem with some of the other results in the literature since most of the adducts of this type prepared by Ouchi utilized the triethylamine catalyst.[8-13]

Why does the tin-based catalyst system work while that based on triethylamine does not? There are several possible explanations, but the presence of side reactions or the aborting of the desired reaction by traces of water are clearly major problems. There is one potential side reaction in the synthesis sequence wherein the diisocyanate could react with two 5-FU units. If this double derivative forms, it is possible that this material could be embedded within the poly(saccharide) matrix. While we cannot state a *priori* that this double 5-FU:diisocyanate product will release the 5-FU more rapidly, other research from our laboratories does show that an ordinary 5-FU:isocyanate derivative, which is a 5-FU prodrug, releases 5-FU very quickly.[15] If a prodrug of this type forms here, then the release rates should be rapid, as observed. In fact, this side reaction would bind the 5-FU with only half of the diisocyanate, leaving the rest available to react with the poly(saccharide) and cause it to crosslink. Since the 5-FU:diisocyanate product would hydrolyze and release from this matrix fairly readily, probably promoted by the high hydrophilic nature of the poly(saccharide, these adducts should release the anticancer drug rapidly. Naturally, they would give positive results in animal studies since 5-FU is a known anticancer agent.

Contamination of the reaction with traces of water would, of course, destroy the diisocyanate and prevent coupling to the poly(saccharide). Some of the cases reported by Ouchi, specifically those using chitosan, utilize an aqueous LiCl media to dissolve the chitosan.[9,10] Whether this also destroyed all possibility of coupling is unknown, however.

If either of the above side reactions actually occurs, then these 5-FU;poly(saccharide) adducts are actually mixtures of two materials, one of which might show very rapid release rates.

The third possible scenario, that 5-FU chemically bonded to the poly(saccharides) by a carbamate linkage is highly labile, seems unlikely since the results reported in Figures 9 and 10 indicate that some of these adducts will show extended release times. Further studies are necessary to delineate this problem clearly, and to develop the full potential utility of the poly(saccharides) for the controlled release of 5-FU and other anticancer agents.

ACKNOWLEDGEMENT

Portions of this research was abstracted from the undergraduate thesis of DW, KM and TS, submitted in partial fulfillment of the requirements for the Bachelor of Engineering. The xylan adducts prepared using the tin catalyst, and their controlled release studies, were made by Donna Gardner and Tonya Ellis.

REFERENCES

1. C. G. Gebelein, Proc. Polym. Mat. Sci. Eng., **51**, 127-131 (1984).
2. R. R. Hartsough & C. G. Gebelein, Proc. Polym. Mat. Sci. Eng., **51**, 131-135 (1984).
3. R. R. Hartsough & C. G. Gebelein in: "*Polymeric Materials in Medication*," C. G. Gebelein & C. E. Carraher, Jr., Eds., Plenum Publ. Corp., New York, 1985, pp. 115-124.
4. C. G. Gebelein & R. R. Hartsough in: "*Controlled Release of Bioactive Materials, 11th International Symposium*," W. E. Meyers & R. C. Dunn, Eds., Controlled Release Society, Lincolnshire, IL, 1984, pp. 65-66.
5. C. G. Gebelein, T. Mirza & R. R. Hartsough in: "*Controlled Release Technology, Pharmaceutical Applications*," P. I. Lee and W. R. Good, Eds., Symp. Series #348, American Chemical Society, Washington, DC, 1987, pp. 120-126.
6. C. G. Gebelein, M. Chapman & T. Mirza in: "*Applied Bioactive Polymeric Systems*," C. G. Gebelein, C. E. Carraher & V. Foster, Eds., Plenum Publ., New York, 1988, pp. 151-163.
7. C. G. Gebelein, M. Davison, T. Gober & M. Chapman, Proc. Polym. Mat. Sci. Eng., **59**, 798-802 (1988).
8. T. Ouchi, T. Banba, M. Fujimoto and S. Hamamoto, Makrol. Chem., **190**, 1817-1825 (1989).
9. T. Ouchi, Poly. Mater. Sci. Eng., **62**, 412-415 (1990).
10. T. Ouchi, T. Banba, T. Z. Huang and Y. Ohya, Polymer Preprints, **31** (2), 202-3 (1990).
11. T. Ouchi, T. Banba, T. Matsumoto, S. Suzuki and M. Suzuki, J. Bioactive and Compatible Polymers, **4**, 362-371 (1990).
12. T. Ouchi, T. Banba, T. Matsumoto, S. Suzuki and M. Suzuki, Drug Design and Delivery, **6**, 281-287 (1990).
13. T. Ouchi, T. Banba, T. Z. Huang and Y. Ohya, in: "*Polymeric Drugs and Drug Delivery Systems*," R. M. Ottenbrite and R. L. Dunn, Eds., A.C.S. Symposium Series No. 469, 1991.
14. T. Higuchi, J. Pharm. Sci., **59**, 353 (1961).
15. C. G. Gebelein, D. Gardner and T. Ellis, This book.

RELEASE BEHAVIOR OF 5-FLUOROURACIL FROM CHITOSAN-GEL MICROSPHERES

MODIFIED CHEMICALLY AND THEIR ANTITUMOR ACTIVITIES

T. Ouchi, M. Shiratani, H. Kobayashi, T. Takei
and Y. Ohya

Department of Applied Chemistry
Faculty of Engineering
Kansai University
Suita, Osaka 564, Japan

In order to provide a device which releases 5-fluorouracil (5FU) in a controlled manner and has targetability to the specific organ cells, chitosan-gel microspheres immobilizing 5FU derivatives (aminopentyl-carbamoyl-5FU, aminopentyl-ester-methylene-5FU) coated with polysaccharides or lipid multilayers were prepared. The chitosan-gel microspheres cross-linked with glutaraldehyde (MS(CM)) were obtained by applying emulsion method using an ultrasonicator. The MS(CM)s were coated with polyanionic polysaccharides, such as CM-N-acetyl-α-1,4-polygalactosamine, CM-chitin and hyaluronic acid, by formation of polyelectrolyte complex membrane to give MS(CMG), MS(CMC) and MS(CMH), respectively. Moreover, MS(CML) was obtained by coating MS(CM) with dipalmitoyl phosphatidylcholine (DPPC) multilayer. The release rate of 5FU from the MS(CM) was depressed by immobilization of 5FU derivatives into MS(CM) via covalent bonds and by coating with polysaccharide or DPPC multilayer at 37°C. The temperature-sensitive release behavior of 5FU from MS(CM) was achieved between 37°C and 42°C by coating with DPPC multilayer. Moreover, MS(CML-CM-Poly(GalNAc)) and MS(CML-Lac), MS(CMG) immobilizing 5FU derivatives showed the cell specific cytotoxicities against SK-Hep-1 *human hepatoma* cells and HLE *human hepatoma* cells *in vitro*, respectively.

INTRODUCTION

Chitosan is low toxic, non-immunogenetic, biodegradable polysaccharide. Partially N-acetylated chitosan was reported to be collected into L1210 *leukemia* cells and to inhibit the growth of tumor cells.[1,2] Therefore, chitosan is expected to be used as a carrier in a drug delivery system (DDS). 5-Fluorouracil (5FU) has a remarkable antitumor activity which is accompanied, however, by undesirable side-effects. In order to control the release behavior of 5FU, to reduce the side-effects of 5FU, to achieve the cell specific transport of 5FU to the target tumor cells and to exhibit a high antitumor activity, the present paper is

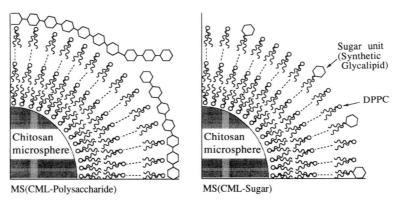

Figure 1. Estimated structures of MSs prepared by emulsion method.

LEGEND

MS	Chitosan gel MS	Lipid	Coating polysaccharide Cationic	Anionic
MS(CM)	O			
MS(CML)	O	ⓐ		
MS(CMA)	O			O
MS(CMLA)	O	ⓐ		O
MS(CMLC)	O	ⓐ	O	
MS(CML-Sugar)	O	ⓑ		

ⓐ:DPPC, ⓑ:Glycolipid

concerned with the fundamental study on the controlled release of 5FU from some kinds of chitosan gel microspheres (MS) coated with polysaccharide chains recognizing tumor cells; our research objects are (1) the construction of chitosan gel microspheres (MS) immobilizing 5FU coated with lipid multilayer and polysaccharide chains, (2) the temperature-sensitive controlled release of 5FU from the MS, and (3) the possibility of target of 5FU to the objective cells by coating the MS with polysaccharide chains.

EXPERIMENTAL

1. CONSTRUCTION OF CHITOSAN GEL MICROSPHERES CONTAINING 5FU

Figure 1 illustrates schematically the structures of the microspheres prepared. MS(CMLS) consisted of chitosan gel microsphere core crosslinked by glualaldelyde, lipid multilayer and one polyelectrolyte complex membrane.

As an immobilized drug, a temperature-sensitive barrier and the

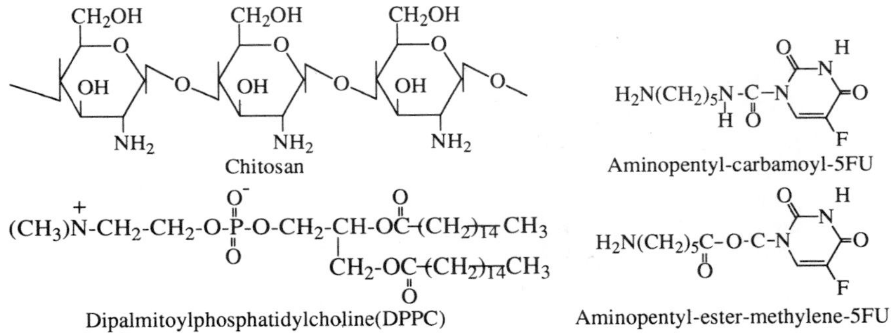

CH₂OH structures of Chitosan, Aminopentyl-carbamoyl-5FU, Dipalmitoylphosphatidylcholine (DPPC), and Aminopentyl-ester-methylene-5FU

Chitosan

Aminopentyl-carbamoyl-5FU

$H_2N(CH_2)_5N-C-N$ ... (Aminopentyl-carbamoyl-5FU)

$(CH_3)N-CH_2-CH_2-O-P-O-CH_2-CH-OC(CH_2)_{14}CH_3$
$CH_2-OC(CH_2)_{14}CH_3$

Dipalmitoylphosphatidylcholine(DPPC)

$H_2N(CH_2)_5C-O-C-N$ ... (Aminopentyl-ester-methylene-5FU)

Aminopentyl-ester-methylene-5FU

Structures of materials used.

coating materials, aminopentyl-carbamoyl-5FU hydrochloride, amino-entyl-ester-methylene-5FU hydrochloride, dipalmytoyl phoshpatidylcholine (DPPC) and cationic or anionic polysaccharide were used, respectively. (These structures are shown above.) Chitosan were cross-linked and 5FU amino derivatives were immobilized chemically with glutaraldehyde to give MS(CM) containing 5FU. MS(CML) and MS(CMLA), MS(CMLC) were prepared by the techniques of lipid multilayer construction and polyelectrolyte complex membrane formation. We prepared small size chitosan gel microspheres coated with DPPC immobilizing 5FU by such W/O emulsion and ultrasonication techniques. It was confirmed by SEM observation view that the diameters of the obtained MSs could be regulated to be about 0.3-0.5 μm (Figure 2). The MS(CML), MS(CMLA) and MS(CMLC) treated with DPPC were confirmed to be coated with a lipid multilayer of DPPC having gel/liquid-crystalline phase transition temperature at 41.4°C, similar to DPPC liposome by measurement of DSC. The location of 6-CM-N-acetyl-α-1,4-polygalactosamine and galactose chains on the surface of the obtained MS(CML-CM-Poly(GalNAc)) and MS(CML-Lac) was checked by the phenomena of lectin mediated specific aggregation of the MSs; the transmittance of PBS suspension of the MS(CML-CM-Poly(GalNAc))s and MS(CML-Lac)s varied dramatically by addition of APA and RCA 120 lectins, respectively.

Moreover, by introduction of lipid multilayer, the good dispersibility of microspheres was found to be achieved in pH 7.4 phosphate buffer solution (Figure 3).

The cytotoxicities of MSs immobilizing aminopentyl-carbamoyl-5FU and aminopentyl-ester-methylene-5FU coated with kinds of anion charged polysaccharide or glycolipid against SK-Hep-1 or HLE *human hepatoma* cells were evaluated *in vitro* by Scheme I.

RESULTS AND DISCUSSION

1. SLOW RELEASE OF 5FU FROM MS(CM)

The release behavior of 5FU from the obtained MS immobilizing the 5FU derivative in physiological saline at 37°C was investigated. Firstly, 5FU derivative immobilized in chitosan gel MS was hydrolyzed to give free 5FU itself and to afford no 5FU derivative. Secondly, free 5FU was permeated from MS through hydrogel, lipid multilayer and polyelectrolyte complex membrane. As an example, the results of release rate of 5FU from

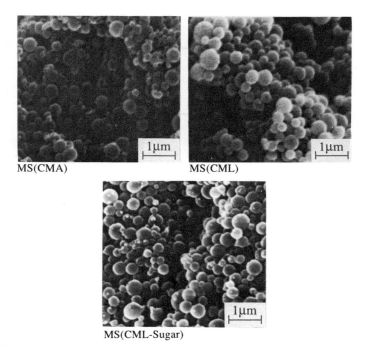

MS(CMA)   MS(CML)

MS(CML-Sugar)

Figure 2. SEM views of MS(CMA), MS(CML) and MS(CML-sugar).

MS(CM) containing free 5FU or immobilizing 5FU derivatives in physiological saline at 37°C are shown in Figure 4. The release rate of 5FU from
the MS was depressed by immobilization of 5FU derivatives into MS(CM) via
covalent bonds and by formation of polyelectrolyte complex membrane
(Figure 5).

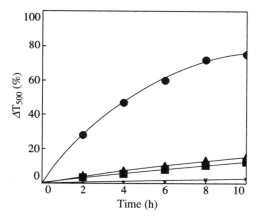

Figure 3. Effect of lipid multilayer on the dispersibility of
MSs in 1/15 M-phosphate buffer solution (pH 7.4).

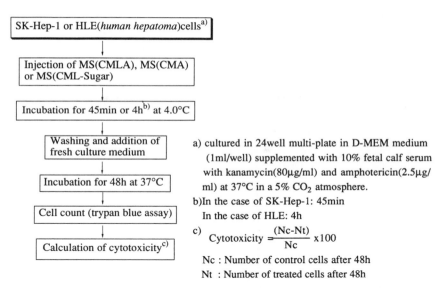

SK-Hep-1 or HLE(*human hepatoma*)cells[a]

↓

Injection of MS(CMLA), MS(CMA) or MS(CML-Sugar)

↓

Incubation for 45min or 4h[b] at 4.0°C

↓

Washing and addition of fresh culture medium

↓

Incubation for 48h at 37°C

↓

Cell count (trypan blue assay)

↓

Calculation of cytotoxicity[c]

a) cultured in 24well multi-plate in D-MEM medium (1ml/well) supplemented with 10% fetal calf serum with kanamycin(80μg/ml) and amphotericin(2.5μg/ml) at 37°C in a 5% CO_2 atmosphere.

b)In the case of SK-Hep-1: 45min
  In the case of HLE: 4h

c)
$$\text{Cytotoxicity} = \frac{(Nc-Nt)}{Nc} \times 100$$

Nc : Number of control cells after 48h
Nt : Number of treated cells after 48h

Scheme 1. Procedure of measurement of cytotoxicity by MSs against *human hepatoma* cells.

2. TEMPERATURE-SENSITIVE RELEASE OF 5FU FROM MS(CML)

The temperature-sensitive release of 5FU from MS(CML) coated with lipid multilayer of DPPC having liquid-crystalline phase transition temperature (T_c = 41.4°C) was tested in physiological saline. The release rate of 5FU increased significantly with temperature change from 37°C to 42°C, as shown in Figure 6. It was thought that the barrier ability of

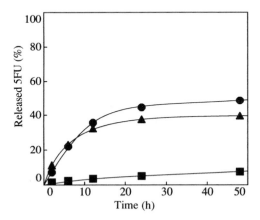

Figure 4. Release rate of 5FU from MS(CM) containing free 5FU or immobilizing 5FU derivatives in physiological saline at 37°C. ● = Free 5FU, ▲ = Aminopentyl-carbamoyl-5FU, ■ = Aminopentyl-ester-methylene-5FU.

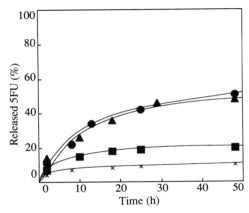

Figure 5. Release rate of 5FU from MS(CM), MS(CMChi), MS(CML) and MS(CMLChi) immobilizing aminopentyl-carbamoyl-5FU in physiological saline at 37°C.

DPPC multilayer on the release of 5FU was dropped by liquid-crystalline phase transition: The ON-OFF controlled release of 5FU was found to be achieved by construction DPPC lipid multilayer in the MS(CM).

3. CELL SPECIFIC ACTIVE TARGETING OF 5FU TO HEPATOCYTO BY USING GALACTOSE RESIDUE AS COATED SACCHARIDE CHAIN

The possibility of targeting of 5FU to *hepatoma* cells by using chitosan gel microspheres coated with polysaccharide or glycolipid was investigated. The MS(CM) coated with N-acetyl – α-1,4-polygalactosamine chains was found to exhibit the higher cytotoxicities against SK-Hep-1

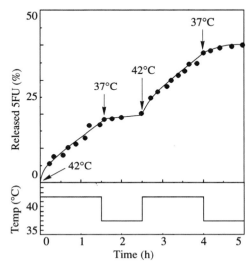

Figure 6. Effect of step-wise temperature change between 37°C and 42°C on the release rate of 5FU from MS(CML) immobilizing aminopentyl-carbamoyl-5FU after incubation for 10 h in physiological saline at 37°C.

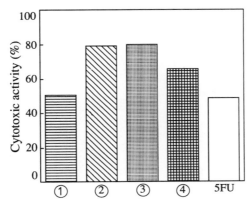

Figure 7. Cytotoxicity of MS(CMLA) immobilizing aminopentyl-
carbamoyl-5FU coated with different kinds of
anionic polysaccharide against SK-Hep-1 *human
hepatoma* cells *in vitro*. Coated anionic
polysaccharide: 1 = Heparin/Na, 2 = CM-chitin/Na,
3 = CM-N-acetyl-1,4- poly-galactosamine/Na, 4 =
Alginic acid/Na. 5FU concentration = 50 μg/well.

and HLE *hepatoma* cells than the MS(CM) coated with other polysaccharides
(Figures 7 and 8).

The MSs coated with galactolipid immobilizing 5FU also exhibited the
cell specific cytotoxic activity against HLE *human hepatoma* cells (Figure
9). These results can be explained by the smooth uptake of the MSs coated
with N-acetyl α-1,4 polygalactosamine or galactose into *hepatoma* cells
via galactose-receptor mediated endocytosis. It was observed by the SEM
views that the MS-CM-CM-Poly(GalNAc))s were easily uptaken into SK-Hep-1
and HLE *hepatoma* cells.

These microspheres were found to be disintegrated and decomposed
with the passage of long time after achievement of release of 5FU. Therefore, such type chitosan gel MSs could be concluded to be available as
the DDS technique.

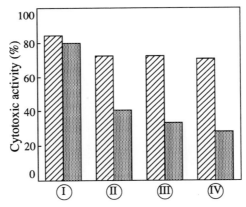

Figure 8. Cytotoxicity of MS(CM) and MS(CM)-Polysaccharide
immobilizing 5FU derivatives against HLE human
hepatoma cells *in vitro*. ▨ = aminopentyl-ester-
methylene-5FU, ▨ = aminopentyl-carbamoyl-5FU. I =
MS(CMG), II = MS(CMH), III = MS(CMC), IV = MS(CM).

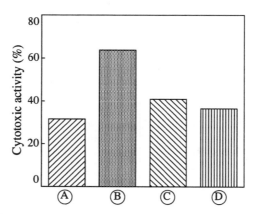

Figure 9. Cytotoxicities of MS(CML) and MS(CML-sugar) immobilizing aminopentyl-carbamoyl-5FU coated with some kinds of glycolipid against HLE *human hepatoma* cells *in vitro*. A = MS(CML), B = MS(CML-Lac), C = MS(CML-Glu), D = MS(CML-Man). 5FU concentration = 30 μg/well.

REFERENCES

1. A. E. Sirca, R. J. Woodman, J. Natl. Cancer Inst., **47**, 377 (1971).
2. A. E. Sirca, R. J. Woodman, Pharmacology, **29**, 681 (1970).

BIOLOGICALLY ACTIVE POLYPHOSPHATE AND POLYPHOSPHONATE ESTERS - NUCLEIC ACID ANALOGS

Charles E. Carraher, Jr., Daniel S. Powers and Bhoomin Pandya

Florida Atlantic University
Department of Chemistry
Boca Raton, FL 33431
and
Wright State University
Department of Chemistry
Dayton, Ohio 45435

A number of polyphosphate and polyphosphonate esters derived from biologically active diols were synthesized. The diols included androstendiol, amcinafal, dienestrol, 5-iodo-2-deoxyuridine and diethylstilbestrol. The products showed a wide range of activities towards the bacteria tested and generally decent to good activity towards the cancer cell lines tested.

INTRODUCTION

Polyphosphate esters, in the form of nucleic acids, have been synthesized by nature since life began in any complex form. The first non-biological synthesis of polyphosphonate and polyphosphate esters was reported by Arvin in 1936.[1] He reported the formation of a resinous material by refluxing phosphorus oxychloride in the presence of 2,2-bis-(4-hydroxyphenyl)propane. Zentfman and Wright,[2] in 1952, reported the synthesis of polyphospate and polyphosphonate esters, again employing the melt condensation technique producing mainly glassy solids. Sander and Steininger,[3] in 1967, reviewed the synthesis of phosphorus-containing polymers including polyphosphonate and polyphosphate esters. Since then, extensive work has been done employing the use of interfacial techniques in the production of polymers.[4] Related to this is the synthesis of phosphorus-containing polymers by solution techniques.[4,5] The synthesis of these esters has been recently reviewed by Carraher, Millich, and co-workers.[4,6,7]

The first reported use of the interfacial technique in the synthesis of polyphosphonate esters was by Aufderhaar,[8] and Kuznetsov et al.,[9] in 1961. The first reported synthesis of polyphosphate esters by the interfacial technique was made by Rabek and Prot,[10] in 1963. Recently, new synthetic condensation techniques have been developed by Carraher and co-workers for the synthesis of condensation organometallic polymers, and

Biotechnology and Bioactive Polymers, Edited by C. Gebelein
and C. Carraher, Plenum Press, New York, 1994

possibly suitable for use in the synthesis of poly-phosphonate and polyphosphate esters.[11-17]

With the exception of the work of Carraher, Millich and co-workers,[4,6,7] the emphasis in the synthesis of polyphosphonate and polyphosphate esters has been the production of industrially acceptable flame retardants, specialty adhesives, etc., for instance references 3-7.

Carraher, Millich, and co-workers have synthesized a number of biologically active phosphorus-containing polymers, but did not emphasize the use of biologically active Lewis bases, but mainly relied on the general biological activity of the phosphorus-containing moiety to impart the desired biological activity.[4,6,7] More recently Carraher and co-workers have found that the combining of the biologically active moieties within the same polymer may result in a synergistic effect whereby the biological activity, but not general toxicity, of the combined polymer is greater than that of either monomer alone. Thus, the polymer produced from the reaction of dipotassium tetrachloroplatinate and methotrexate is more active against selected cancers, but less toxic to normal cells than is the tetrachloroplatinate or methotrexate (C. Carraher, D. Giron, unpublished results). Therefore, the present research focused on the use of biologically active diols. Thus, by incorporating biologically active diols into the backbone of these phosphorus-containing polymers, it might prove helpful to the development of new drugs for medicinal purposes.

EXPERIMENTAL

1. CHEMICALS

The following chemicals were used without further purification: Phenylphosphonic dichloride, Eastman Organic Chemical (Rochester, NY). Phenyl dichlorophosphate, Aldrich Chemical Co., Inc. (Milwaukee, WI). Phenyl phosphonothioic dichloride, Aldrich Chemical Co., Inc. (Milwaukee, WI). Diethylstilbestrol, Aldrich Chemical Co., Inc. (Milwaukee, WI). Androstendiol, Aldrich Chemical Co., Inc. (Milwaukee, WI). 5-Iodo-2'-deoxyuridine, Aldrich Chemical Co., Inc. (Milwaukee,WI). Dienestrol, Sigma Chemical Co. (St. Louis, MO). Amcinafal, Batch #8B, U.S. Biochemical Corp. (Cleveland, OH).

2. GENERAL SYNTHESIS

Polymerizations were carried out in a 1-pt. Kimax emulsifying mill jar, which was placed on a Waring Blendor (model 1120). The rotor speed was listed at 15,000 rpm. The jar was fitted with a metal screw-type lid which had a 3/4 inch hole in it. A large mouthed funnel was placed through the hole to allow large volumes of solution to be delivered in a short time (each 100 mL within three seconds). This was to help eliminate errors due to rate of addition.

Acid chloride (0.66 mL, 4 mmole) in 50 mL carbon tetrachloride was quickly added, with stirring, to the reaction jar containing diol (1072 mg, 4 mmole) and 50 mL 0.08N sodium hydroxide, both dissolved in 50 mL water. After a few seconds of rapid stirring, a solid begins to precipitate from the reaction mixture. After 1 minute of blender stirring time at 25°C, 50 mL of 0.1N hydrochloric acid is added to "kill" (i.e., neutralize excess sodium hydroxide) the reaction.

It should be noted that product yield is lowered if the sodium hydroxide is not rapidly (with 10 seconds) neutralized by the addition of dilute hydrochloric acid. The reason for this is because polyphosphonate and polyphosphate esters are unstable in strong basic solutions and undergoes degradation. Once the dilute hydrochloric acid has been added, the blender is turned off.

3. INSTRUMENTATION FOR CHEMICAL CHARACTERIZATION

The weight average molecular weight (M_w) determinations were accomplished using a Brice-Phoenix Universal Light Scattering Photometer, Model BP-3000. These determinations were carried out in the usual manner using 1% polymer solutions with serial dilutions in dimethyl sulfoxide (DMSO). The refractive index increments, dn/dc, were determined using a Bausch and Lomb Abbe Refractometer, Model #3-L.

Pyroprobe-mass spectroscopy and elemental analysis (for carbon, hydrogen and phosphorus) were done by the USAF Material Laboratory, Analytical Department at Wright-Patterson AFB, Dayton, Ohio. The mass spectral analyses were performed using a DuPont Mass Spectrograph, Model #21-941 coupled with a Hewlett-Packard Computer, Model HP-2216C. The elemental analyses were performed using a Perkin-Elmer Elemental Analyzer, Model #240.

For identification purposes, a Perkin-Elmer Grating Spectro-photometer, Model #457, was used to obtain infrared spectra. The polymer was embedded into potassium bromide, (KBr) pellets for this analysis.

4. PRELIMINARY BIOLOGICAL CHARACTERIZATION

Selected types of polymers were utilized to perform toxicological experiments. A wide variety of bacterial cells, along with certain types of tissue culture cells, were chosen for preliminary biological studies. The results of these preliminary investigations could prove to be vital for potential drug applications.

The bacterial tests were conducted in the usual fashion. Tryptic Soy Agar (TSA) plates were seeded with suspensions of the bacterial test organism to produce an acceptable lawn of test organism. After 24 hours of incubation at 37°C, the polymeric compounds (0.1 mg) were deposited as solids on the plates, as small round spots. The plates were incubated again at 37°C for 24 hours, removed and examined visually for zones of inhibition. These zones are indicated by a clear area around the deposits of polymeric solids. The plates were placed back into the incubator for an additional 24-hour period and observed again for zones of inhibition.

Three cell lines were used for tissue cell studies. These cell lines were L929 (a mouse connective tissue tumor cell line), HeLa (a human cervical cancer cell line), and BHK-21 (baby hamster kidney cells). The cells were grown to confluency in a T-75 tissue culture flask. The growth medium is Dulbecco's Modified Eagle Medium (DMEM), which is a mixture of 450 mL of Dulbecco's solution, 50 mL of fetal calf serum, and 5 ml of antibiotic Penicillin-streptomycin.[28] The cell lines were then trypsinized, the cells suspended and counted (using a Coulter Counter, Model Z_{BI}), then plated onto Corning 1 mL well plates. The plates were incubated overnight in a 100% humidity, 5% carbon dioxide incubator at 37°C. This was to allow the cells to form a monolayer on the bottom of the plates. The next day the DMEM was aspirated off and substituted with DMEM that contained various microgram quantities of the polymers. This polymer-treated media was incubated overnight at 37°C. The DNEM-polymer solution was aspirated off and replaced with 1 mL of a solution of DMEM and Neutral Red in 0.85% sodium chloride. The plates were placed back into the incubator for two hours to allow the cells to take up the dye. The plates were removed after the 2-hour period and the media was aspirated off. The cells were then washed twice with 1 mL aliquots of 1:1 Sorensen Citrate Buffer/Ethanol. (This solution was used to develop the full acid color to the dye). The cells were then visually inspected to determine the range of cell death. After the cell death had been rated, the plates were then placed back into the incubator for a final 24-hour period. When this final incubation period had expired, the plates were removed and once again, visually examined for cell death.

The cell death is rated according to the amount of dye taken up by the cells. Viable cells will take up the dye, whereas, dead cells will not. Cell death is rated on a scale from 0-4. A rating of 4, indicates 100% cell death. A rating of 2, indicates 50% cell death and so on.

RESULTS AND DISCUSSION

1. BIOLOGICALLY ACTIVE DIOLS

There are a large number of diols commercially available; of particular importance are those diols which are noted for their biological activities. One diol that was emphasized in the synthesis of polyphosphonate and polyphosphate esters was DES, or diethylstilbestrol. DES is a synthetic estrogenic hormone. This chemical was used as a growth hormone in cattle feeds. It was banned by the U.S. Food and Drug Administration in 1972 after it was shown to be cancer-producing in some animals.[18] Formerly, DES was prescribed in treating menopausal symptoms as well as for the prevention of miscarriages. Studies have shown a high incidence of vaginal cancer in the teenage daughters of women who took the drug during pregnancy.[19]

DES is known to be an effective post-coital contraceptive, even when taken as long as 72 hours after intercourse. Its use is now limited to situations involving rape, incest and other emergencies. The U.S. Food and Drug Administration considers DES a potentially dangerous drug and warns against using it routinely as a contraceptive.

Although this chemical does have its drawbacks, it might be useful when placed in the backbone of polymeric compounds. By tieing up the chemical in the polymer, it might be possible to reduce its harmful effects, yet permit desired biological activities to occur. At any rate, it might prove effective in the study of different reaction systems for other medicinal chemicals.

Dienestrol, another biologically active diol, is an estrogenic hormone like DES and is similar in structure. It has been used as a drug against virus and interferon production. The structural similarities can be noted in the two figures below.

Structure 1. Diethylstilbestrol (DES).

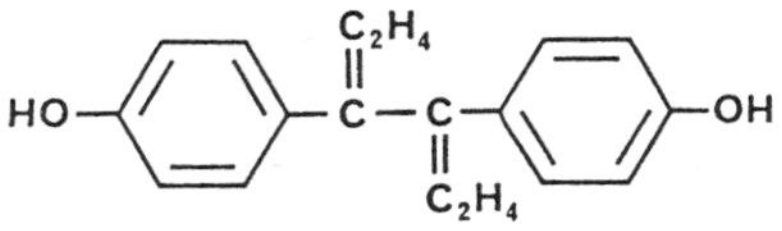

Structure 2. Dienestrol.

Nucleosides are known for their biological importance since they are incorporated into the backbone of nucleic acids. The gene that carries genetic information is composed of deoxyribonucleic acid, or DNA. The sugar moieties, which are incorporated into the nucleoside of the DNA structure, contain hydroxyl groups. Therefore, certain nucleosides, such as, 5-Iodo-2'-deoxyuridine (5-IDU), could be used in the present study as biologically active diols. If 5-IDU can be substituted into the structure of a polyphosphate or poly-phosphorate ester, there might be a chance that it can be used to treat certain cancers or herpes viruses since 5-IDU itself is active against selected cancers and viruses.[20] The structure of 5-IDU is shown below.

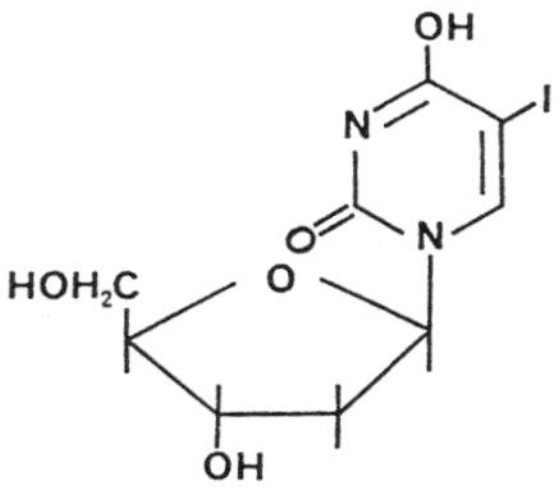

Structure 3. 5-Iodo-2'-deoxyuridine.

Certain compounds have structures analogous to the cholesterols. Cholesterols are used as sources of energy during biological reaction cycles. Structures like Androstendiol and Amcinafal are two compounds which have similar structures to cholesterols. Androstendiol has some effect on virus and interferon production. Amcinafal is used against virus replication and interferon production.

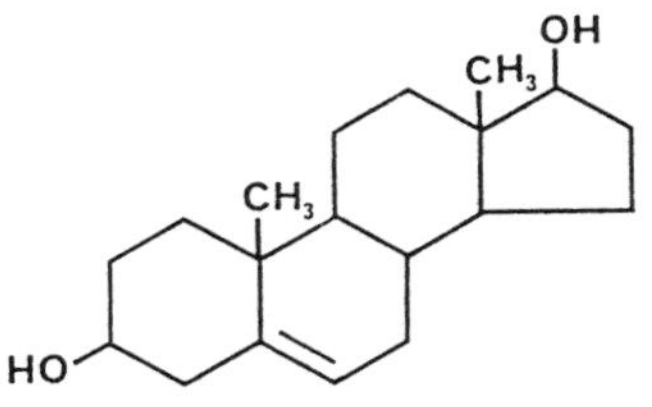

Structure 4. Androstendiol.

Structure 5. Amcinafal.

2. PRODUCT PHYSICAL RESULTS

The precise structural results are given elsewhere.[12] Briefly, the products are polyesters with chain lengths of about 100. Exact chain length data is given in Table 1.

3. BIOLOGICAL RESULTS

The treatment of cancer has led to the use of chemotherapeutic substances. Chemotherapy is usually used when the tumor has spread over the body or if the tumor cannot be removed by surgery or destroyed by radiation. A cancer drug may reach all parts of the body and should adhere to reactive sites without causing extensive lethal damage to the living host. The problem with chemotherapy, which now exists, is that the chemical doesn't distinguish between malignant and normal cells of the body. This has led to the search for compounds which have the ability to reduce or stop the growth rate of tumors while minimizing unwanted secondary actions.

Table 1. Refractive index increments (dn/dc), weight average molecular weight (M_w) and the degree of polymerization (DP) of phosphorus-containing polyesters used in biological testing.

Phosphorus acid chloride*	Biological diol**	dn/dc	M_w	DP
PPD	DES	2.97×10^{-1}	5.00×10^4	1.28×10^2
PPD	DEL	2.97×10^{-1}	5.00×10^4	1.28×10^2
PPD	ASD	7.78×10^{-2}	7.04×10^4	1.71×10^2
PDP	ACB	1.40×10^{-1}	5.39×10^4	9.46×10^1
PTD	DES	1.20×10^{-1}	5.03×10^4	1.24×10^2
PTD	DES	1.10×10^{-1}	5.38×10^4	1.32×10^2

*	PPD = Phenylphosphonic dichloride
	PDP = Phenyl dichlorophosphate
	PTD = Phenyl phosphonothioic dichloride
**	DES = Diethylstilbestrol
	ACB = Amcinafal
	DEL = Dienestrol
	ASD = Androstendiol

Polymers may have several advantages over low molecular weight drugs. One possible advantage is that fragments from large molecular weight polymers can be released slowly into the biological environment. This will allow an effective concentration to be present over a period of time. The result might be to lower the toxic effect of the drug which previously was known to be toxic when administered by itself in large single dosages.[21] Another advantage may be that polymeric materials may have a mechanism of action which is different from that of the reactants themselves.[22] Finally, because of its size, its movement may be limited allowing certain undesired toxicities to be minimized.

Research has shown that incorporating different drugs into a single polymer can lead to synergistic effects. Drugs which have been used to treat cancer initially have some positive effect in reducing or slowing cancerous growth, however, the result tends to dissipate with time. But when different drugs are incorporated into polymers the results may be additive or even logarithmic in the treatment of malignant cells. Polymers can offer a wider variety of possibilities of compositions, offering the opportunity of optimizing the individual factors of solubility, electron effect, and stereo arrangements that are important in the design of good chemotherapeutic agents.[23]

The testing of polymers for antitumor activity is difficult since there is no universally accepted animal model for cancer and live animal tests are typically costly. Consequently, a large number of tumor cell lines are used for this purpose. For a preliminary evaluation, three types of tissue cells were evaluated. These lines were L929 (mouse connective tissue tumor line), BHK-21 (baby hamster kidney), and HeLa

Table 2. Visual observation results of the effect of select phosphorus-containing polyesters on the tissue cell line L929 after 24 hours incubation.

Phosphorus acid chloride[*]	Biological diol[**]	Concentrations (µg/mL)					
		8	16	40	100	150	Control
PPD	IDU	0	1	2	4	4	0
PPD	DEL	1	2	2	3	4	0
PPD	ACB	1	1	2	2	3	0
PPD	ASD	0	0	1	1	2	0
PTD	DES	1	1	2	2	4	0
DMSO Control		0	0	0	0	0	0

[*] PPD = Phenylphosphonic dichloride
    PTD = Phenyl phosphonothioic dichloride

[**] DES = Diethylstilbestrol
     IDU = 5-Iodo-2'-deoxyuridine
     DEL = Dienestrol
     ASD = Androstendiol
     ACB = Amcinafal

Ratings:
0 =   0% cell death
1 =   25% cell death
2 =   50% cell death
3 =   75% cell death
4 = 100% cell death

Table 3. Eosin stained results of the effect of select
phosphorus-containing polyesters on the tissue cell
line HeLa after 48 hours incubation.

Phosphorus acid chloride[*]	Biological diol[**]	Concentrations (μ g/mL)					
		8	16	40	100	150	Control
PPD	IDU	0	2	2	4	4	0
PPD	DEL	1	1	2	3	4	0
PPD	ACB	1	1	2	3	4	0
PPD	ASD	1	1	1	1	2	0
PTD	DES	1	2	2	4	4	0
DMSO Control		0	0	0	0	0	0

*   See Table 2.
** See Table 2.

(dpitheloid carcinoma, cervix, human). The results of these cell lines
are tabulated in terms of percentage of cells that survived (Tables 2-5).

It should be noted that two polymers were picked at random and
tested at two concentrations (100 and 150 μg/mL) with the tissue cell,
WI-38, which is a normal tissue cell. The results of the two polymers
(phenylphosphonic dichloride with dienestrol and phenylphosphonic
dichloride with androstendiol) indicated that the polymers had the same
effect on the normal cells as with the tumor cells (L929 and HeLa). This
indicates that the two polymers did not differentiate between these
"normal" and "cancer" cells.

Some synthetic and natural polymers have the ability to stimulate a
non-specific immune response against certain types of bacteria. To
evaluate this, the polymers were tested against a wide range of bacteria.
These bacteria are known to effect various regions or parts of the human
body. The bacteria studies, the effects they elicit, and the regions of
the body they effect are given in Table 6. The idea was to perform a spot
check on these selected bacteria to determine if any immune response did

Table 4. Visual observation results of the effect of select
phosphorus-containing polyesters on the tissue cell
line HeLa after 24 hours incubation.

Phosphorus acid chloride[*]	Biological diol[**]	Concentrations (μg/mL					
		8	16	40	100	150	control
PPD	DES	0	0	1	2	3	0
PPD	IDU	1	1	2	2	2	0
PPD	DEL	1	2	2	3	3	0
PPD	ACB	0	0	1	1	1	0
PPD	ASD	0	0	1	1	1	0
PDP	DES	1	1	1	2	2	0
PTD	DES	2	3	3	3	4	0
DMSO control		0	0	0	0	0	0

*   See Table 2.
** See Table 2.

Table 5. Eosin stained results of the effect of select
         phosphorus-containing polyesters on the tissue cell
         line HeLa after 48 hours incubation.

Phosphorus acid chloride*	Biological diol**	Concentrations					
		8	16	40	100	150	Control
PPD	DES	1	1	2	3	3	0
PPD	IDU	1	2	2	2	3	0
PPD	DEL	2	3	3	4	4	0
PPD	ACB	0	1	1	1	2	0
PPD	ASD	0	1	2	2	2	0
PDP	DES	1	1	1	2	3	0
PTD	DES	2	3	3	4	4	0
DMSO Control		0	0	0	0	0	0

 * See Table 2.
** See Table 2.

in fact occur. The results given in Table 7 indicate that all exhibited
some inhibition of test organisms. Inhibition was also selective. These
results indicated that further study is merited to see if, in fact, some
of the polymers may be useful as selective agents of inhibition.

These biological analyses were meant only to briefly evaluate the
potential inhibitory nature of the compounds. Further detailed work would
be required before specific applications should be considered.

Table 6. Types of bacteria studied, the regions of the body
         they effect, and the diseases they can cause.

(1) *Actinobacter calcoacetius*

gram-negative rod
can be an opportunistic pathogen but is part of the normal
          flora of the skin and mouth
associated with conjunctivitis, keratitus, and chronic ear
          infections.

(2) *Enterobacter aerogenes*

gram-negative rods
part of the normal flora of the intestinal tract
has caused urinary tract infections, pneumonia, endocarditis,
          and bacteremia.

(3) *Klebsiella pneumonias*

gram-negative rod
part of the normal flora of the nose, mouth and intestines
may cause lesions in various parts of the body, pneumonia,
chronic lung abscess, sinusitis and upper respiratory
          infections.

Table 6. Continued.

---

(4) *Staphylococcus epidermidis*

---

gram-positive cocci
generally causes mild infections but has caused septicemia,
        bacterial endocarditis, and urinary tract
        infections.

---

(5) *Staphylococcus aureus*

---

gram-positive cocci
causes acne, abscesses, impetigo, wound infections, pyelitis,
        cystitis, "food poisoning", pneumonia,
        meningitis, and enteritis.

---

(6) *Pseudomonas aeruginosa*

---

gram-negative rod
common inhabitant of soil
frequently found as part of normal flora of the intestines
        and skin
opportunity pathogen, may infect wounds, cause urinary tract
        infection, eye infection, meningitis,
        contaminates burns, draining sinuses and
        decubitis ulcers.

---

(7)*Escherichia coli*

---

gram-negative rod
part of the normal intestinal flora
most frequent cause of urinary tract infections
may cause cholecystitus, appendicitis, peritonitis,
        sinusitis, and summer diarrhea.

---

(8)*Alcaligenes faecalis*

---

gram-negative rod
part of the normal intestinal flora
has been found to cause urinary tract infections and may
        cause indebilitated individuals septicemia or
        meningitis.

Table 7. Bacterialogical results of select phosphorus-containing polyesters.

Acid chloride*	Diol**	*Actin*	*Entereo*	*Kleb*	*StaphE*	*StaphA*	*Pseudo*	*E.coli*	*Alcal*
PPD	DES	P	N	N	N	C	C	N	N
PPD	IDU	N	C	N	N	N	N	N	N
PPD	DEL	C	P	C	C	C	P	C	N
PPD	ACB	N	P	P	N	N	C	N	N
PPD	ASD	N	C	C	N	N	N	C	N
PDP	DES	C	C	C	C	C	P	C	N
PTD	DES	N	N	P	P	C	N	N	N
Saline Control		N	N	N	N	N	N	N	N

N = no inhibition   P = partial inhibition   C = complete inhibition, clear zone.

* PPD = Phenylphosphonic dichloride,
  PTD = Phenyl phosphonothioic dichloride
  PDP = Phenyl dichlorophosphate
** DES = Diethylstilbestrol
  DEL = Dienestrol
  IDU = 5-Iodo-2'-deoxyuridine
  ASD = Androstendiol
  ACB = Amcinafal

REFERENCES

1. J. Arvin, U.S. Patent 2,058,394 (1936).
2. H. Zentfman and H. R. Wright, Br. Plastics, **25**, 374 (1952).
3. M. Sander and E. Steininger, J. Macrol. Sci., **C1**, 91 (1967).
4. (a) P. W. Morgan, "*Condensation Polymers: By Interfacial and Solution Methods,*" Wiley, New York, 1965. (b) Vol. II, Dekker, New York, 1978. (c) C. Carraher and J. Preston, "*Interfacial Synthesis, Vol. III,*" Dekker, New York, 1982.
5. Y. L. Gefter, "*Organophosphorus Monomers and Polymers,*" Pergamon New York, 1962.
6. C. Carraher, Inorganic Macromolecules Revs., **1**, 271 (1972).
7. C. Carraher, in: "*Interfacial Synthesis, Vol. II,*" F. Millich and C. Carraher, Eds., Dekker, New York, 1978, Chapter 20.
8. E. Aufderhaar, Dissertation, Bonn (1961).
9. E. V. Kuznetsov, I. M. Shermergorn and V. A. Belyaeva, U.S.S.R. Patent 137,673 (1961).
10. T. Rabek and T. Prot, Roczniki Chem., **37**, 747 (1963).
11. C. Carraher and G, Scherubel, Makromol. Chemie., **160**, 259 (1972) and **152**, 61 (1972).
12. C. Carraher, Inorganic Macromolecules Revs., **1**, 287 (1972) and C. Carraher, D. Powers and B. Pandya, Polymer Materials-Sci. & Eng.,**65**, 32 (1991).
13. C. Carraher and S. Bajah, Polymer, **15**, 9 (1974).
14. C. Carraher, M. Naas, D. Giron and D. R. Cerutis, J. Macromol. Sci-Chem., **A19**, 1101 (1983).
15. C. Carraher and C. Deremo-Reese, in: "*Metallorganic Polymers,*" C. Carraher, J. Sheets and C. Pittman, Eds., MER Press, Moscow, 1981, Chapter 10, p. 115.
16. C. Carraher, J. Chem. Ed., **58**, C111, 921 (1981).
17. C. Carraher and S. Bajah, British Polymer J., **7**, 155 (1975).
18. J. W. Long, M. D., "*The Essential Guide to Prescription Drugs,*" Harper and Row, New York, 1980, p. 221-227.
19. Teen, **28**, 52-54 (1984).
20. D. Silburt, MaCleans, **97**, 58 (1984).
21. A. Albert, Selective Toxicity, 5th ed., Chapman and Hall, London, 1973.
22. E. M. Hodnett and J. Tien Hai Tai, J. Med. Chem., **17**, 1335 (1974).
23. E. M. Hodnett, Polymer News, **8**, 323-328 (1983).

THE CONTROLLED RELEASE OF A 5-FLUOROURACIL PRODRUG FROM A BIODEGRADABLE MATRIX

Charles G. Gebelein,[1,3,4*], Donna Gardner[2] and
Tonya Ellis[2]

(1) Department of Chemistry
    Youngstown State University
    Youngstown, OH 44555
(2) Department of Chemical Engineering
    Youngstown State University
    Youngstown, OH 44555
(3) Department of Chemistry
    Florida Atlantic University
    Boca Raton, FL 33431
(4) LIONFIRE, Inc.
    Edgewater, FL 32132

A prodrug of 5-fluorouracil [5-FU] was prepared by reacting a 5-FU slurry with hexylisocyanate, in dioxane, at room temperature. The resulting solid, called HIFU, was purified and the release rates of 5-FU were determined, in distilled water, and compared against 5-FU using the gas dispersion tube technique previously described, at both room temperature and at 37°C. HIFU was studied in the form of a powder and as a monolithic dispersion in poly(caprolactone), The prodrug releases 5-FU very rapidly in the powdered form and both HIFU and 5-FU gave the typical monolithic release profiles, which were linear when plotted against the square root of time.

INTRODUCTION

In previous reports from our laboratory, we have shown that the release of 5-fluorouracil (5-FU) from a copolymer system in which this anticancer agent is attached to a methacrylate-type monomer unit, called [EMCF], follows a zero-order release profile.[1-10] On the other hand, when 5-FU is dispersed, or encapsulated, within a monolithic matrix of poly(caprolactone) the release pattern is linear with the normal square root of time; i.e., the Higuchi pattern.[11] While these [EMCF] copolymer systems are nominally monolithic, the form of the 5-FU itself resembles a

---

* To whom requests should be directed at the following address:
    Dr. Charles G. Gebelein, LIONFIRE, Inc., 1730 Umbrella Tree
    Drive, Edgewater, FL 32132-3111.

Biotechnology and Bioactive Polymers, Edited by C. Gebelein
and C. Carraher, Plenum Press, New York, 1994

prodrug. In this paper, we will report on the release behavior of 5-FU from a low-molecular-weight, non-polymerizable 5-FU prodrug encapsulated within a poly(caprolactone) matrix.

EXPERIMENTAL

1. THE SYNTHESIS OF L-(N-HEXYLCARBAMOYL)-5-FLUOROURACIL [HIFU]

The prodrug chosen for this study was prepared by stirring a slurry of an excess of 5-FU with n-hexylisocyanate using dioxane as the solvent, following the general procedure outlined for [EMCF] monomer.[3] A small quantity of dibutyltindilaurate was added as a catalyst, instead of the triethylamine. The reaction was run at room temperature until the IR spectra showed disappearance of the -N=C=O peak. The dioxane layer containing the soluble [HIFU] sample was separated and evaporated to dryness. The solid [HIFU] was then recrystallized from dioxane to give a white solid which melted at 63-66°C. This synthetic procedure is outlined in Equation 1.

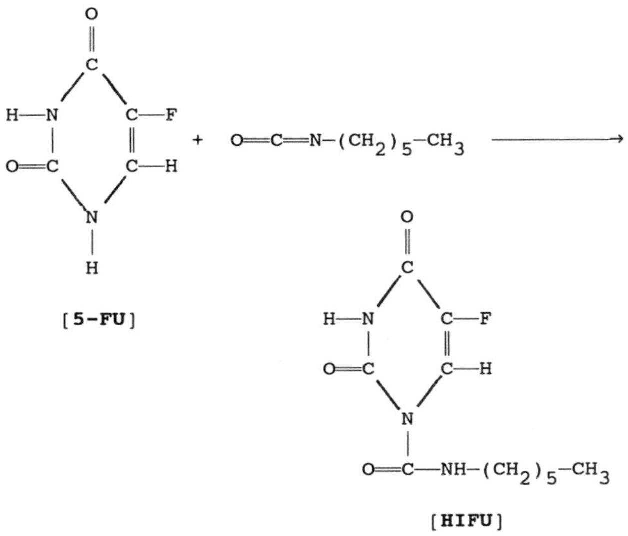

(Equation 1)

2. PREPARATION OF [HIFU]:POLY(CAPROLACTONE), (PCL), MONOLITHIC SAMPLES

Unlike 5-FU, HIFU melts readily and this molten material could be mixed with the molten poly(caprolactone) [PCL] to obtain good, fairly clear disks of the prodrug:polymer blend. Samples were prepared at 5, 10, 15 and 20% HIFU. These disks were prepared at 60+°C and are, therefore, annealed samples.[9,10]

3. PREPARATION OF [FU]:POLY(CAPROLACTONE), (PCL), MONOLITHIC SAMPLES

Monolithic dispersions of 5-FU in poly(caprolactone) were prepared as previously described.[9,10] While these disks are more difficult to prepare in a homogenous fashion than the HIFU monolithic dispersions, the 5-FU will disperse evenly if stirred slightly while cooling. Care is necessary to avert entrapment of air, however. These samples are also annealed. The final disks are cloudy in appearance.

4. CONTROLLED RELEASE STUDIES

The controlled release of the 5-FU from the [HIFU]:PCL disks was run using the fritted glass tube method (gas dispersion tubes) previously described.[4-7] The release measurements were made at room temperature (about 23°C) and at 37°C. The 5-FU level was measured by periodically removing an aliquot from the system and determining the concentration using a Hewlett-Packard Diode Array Spectrophotometer at 266 nm, as previously described.[10]

RESULTS AND DISCUSSION

Although most monolithic dispersions of drugs release the active agent as a function of the square root of time, a release profile which is linear with time should normally be more useful in medication. Our previous studies showed that copolymers of [EMCF] gave this zero-order release, even in the form of a powdered sample.[2-5] Nominally polymers and copolymers are monolithic systems, but the use of [EMCF] effectively converts the 5-FU into a prodrug which is attached to the polymeric backbone. Some typical results are shown below, in Figure 1, for copolymers of [EMCF] with methyl acrylate (MA). It is immediately evident that the release rates obey zero order kinetics.

Other previous studies from our laboratories clearly showed that 5-FU was released from poly(caprolactone), PCL, monolithic systems according to the Higuchi pattern (i.e., linear with the square root of time). This observation agrees with work from other laboratories using different types of polymer matrices.[11-16] Some typical results are shown in Figure 2 for 5%, 10% and 15% 5-FU in PCL.

Earlier we proposed that the EMCF copolymers followed zero-order kinetics because the rate controlling step in the release was the hydrolysis of the EMCF to release the 5-FU.[10] Since the polymer was insoluble in water, and the water was present in large excess, the kinetic order defaulted to zero. Similar results have been reported for

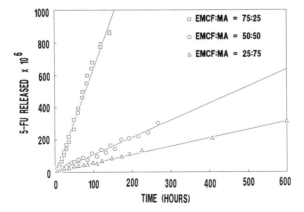

Figure 1. Zero order release of 5-fluorouracil from EMCF:MA copolymers.[3]

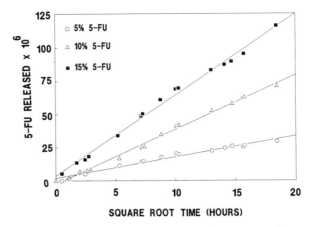

Figure 2. The release of 5-fluorouracil from poly(caprolactone) matrices.[10]

other systems in which the active (or drug) unit was attached to a
polymeric change via a linkage that could undergo hydrolytic cleavage.
For example, Ghosh has reported that the release of nalidixic acid from
some condensation polymers followed zero-order kinetics.[17] In each case,
the drug was released, via hydrolysis, from its attachment to a polymer,
which served as a prodrug for the agent.

Accordingly, we prepared HIFU, which is another 5-FU prodrug, in
order to learn if this material might also give a zero-order release
profile, in the same manner as the [EMCF] copolymers (Figure 1), when it
is encapsulated within a monolithic system, or whether it would follow
the normal Higuchi style pattern. Figure 3 shows the release of powdered

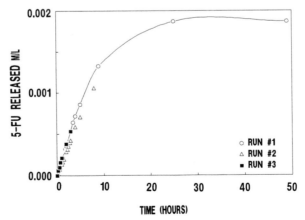

Figure 3. The release of 5-fluorouracil from the sample
tubes. The results of three separate trials are
shown.

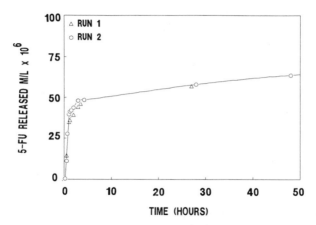

Figure 4. The release of 5-fluorouracil via the hydrolysis of
HIFU. The results of two separate trials are
shown.

5-FU from the fritted glass tubes, at 37°C, as a function of time. Almost
all the 5-FU is released in about 10-15 hours. In addition, it can be
seen that these results are reproducible. Powdered [HIFU] hydrolyzed very
rapidly in water and was released from the fritted glass tubes at
approximately the same rate, or faster, than the powdered 5-FU (Figure
4). These results are also reproducible. The [HIFU], however, only
contains about 50% 5-FU. This release rate is much faster than we
previously observed with [EMCF] monomer.[2-4]

The release of 5-FU from HIFU dispersed monolithically in poly-
(caprolactone) is shown in Figure 5. The results do not follow zero-order
kinetics, as desired. In fact, when this release of 5-FU data is plotted
against the square root of time, the Higuchi pattern, we obtain fairly
good straight lines (Figure 6). The fact the HIFU is a prodrug is not
sufficient to transform the release pattern from the Higuchi style into
the zero-order profile. Possibly the hydrolysis of HIFU is too rapid, as
noted in Figure 4, and, therefore, does not significantly alter the
migration of the 5-FU out of the matrix. Alternatively, it may be
necessary for the prodrug to be bound to the polymer in order to achieve
zero-order release. In any event, the simple monolithic dispersion of
HIFU in PCL does not form a zero-order release system.

Unsurprisingly, the release of 5-FU from the HIFU:PCL systems is
more rapid at higher temperatures. Figure 7 illustrates this for a system
containing 10% HIFU. Similar results were obtained with the other levels
of HIFU in PCL.

Figure 8 shows the release of 5-FU from monolithic systems in which
either 20% [HIFU] or 20% 5-FU was dispersed in PCL. Both systems clearly
follow the Higuchi kinetic pattern, demonstrating again that the HIFU
prodrug does not lead to zero-order kinetics. The release rates are about
the same as with 5-FU, when adjusted for the lower concentration of this
drug in HIFU.

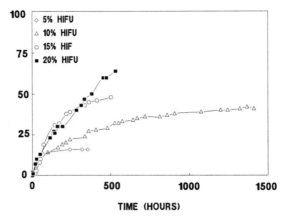

Figure 5. The release of 5-fluorouracil from poly(caprolactone) matrices containing various levels of HIFU.

Although [HIFU] did not give the hoped for release pattern, it is
much easier to prepare the monolithic PCL matrices with this analog than
with 5-FU itself, since HIFU melts at about the same temperature as the
PCL. From a practical, manipulative aspect, this could be a big advantage
for preparing monolithic dispersions of 5-FU. In addition, HIFU is also
soluble in many simple organic solvents, unlike 5-FU which is insoluble
in essentially all solvents, except water. (Even in water, the solubility
is marginal.) These features could make HIFU, or other 5-FU analogs, a
useful prodrug.

The reason why [HIFU] did not behave in the same manner as [EMCF],
and form zero-order release systems, is not completely clear. Several

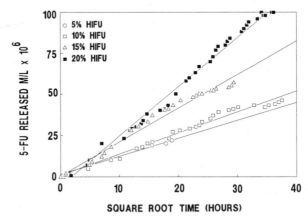

Figure 6. The data of Figure 5 plotted against the square
root of time.

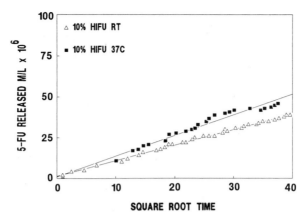

Figure 7. The release of 5-fluorouracil from a poly(caprolactone) matrix containing 10% HIFU at room temperature and at 37°C.

factors may be involved, but the fact that the HIFU prodrug hydrolyzes very rapidly in water, to release 5-FU, may be a major contributing factor. Whether a slower releasing prodrug would work more satisfactorily in these PCL matrices is unknown, however, but some other systems are under investigation and the results will be reported at a later date.

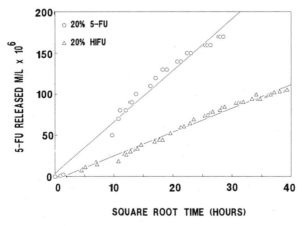

Figure 8. Comparison of the release of 5-fluorouracil from poly(caprolactone) matrices containing either 20% 5-fluorouracil or 20% HIFU.

The alternative explanation, that the prodrug, or the drug itself, must be attached to a polymer chain has, however, been tested, and these results are reported elsewhere in this book.[18] Those results suggest that the nature of the polymer backbone is more crucial than the mere fact of drug attachment. If zero-order release systems are desired for 5-FU, the best candidate remains the EMCF copolymers, which can give constant release for a fairly long period of time, and at widely variable release rates. This should be advantageous in cancer chemotherapy.

ACKNOWLEDGEMENT

Portions of this research was abstracted from the undergraduate thesis of DG and TE, submitted in partial fulfillment of the requirements for the Bachelor of Engineering. They also prepared the xylan adducts, and the controlled release studies on them, which were reported in the paper by Gebelein, Williams, et al, elsewhere in this book.

REFERENCES

1. C. G. Gebelein, Proc. Polym. Mat. Sci. Eng., **51**, 127-131 (1984).
2. R. R. Hartsough & C. G. Gebelein, Proc. Polym. Mat. Sci. Eng., **51**, 131-135 (1984).
3. R. R. Hartsough & C. G. Gebelein in: "*Polymeric Materials in Medication*," C. G. Gebelein & C. E. Carraher, Jr., Eds., Plenum Publ. Corp., New York, 1985, pp. 115-124.
4. C. G. Gebelein & R. R. Hartsough in: "*Controlled Release of Bioactive Materials, 11th International Symposium*," W. E. Meyers & R. C. Dunn, Eds., Controlled Release Society, Lincolnshire, IL, 1984, pp. 65-66.
5. C. G. Gebelein, T. Mirza & M. Chapman, Proc. Polym. Mat. Sci. Eng., **57**, 413-416 (1987).
6. C. G. Gebelein, R. R. Hartsough & T. Mirza in: "*Controlled Release of Bioactive Materials, 13th International Symposium*," I. A. Chaudry & C. Thies, Eds., Controlled Release Society, Lincolnshire, IL, 1986, pp. 188-189.
7. C. G. Gebelein, T. Mirza & R. R. Hartsough in: "*Controlled Release Technology, Pharmaceutical Applications*," P. I. Lee and W. R. Good, Eds., Symp. Series #348, American Chemical Society, Washington, DC, 1987, pp. 120-126.
8. C. G. Gebelein, M. Chapman & T. Mirza in: "*Applied Bioactive Polymeric Systems*," C. G. Gebelein, C. E. Carraher & V. Foster, Eds., Plenum Publ., New York, 1988, pp. 151-163.
9. C. G. Gebelein, M. Davison, T. Gober & M. Chapman, Proc. Polym. Mat. Sci. Eng., **59**, 798-802 (1988).
10. C. G. Gebelein, M. Chapman, M. Davison & T. Gober, in: "*Progress in Biomedical Polymers*," C. G. Gebelein & R. L. Dunn, Eds., Plenum Publ. Corp., New York, 1990, pp. 321-334.
11. T. Higuchi, J. Pharm. Sci.,**59**, 353 (1961).
12. M. Yoshida, M. Kumakura & I. Kaetsu, Polymer, **19**, 1375 (1978).
14. M. Yoshida, M. Kumakura & I. Kaetsu, Polymer, **11**, 775 (1979).
15. I. Kaetsu, M. Yoshida, M. Kumakura, A. Yamada & Y. Sakurai, Biomaterials, **1**, 17 (1980).
16. I. Kaetsu, M. Yoshida & A. Yamada, J. Biomed. Mater. Res., **14**, 185 (1980).
17. M. Ghosh in: "*Progress in Biomedical Polymers*," C. G. Gebelein & R. L. Dunn, Eds., Plenum Publ. Corp., New York, 1990, pp. 335-345.
18. C. G. Gebelein, D. Williams, K. Marshall and T. Slaven, This book.

NOVEL ANTIINFECTIVE BIOMATERIALS BY POLYMER MODIFICATION

W. Kohnen, B. Jansen, D. Ruiten, H. Steinhauser and
G. Pulverer

Institute of Medical Microbiology and Hygiene
University of Cologne
Goldenfelsstr. 19-21
D-5000 Cologne 41, Germany
and
Institute of Physical Chemistry
University of Cologne
Luxemburger Str. 116
D-5000 Cologne 41, Germany

The first significant step in the pathogenesis of poly-
mer-associated infections (foreign-body infections) is the
adhesion of bacteria to the synthetic material. The development of an anti-adhesive synthetic material seems to be a
promising method to prevent that kind of infection. A possible approach is to modify the polymer surface without affect-
ing the bulk properties (e.g. mechanical stability, elasticity). An elegant and versatile method to achieve this is the
glow discharge technique (NT-plasma), where the surface of a
synthetic material is exposed to a glow discharge under
reduced pressure.

In this study, polyurethane films were grafted with
acrylic acid using a glow discharge treatment. The influence
of different experimental parameters - discharge power,
pressure, gas flow, and posttreatment pressure and time - was
examined by measuring the amount of a cationic dye adsorbed
by the carboxyl groups of the newly formed surface coating.
Further, the relation between the *in vitro* adhesion of
Staphylococcus epidermidis KH 6 and acrylic acid grafting
yield, determined by the amount of dye adsorbed, has been
investigated. Bacterial adhesion on the standard sample after
pretreatment was about 15%. After grafting with acrylic acid,
adhesion was reduced to about 1%, but no dependence on the
grafting yield could be observed.

INTRODUCTION

Today synthetic materials are used in many fields of medicine. Their
benefits for the patient are beyond doubt; however, their use may result

in undesirable side effects. One of these complications is the infection of the synthetic material.[1] The main causative organisms of these so-called foreign-body infections are coagulase-negative staphylococci. In most cases antimicrobial therapy alone cannot cure the infection, and the removal of the infected device becomes necessary. Therapy is difficult since many bacterial strains - after settling on the polymer surface - are able to produce extracellular substrates protecting them against host defense mechanisms and antibiotics.[2]

The adhesion of bacteria to synthetic materials is the first step in the pathogenesis of foreign-body infections. Therefore, preventing bacteria from adhering to the material seems to be the most effective way of avoiding such infections. This may be achieved by developing polymers with anti-adhesive properties. Such polymers could either be produced by developing entirely new synthetic materials or by modifying the surface of common polymers.

Surface modification makes it possible to retain the bulk properties of synthetic materials (e.g., mechanical stability, elasticity) and it is therefore more advantageous. The surface of most polymers can be modified in a great variety using the glow discharge technique (NT-plasma) under reduced pressure.[3] The experimental arrangement used in the present study is shown in Figure 1.

The 'plasma' (generated through the glow discharge) is a gaseous complex composed of ionized gas, radicals, photons, and excited and non-excited gas molecules. This reactive plasma modifies a synthetic material which is exposed to the discharge.

Depending on the nature of the gas used, the polymer surface may be modified in different ways: modifications include etching, crosslinking, functionalization, plasma polymerization and plasma-induced grafting (Figure 2). When non-polymerizable gases (e.g. oxygen, noble gases) are used, etching or crosslinking of the surface will occur. Functionalization with nitrogen or water-vapor will cause the formation of new chemical groups in the surface. Treatment with polymerizable gases (e.g. organic compounds) results in the formation of a dense coating of a plasma-polymer on the polymer surface, i.e. layers with new properties are generated. Plasma-induced grafting, on the other hand, will lead to the formation of a coating consisting of mobile polymer chains. In this process glow discharge is used to generate radicals inside the surface of the synthetic material. Following the glow discharge reaction a polymer-izable monomer is transferred into the reaction system and will react with the polymer, thereby creating graft chains. This method also generates surfaces with new properties that may have improved biocompatibility.[4]

In the present study acrylic acid was grafted onto polyurethane films which were previously treated in a glow discharge with oxygen. The parameters varied in the glow discharge treatment phase included discharge power, pressure and gas flow rate; the parameters varied in the post treatment phase (grafting with acrylic acid) were pressure and time.

The effects of the parameter variations on surface modification was assessed by measuring the amount of cationic dye adsorbed by the carboxyl groups of the newly formed coating. Further, bacterial adhesion as a function of the amount of dye the modified surface can adsorb was measured.

318

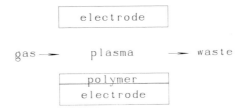

Figure 1. Experimental arrangement of glow discharge treat-
　　　　　 ment: gas passing into the reaction tube forms a
　　　　　 plasma between the electrodes where the glow
　　　　　 discharge can take place; waste products are
　　　　　 removed by a pump.

EXPERIMENTAL

1. MODIFICATION OF POLYMER SURFACES

Poly(etherurethane) Walopur 2201 U (Wolff, Walsrode, Germany) was
extracted in ethanol prior to use. The liquid monomer acrylic acid was
distilled under reduced pressure. Oxygen for a glow discharge treatment
was used without further purification.

The plasma reactor (Softal, Hamburg, Germany) was designed for the
treatment of films with a size of 100 mm x 100 mm. The high frequency
generator used for glow discharge operates at a frequency of 20 - 30 kHz.
The standard conditions for the glow discharge with oxygen for 600 sec
were: power = 40 watts, pressure = 40 Pa, gas flow rate = 20 mL(STP)/min.
After glow discharge, the reaction vessel was evacuated and subsequently
the tube connection of the reaction vessel with supply vessel containing
acrylic acid was opened. During glow discharge and the following grafting

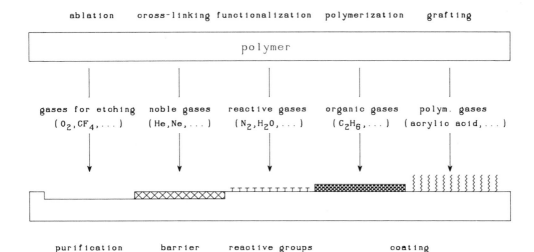

Figure 2. Polymer surface modifications with the glow dis-
　　　　　 charge technique.

process (time = 300 sec and pressure = 500 Pa as standard conditions) the pressure inside the reactor was kept constant by a pressure control device. Temperature of the electrode onto which the sample was mounted was 296°K.

After the grafting process, the film were cleansed by water and finally dried under reduced pressure.

2. ADSORPTION OF DYE

The carboxyl groups of the acrylic acid grafted onto the films were determined in accordance with the method developed by W. Kesting, et al.[5] The cationic dye Astrazon Blue F2RL (Bayer AG, Leverkusen, Germany) was purified by filtration and subsequently removal of the solvent (acetone).

A buffered dye solution (0.05% in PBS) was used to dye the treated surface of the polymer samples (surface about 1 cm^2) at pH 7.2. After rinsing the samples with 50 mL of water in small portions, the adsorbed dye was removed with 5 mL of acetic acid (84%). A photometric measurement of this extract was made at a wavelength of 591 nm using a Zeiss Q II photometer.

3. BACTERIAL ADHESION TO MODIFIED FILMS

Bacterial adhesion to the surface-modified films was measured using the bioluminescence assay described by Ludwicka, et al.[6] The modified surface of a polymer disc (d = 12 mm) was incubated in PBS (pH = 7.2) usually containing 10^8 colony forming units (cfu) at room temperature. After an adhesion period of three hours, the polymer disc was washed three times in PBS and then exposed to 100 µl 2% trichloroacetic acid to extract bacterial ATP. Next, the extract was diluted with 400 µl TRIS-EDTA buffer (pH 7.75). An aliquot of 50 µl was taken to which 400 µl TRIS-EDTA buffer and 200 µl ATP monitoring reagent (LKB Wallac, Finland) were added. The ATP monitoring reagent converts ATP to AMP and light, which was measured in a bioluminometer (LKB Wallec, Finland). The light emission is proportional to the ATP-concentration, and by establishing a standard curve bacterial concentration vs. ATP-content, the percentage of adhered bacteria (in relation to the cfu of the initial bacterial solution) per cm^2 surface could be calculated. The mean value of nine measurements per sample was taken as the final result.

RESULTS AND DISCUSSION

1. INFLUENCE OF DISCHARGE PARAMETERS ON ACRYLIC ACID GRAFTING

1A. Discharge Power

Power of the generator was varied in the range of 0 to 50 Watts. It was found that with increasing power the dye adsorption by the grafted acrylic acid layers also increased. (Figure 3).

This can be explained by the fact that by increasing the power there will be a growing number of gas molecules having enough energy to generate radicals inside the film surface. A growing number of radicals will raise the amount of acrylic acid grafted onto the film, which is confirmed by the higher amounts of dye adsorbed to the surface carboxylate groups.

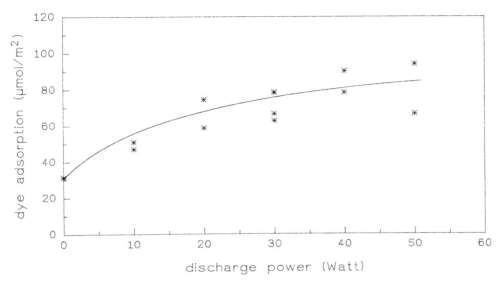

Figure 3. Influence of discharge power on acrylic acid grafting yield, determined by the amount of dye adsorbed.

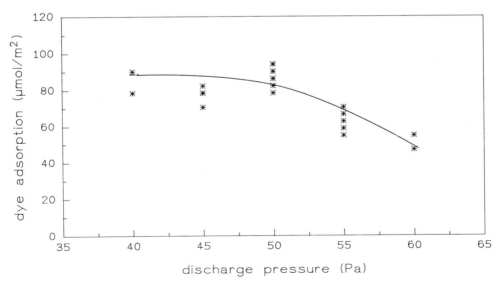

Figure 4. Influence of discharge pressure on acrylic acid
grafting yield, determined by the amount of dye
adsorbed.

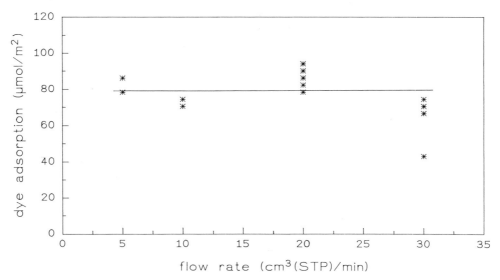

Figure 5. Influence of gas flow rate on acrylic acid grafting
yield, determined by the amount of dye adsorbed.

1B. Discharge Pressure

Varying the discharge pressure from 40 up to 60 Pa leads to reduced
dye adsorption at higher pressures (Figure 4). With increasing pressure
the mean free path of gas molecules will become shorter and, consequently, the active species will lose energy when colliding with other gas
molecules before meeting the surface. Therefore, fewer radicals will be
generated inside the surface at higher pressures, resulting in lower
amounts of acrylic acid grafted onto the surface in the post-treatment
phase. This is demonstrated by the reduction of the adsorbed dye amount.

1C. Gas Flow Rate

Due to the chosen discharge pressure of 50 Pa, the gas flow rate
could be varied in the range of 5 cm^3(STP)/min to 30 cm^3(STP)/min. In
this range, variation of the gas flow rate obviously had no influence on
the subsequent grafting process (Figure 5), as was shown by the same
levels of adsorbed dye for all flow rates.

2. INFLUENCE OF POST TREATMENT PARAMETERS ON ACRYLIC ACID GRAFTING

2A. Monomer Pressure

The pressure of acrylic acid was varied in the range of 0 to 500 Pa
(Figure 6). At 200 Pa acrylic acid begins to condense on the polymer
surface, therefore more monomer is offered to the surface and, thus, the
grafting yield increases. At 400 Pa the surface of the synthetic material
is completely covered with a liquid monomer coating, so that a maximum in
grafting yield is reached.

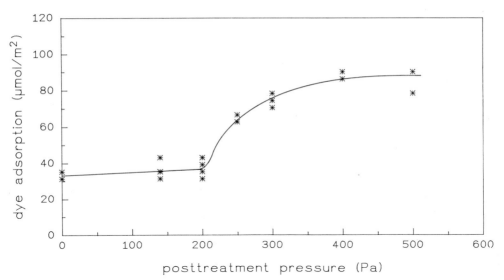

Figure 6. Influence of acrylic acid pressure on acrylic acid grafting yield, determined by the amount of dye adsorbed.

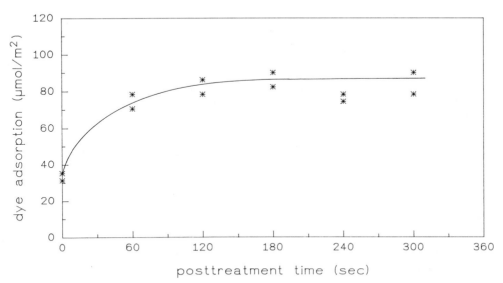

Figure 7. Influence of grafting period on acrylic acid grafting yield, determined by the amount of dye adsorbed.

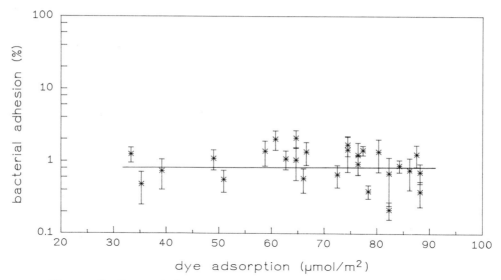

Figure 8. Bacterial adhesion to acrylic acid-grafted poly-
(urethane) films as a function of the amount of
dye the film can adsorb. Bacterial adhesion on the
standard sample after pretreatment was about 15%.

2B. Grafting Time

The grafting period was varied in the range of 0 to 300 sec (Figure
7). After about 120 sec the grafting reaction was already terminated, and
saturation was reached.

3. BACTERIAL ADHESION TO ACRYLIC ACID GRAFTED POLYURETHANE FILMS

Bacterial adhesion onto the polyurethane films activated with oxygen, under standard conditions, was about 15% (related to the initial
bacterial inoculum). After grafting with acrylic acid the adhesion was
decreased (about 1%). The grafted polymer surface obviously reduces
adhesion of bacteria, most probably by decreasing hydrophobic interactions through the newly created hydrophilic surface.

Further, the relative bacterial adhesion as a function of the amount
of adsorbed dye on the polymer surface (determined in a separate test)
was measured. It is assumed that the amount of adsorbed cationic dye
represents the extent of grafting and is proportional to the density of
negative charges in the newly formed coating. We found that bacterial
adhesion in the range examined is independent of charge density and
grafting yield (Figure 8).

CONCLUSIONS

In this study, a method was evaluated by which acrylic acid can be
grafted onto a polymer surface. A reduction of bacterial adhesion by

introducing these hydrophilic groups was observed, but no influence of the grafting yield on adhesion could be detected.

In the next step, protein adhesion onto the modified polymer surfaces and the resulting influence on bacterial adhesion will be examined, as bacterial adhesion is influenced by the nature of the protein layer.[7] Also, the influence of different grafting monomers on bacterial adhesion will be investigated. Further, we want to link - by using the potential for coupling reactions offered by acrylic acid - antimicrobial agents to the surface.

REFERENCES

1. B. Sugarman & E. J. Young, *"Infections Associated with Prosthetic Devices,"* CRC Press, Boca Raton, 1984.
2. B. Jansen, F. Schumacher-Perdreau, G. Peters, G. Pulverer, J. Invest. Surg., **2**, 361 (1989).
3. W. R. Gombotz & A. S. Hoffman, in: *"CRC Critical Reviews in Biocompatibility,"* Vol. 4, Issue 1, CRC Press, Boca Raton, 1986.
4. B. Jansen, H. Steinhauser & W. Prohaska, Angew. Makromol. Chem., **164**, 115 (1988).
5. W. Kesting, D. Knittel & E. Schollmeyer, Angew, Makromol. Chem., **182**, 187 (1990).
6. A. Ludwicka, L. M. Switalski, A. Lundin, G. Pulverer & T. Wadstrom, J. Microbiol. Meth., **4**, 169 (1985).
7. B. Jansen, G. Peters, S. Schareina, H. Steinhauser, F. Schumacher-Perdreau & G. Pulverer, in: *"Applied Bioactive Polymeric Materials,"* C. E. Carraher, C. G. Gebelein & V. R. Foster, Eds., Plenum Publ., New York, 1988.

CONTRIBUTORS (Current addresses where known)

B. C. Adelmann-Grill,   127
Institut für Experimentelle Chirurgie der Technischen,
Universität München,
Ismaninger Straße 22, 8000 München 80

R. Ascherl,   127
Institut für Experimentelle Chirurgie der Technischen,
Universität München,
Ismaninger Straße 22, 8000 München 80

A. F. Azhar,   177
Boehringer Mannheim Corporation,
9115 Hague Road,
Indianapolis, IN 46250

Yoshinari Baba,   35
Department of Applied Chemistry,
Saga University,
Honjo-machi, Saga 840, Japan

F. Bader,   17, 127
Institut für Experimentelle Chirurgie der Technischen,
Universität München,
Ismaninger Straße 22, 8000 München 80

Endre A. Balazs,   25
Departments of Biochemistry, Histology and Chemistry,
Biomatrix, Inc.,
Ridgefield, New Jersey

L. W. Barrett,   147
Chemical Engineering Department,
Materials Science and Engineering Department,
Materials Research Center,
Center for Polymer Science and Engineering,
Whitaker Laboratory 5,
Lehigh University,
Bethlehem, PA 18015-3194

G. Blümel,   17, 127
Institut für Experimentelle Chirurgie der Technischen,
Universität München,
Ismaninger Straße 22, 8000 München 80

A. D. Burke,   177, 195
Boehringer Mannheim Corporation,
9115 Hague Road,
Indianapolis, IN 46250

Cynthia Butler,   1, 71
Florida Atlantic University,
Departments of Chemistry and
Biological Sciences,
Boca Raton, FL 33431

Charles E. Carraher, Jr.,   1, 71, 297
Florida Atlantic University,
Department of Chemistry,
Boca Raton, FL 33431

Stoil Dirlikov,   79
Coatings Research Institute,
Eastern Michigan University,
122 Sill Hall,
Ypsilanti, MI 48197

J. E. DuBois,   177
Boehringer Mannheim Corporation,
9115 Hague Road,
Indianapolis, IN 46250

Tonya Ellis,   309
Department of Chemical Engineering,
Youngstown State University,
Youngstown, OH 44555

W. Erhardt,   127
Institut für Experimentelle Chirurgie der Technischen,
Universität München,
Ismaninger Straße 22, 8000 München 80

D. R. Flanagan,   249
Division of Pharmaceutics and
Division of Medicinal and Natural Products Chemistry,
College of Pharmacy,
University of Iowa,
Iowa City, Iowa 52242

Donna Gardner,   309
Department of Chemical Engineering,
Youngstown State University,
Youngstown, OH 44555

Charles G. Gebelein,   277, 309
LIONFIRE, Inc.,
1730 Umbrella Tree Drive,
Edgewater, FL 32132

Malay Ghosh,   239
Alcon Laboratories,
Preformulation: Product Design Group,
6201 South Freeway,
Ft. Worth, TX 76134

Ina Goldberg,   115
Allied-Signal Inc.,
101 Columbia Road,
Morristown, NJ 07962-1021

Gershon Golomb,   259
School of Pharmacy,
The Hebrew University of Jerusalem,
Jerusalem, Israel

D. Channe Gowda,   95
Laboratory of Molecular Biophysics,
School of Medicine,
The University of Alabama at Birmingham, VH300,
Birmingham, Alabama 35294-0019

S. Haas,   17
Institut für Experimentelle Chirurgie der Technischen,
Universität München,
Ismaninger Straße 22, 8000 München 80

Shigehiro Hirano,   43
Department of Agricultural Biochemistry and Biotechnology,
Tottori University,
Tottori, Japan 680

Yoshiaki Inaki,   207
Department of Applied Fine Chemistry,
Faculty of Engineering,
Osaka University,
Suita, Osaka 565, Japan

Katsutoshi Inoue,   35
Department of Applied Chemistry,
Saga University,
Honjo-machi, Saga 840, Japan

Hiroshi Inui,   43
Department of Agricultural Biochemistry and Biotechnology,
Tottori University,
Tottori, Japan 680

Balkrishna S. Jadhav,   169
Johnson & Johnson Orthopaedics,
325 Paramount Drive,
Raynham, Massachusetts 02052-0350

B. Jansen,   317
Institute of Medical Microbiology and Hygiene,
University of Cologne,
Goldenfelsstr. 19-21,
D-5000 Cologne 41, Germany,

J. Jeckle,   127
Institut für Experimentelle Chirurgie der Technischen,
Universität München,
Ismaninger Straße 22, 8000 München 30

Nariyoshi Kawabata,   201
Department of Chemistry and Materials,
Faculty of Engineering and Design,
Kyoto Institute of Technology,
Matugasaki, Sakyo-ku, Kyoto, 606 Japan

J. E. Kennamer,   195
Boehringer Mannheim Corporation,
9115 Hague Road,
Indianapolis, IN 46250

John F. Kennedy,   55
Research Laboratory for the Chemistry of Bioactive
Carbohydrates and Proteins,
Department of Chemistry
The University of Birmingham, P O Box 363
Birmingham B15 2TT. England

H. Kobayashi,   289
Department of Applied Chemistry,
Faculty of Engineering,
Kansai University,
Suita, Osaka 564, Japan

W. Kohnen,   317
Institute of Medical Microbiology and Hygiene,
University of Cologne,
Goldenfelsstr. 19-21,
D-5000 Cologne 41, Germany,

Akihiro Kondo,   201
Wakayama Res. Lab.,
Kao Corp.,
1334 Minato, Wakayama, 640 Japan

Hideto Kosaki,   43
Department of Agricultural Biochemistry and Biotechnology,
Tottori University,
Tottori, Japan 680

Vinod Labhasetwar,   259
The University of Michigan Medical School,
Ann Arbor, Michigan

Nancy E. Larsen,   25
Departments of Biochemistry, Histology and Chemistry,
Biomatrix, Inc.,
Ridgefield, New Jersey

Edward Leshchiner,   25
Departments of Biochemistry, Histology and Chemistry,
Biomatrix, Inc.,
Ridgefield, New Jersey

Hilton Levy,   269
NIAIAD,
Bethesda, MD

Robert J. Levy,   259
The University of Michigan Medical School,
Ann Arbor, Michigan

R. J. Linhardt,   249
Division of Pharmaceutics and
Division of Medicinal and Natural Products Chemistry,
College of Pharmacy,
University of Iowa,
Iowa City, Iowa 52242

Michael A. Markowitz,   135
Center for Bio/Molecular Science and Engineering,
Code 6900, Naval Research Laboratory,
Washington, DC 20375

Kathy Marshall,   277
Department of Chemical Engineering,
Youngstown State University,
Youngstown, OH 44555

Kenji Matsukawa,   207
Department of Applied Fine Chemistry,
Faculty of Engineering,
Osaka University,
Suita, Osaka 565, Japan

Javier Morales,   269
Ashford Hospital,
Puerto Rico

Owen St. Clair Morgan,   269
University of West Indies,
Kingston, Jamacia

Victor J. Morris,   9
AFRC Institute of Food Research,
Norwich Laboratory, Norwich Research Park,
Colney Norwich, NR4 7UA, UK

David Muller,   259
The University of Michigan Medical School,
Ann Arbor, Michigan

Yoshinobu Naoshima,   1
Okayama University of Science,
Department of Biochemistry,
Ridaicho, Okayama, 700 Japan

That T. Ngo,   185
20 Sandstone,
Irvine, CA 92714

Alastair Nicol,   95
Laboratory of Molecular Biophysics,
School of Medicine,
The University of Alabama at Birmingham, VH300,
Birmingham, Alabama 35294-0019

Y. Ohya,   289
Department of Applied Chemistry,
Faculty of Engineering,
Kansai University,
Suita, Osaka 564, Japan

T. Ouchi,   289
Department of Applied Chemistry,
Faculty of Engineering,
Kansai University,
Suita, Osaka 564, Japan

Bhoomin Pandya,   297
Florida Atlantic University,
Department of Chemistry,
Boca Raton, FL 33431

Timothy M. Parker,   95
Laboratory of Molecular Biophysics,
School of Medicine,
The University of Alabama at Birmingham, VH300,
Birmingham, Alabama 35294-0019

Marion Paterson,   55
Research Laboratory for the Chemistry of Bioactive
Carbohydrates and Proteins,
Department of Chemistry
The University of Birmingham, P O Box 363
Birmingham B15 2TT. England

Cynthia T. Pollak,   25
Departments of Biochemistry, Histology and Chemistry,
Biomatrix, Inc.,
Ridgefield, New Jersey

Daniel S. Powers   297
Wright State University
Department of Chemistry
Dayton, Ohio 45435

G. Pulverer,   317
Institute of Medical Microbiology and Hygiene,
University of Cologne,
Goldenfelsstr. 19-21,
D-5000 Cologne 41, Germany,

Karen Reiner,   25
Departments of Biochemistry, Histology and Chemistry,
Biomatrix, Inc.,
Ridgefield, New Jersey

Thomas Ridgway,   71
Florida Atlantic University,
Departments of Chemistry and
Biological Sciences,
Boca Raton, FL 33431

D. Ruiten,   317
Institute of Medical Microbiology and Hygiene,
University of Cologne,
Goldenfelsstr. 19-21,
D-5000 Cologne 41, Germany,

Masaru Sakata,   201
Wakayama Res. Lab.,
Kao Corp.,
1334 Minato, Wakayama, 640 Japan

Andres Salazar,   269
Walter Reed Army Hospital,
Washington, DC

Anthony J. Salerno,   115
Allied-Signal Inc.,
101 Columbia Road,
Morristown, NJ 07962-1021

Vincent Saurino,   1
Florida Atlantic University,
Department of Biological Sciences,
Boca Raton, FL 33431

M. A. Scherer,   127
Institut für Experimentelle Chirurgie der Technischen,
Universität München,
Ismaninger Straße 22, 8000 München 80

E. Schmitt,   249
Division of Pharmaceutics and
Division of Medicinal and Natural Products Chemistry,
College of Pharmacy,
University of Iowa,
Iowa City, Iowa 52242

T. Seaber,   127
Duke University Medical Center,
Division of Orthopaedic Surgery,
Durham, North Carolina 27710

M. Shiratani,   289
Department of Applied Chemistry,
Faculty of Engineering,
Kansai University,
Suita, Osaka 564, Japan

Maria P. C. Silva,   55
Research Laboratory for the Chemistry of Bioactive
Carbohydrates and Proteins,
Department of Chemistry
The University of Birmingham, P O Box 363
Birmingham B15 2TT. England

Alok Singh,   135
Center for Bio/Molecular Science and Engineering,
Code 6900, Naval Research Laboratory,
Washington, DC 20375

Terri Slaven,   277
Department of Chemical Engineering,
Youngstown State University,
Youngstown, OH 44555

K. H. Sorg,   127
Institut für Experimentelle Chirurgie der Technischen,
Universität München,
Ismaninger Straße 22, 8000 München 80

L. H. Sperling,   147
Chemical Engineering Department,
Materials Science and Engineering Department,
Materials Research Center,
Center for Polymer Science and Engineering,
Whitaker Laboratory 5,
Lehigh University,
Bethlehem, PA 18015-3194

H. Steinhauser,   317
Institute of Medical Microbiology and Hygiene,
University of Cologne,
Goldenfelsstr. 19-21,
D-5000 Cologne 41, Germany,

A. W. Stemberger,   17, 127
Institut für Experimentelle Chirurgie der Technischen,
Universität München,
Ismaninger Straße 22, 8000 München 80

Dorothy C. Sterling,   1, 71
Florida Atlantic University,
Departments of Chemistry,
Biological Sciences
Boca Raton, FL 33431

M. Stoltz,   127
Institut für Experimentelle Chirurgie der Technischen,
Universität München,
Ismaninger Straße 22, 8000 München 80

Graham Swift,   161
Rohm and Haas Company,
Spring House, PA 19477

T. Takei,   289
Department of Applied Chemistry,
Faculty of Engineering,
Kansai University,
Suita, Osaka 564, Japan

Kiichi Takemoto,   207
Department of Applied Fine Chemistry,
Faculty of Engineering,
Osaka University,
Suita, Osaka 565, Japan

David W. Taylor,   55
Research Laboratory for the Chemistry of Bioactive
Carbohydrates and Proteins,
Department of Chemistry
The University of Birmingham, P O Box 363
Birmingham B15 2TT. England

H. P. Thomi,   127
Institut für Experimentelle Chirurgie der Technischen,
Universität München,
Ismaninger Straße 22, 8000 München 80

Tsuyoshi Toda,   43
Department of Agricultural Biochemistry and Biotechnology,
Tottori University,
Tottori, Japan 680

Eric Topol,   259
The Cleveland Clinic Foundation,
Cleveland, Ohio

Joseph Trachy,   259
The University of Michigan Medical School,
Ann Arbor, Michigan

Li-I. Tsao,   135
Center for Bio/Molecular Science and Engineering,
Code 6900, Naval Research Laboratory,
Washington, DC 20375

Rikio Tsushima,   201
Wakayama Res. Lab.,
Kao Corp.,
1334 Minato, Wakayama, 640 Japan

Deger C. Tunc,   169
Union Carbide Corporation,
P.O. Box 670,
Bound Brook, NJ 08805

Yoshitaka Uno,   43
Department of Agricultural Biochemistry and Biotechnology,
Tottori University,
Tottori, Japan 680

Dan W. Urry,   95
Laboratory of Molecular Biophysics,
School of Medicine,
The University of Alabama at Birmingham, VH300,
Birmingham, Alabama 35294-0019

A. M. Usmani,   177, 195
Dept. of Chemistry,
Kuwait University,
Kuwait

Tyrone L. Vigo,   225
USDA, ARS, SRRC,
1100 R. E. Lee Blvd,
New Orleans, LA 70124

J. M. Walenga,   17
Department of Pathology,
Loyola University,
Maywood, IL 60153

Darrell Williams,   277
Department of Chemical Engineering,
Youngstown State University,
Youngstown, OH 44555

Kazuharu Yoshizuka,   35
Department of Applied Chemistry,
Saga University,
Honjo-machi, Saga 840, Japan

INDEX

DLVO theory, 232–234
  and Gibbs energy, 232, 234
DNA, 207, 208, 210, 267, 301
  and temperature, 214–222
Drug delivery, periadventitial, 266
Dye adsorption, 320–324

Elastin, 95–96
Electrophoresis, 46, 48
EMCF, 278, 280, 283, 309, 316
Endocarditis, bacterial, 261
Enterobacter aerogenes, 305, 307
Enterotoxin B, staphylococcal, 253, 255
Enthalpy, 218, 219
Entropy, 218, 219
Enzymes, 196–199, *see also* separate enzymes
Eosin Y, 252
Escherichia coli, 3, 6, 7, 74, 75, 115–126, 202–204, 306 307
  gene cassette construction, 117–119
  and mussel adhesive protein, synthetic, 115–126
  removal from water, 203
Ethenoanthracene, 189
Ethylenevinylacetate, 250
Exopolysaccharide, 14–16

Factor 4 (antiplatelet), 260
Factor Xa, 18
Fiber, antimicrobial, 226, 229, 230
  "Letilan", 226
  Quinn test, 226
Fibrin, 129
  thrombosis, 260
Fibrinogen, 128
  -collagen compound, 128
Fibroblast, 97
  growth factor, 261
Fibronectin, 95, 97, 106–109, 112
Flow discharge technique, 317–319
5-Fluorouracil, 277–288
  1,6-diisocyanate derivative, 278–279
  n-hexylisocyanate derivative, 310–315
  release, 277–288
    from chitosan gel, 289–296
    controlled, 309–316
  structure, 310
5-Fluorouracil-poly (caprolactone), 312
  preparation, 310
Freund's adjuvant, 249–251
Fusarium sp.

Galactolipid, 295
Galactomannan, 9–13, 16
Gallium, 38
Gel, 10–13
Gene cassette construction, 117–121, 124–125
  cloning, 120–121
  design, 120–121
Gentamicin-collagen compound, 127–133
  results, clinical, 133

Gibbs energy, 232, 234
γ-Globulin, 250
Glucomannan, 9–11, 16
Glucose, 4, 195
  quantification, 177, 182, 195
Glucose oxidase, 177, 178, 196–198
Glutaraldehyde, 290, 291
Glutaric acid, 18, 20
Glycolic acid copolymer, 251, 254–256
Glycoprotein antibody, 260
Glycosaminoglycan, 19
Gonadotrophin hormone, chorionic, bovine, 254
Group, bioactive, antimicrobial, 228–230
  list of fifteen, 228
Growth factor receptor, platelet-derived, 260
Guinea pig, 27–30

Helminthosporium oryzae, 46, 47
Hemostasis and collagen, 127–130
Heparin, 17–23, 260, 261
Heparin ethylenevinylacetate, 261
Hepatoma cell, human, 291–296
Herpes simplex virus, 254
1-(N-Hexylcarbamoyl)-5-fluorouracil, 310, 313
  -poly(caprolactone), 310, 314, 315
    preparation, 310
  structure, 310
n-Hexylisocyanate, 310
Higuchi pattern, 309, 311, 313
Hirudin, 263
  -hirulog, 260
Hirulog, 259, 262–265
  -hirudin, 260
Horseradish peroxidase, 197
HTLV-1 virus, 274
Human
  melanoma cells, malignant, 97, 103, 105, 106, 109
  vein cells, umbilical, 97, 103–108
Hyaluronan, 25, 31
Hydrometallurgy, 35–41
Hydroxyethylcellulose, 163
Hylan gel, 25–33
  and augmentation of soft tissue, 25–33

Immobilization on silica gel, 207–223
Immunoblotting, 117
Immunoglobulin, 56, 186–191
  adsorption, 188–191
Implant in rabbit, 169–176
  periadventitial, 259
  polymeric, controlled, 259–268
Indium, 38
Indulin C, 72
Infiltration, microbial, 230–235
Interferon, 269, 274
Interleukin-1, 261
5-Iodo-2'-deoxyuridine, 301, 303–305, 307
  structure, 301
1-Iodo-4-oxatetradecyne synthesis, 137
Iron, 38

Poly(anhydride), 257
Poly(N-benzyl-4-vinylpyridinium bromide), 201–206
   and activated sludge, 201–206
Poly(caprolactone), 281, 309, 310
Poly(CTTH-iminocarbonate), 251, 252
Polyelectrolytes, 60, 61 *see also* separate com-
        pounds
Poly(ester), 257
Poly(etherurethane), 319
Polyethylene, 162
Poly(ethylene terephthalate), 147, 150, 162
   crystallization, 153–155
Poly-ICLC, *see* Polyinosinic-polycytidylic acid
Poly(iminocarbonate), 257
Polyinosinic-polycytidylic acid, (poly-ICLC), 269–276
   in AIDS treatment, 269–276
Polylactic acid, 250
Polylactic-polyglycolic acid, 266
Poly-L-lactide, 169–176
   adsorption study in rabbit, 169–176
   forms, 170–175
   uses, orthopaedic, 169–170
Polylysine, 269, 270
Polymer
   antimicrobial, 225–237
      group, bioactive, 227–230
      list of fifteen, 228
      use, 229–230
   bacterial, 201–206
      for adsorption, 201–206
      preparation, 201–206
   biodegradable, 249–258
      are they really ? 161–168
      synthetic, 251–255
      and vaccine delivery, 249–258
   medical, 317–320
   network, interpenetrating, 149–150
      synthesis, 150
   non-aqueous, 177–183
Polymerization, 149, *see also* Actin, Tubulin
   equations, 82–86, 89
   method, 298–299
Polyorthoester, 257
Polypentapeptide, 95–113
   structure, 96
Polyphosphate, 297–307
Polyphosphonate ester, 297–307
   as nucleic acid, 297
   technique, interfacial, 297
Polypropylene, 162
Polysaccharide, 1–8
   anionic, 291
   cationic, 291
   derivatives, 278–279, 287
   structure, 291
Polystyrene, 147, 149, 162
Polyurethane, 266, 317
   and acrylic acid, 317, 318
Polyvinyl alcohol fiber, 226
Polyvinyl chloride, 162, 163
Propylhydroxylase, 96

Protein
   adsorption on pyridinium, 185–193
   and actin, 79–94
   bioadhesive, *see* Mussel
   immunodiffusion, 129
   polymerization, 79–94
   purification, 185–193
   recovery from whey, 55–69
   and tubulin, 79–94
Protein kinase, 96
Pseudomonas aeruginosa, 3, 6, 7, 74, 75, 306, 307
   cholesterol esterase, 197
Pullulan, 163
Pyridine, 186
Pyridinium, 185–193
   immobilized, 185–193
   for protein purification, 185–193
   structure, 186–187

Quinn test, 226
Quinolinic acid, 274

Radish seed, 47, 50
REDV amino acid sequence, 95, 96, 101–105
Release, controlled, 227–230, 261, 277–296, 311
Restenosis, arterial
   gene therapy, 266
   implant, polymeric, controlled, 259–268
   pathophysiology is poorly understood, 260
   and tissue plasminogen activator, 266
   treatment, 259–268
      strategies, 260
Rhamnose, 13, 14
Ribonuclease A, 250
Rice callus, 45–52
RNA
   double-stranded, 269
   polymerase T7, 115

Saccharide, 1–8
Sebacic acid, 149
Shrimp chitosan, 56, 57, 64–66
Silanol, 213
Silastic QC-4840, 262, 264, 266
Silica gel, 207–223
Silicone, 25
   rubber, *see* Silastic
Skin, 31
Sludge, activated, 201–206
Sodium benzoate, 155
Spherulite, 154
Staphylococcus aureus, 3, 6, 7, 74, 75, 128, 231,
              233, 235 306, 307
   S. epidermidis, 231, 306, 307, 317
   S. sp., catalse-negative, 318
Starch, 163
   cross-linked, synthetic, 257
Steroid, 261
Streptococcus pyogenes, 231
Streptomyces sp., 197
   cholesterol oxidase, 197